MAINTENANCE AND SPARE PARTS MANAGEMENT

Second Edition

P. Gopalakrishnan
Former Professor of Administrative Staff College of India
Hyderabad
Former U.N. Advisor on Inventory Management

A.K. Banerji
Former Professor of Administrative Staff College of India
Hyderabad

PHI Learning Private Limited
Delhi-110092
2015

₹ 325.00

MAINTENANCE AND SPARE PARTS MANAGEMENT, Second Edition
P. Gopalakrishnan and A.K. Banerji

© 2013 by PHI Learning Private Limited, Delhi. All rights reserved. No part of this book may be reproduced in any form, by mimeograph or any other means, without permission in writing from the publisher.

ISBN-978-81-203-4739-7

The export rights of this book are vested solely with the publisher.

Tenth Printing (Second Edition) ... ... **April, 2015**

Published by Asoke K. Ghosh, PHI Learning Private Limited, Rimjhim House, 111, Patparganj Industrial Estate, Delhi-110092 and Printed by Raj Press, New Delhi-110012.

Dedication

This title is dedicated to the fond memory of the my only son **G. Ganapathy Ram**, 17 years old, 180 cm tall, brilliant, weighing 60 kg who died on 30 July, 1986, due to medical negligence, wrong diagnosis, wrong operation, callous post operative treatment and lack of timely proper medical help. The integrated policy cases have been named after him.

I also dedicate this book to my son-in law **Ranganath Narasimhan**, who passed away on 28 January, 2007 due to cancer. He persuaded me to update this book, constantly reminding me that my six decades experience should be passed on to the younger generation. He was one of the popular active person in Coimbatore region, associated with several NGOs, blood banks, humour club, Rotary club and several non profitable and religious forms. He was the founder Chairman of ADWISE Advertising—an accredited advertising company.

P. Gopalakrishnan

Contents

Preface xix
Preface to the First Edition xxi
Acknowledgements xxiii

SECTION I   CORE MAINTENANCE

1. Maintenance Management and Challenges 3–15

    Relevance of Maintenance   3
    Asset Management   4
    The Role of Railways, Coal and Power   4
    Failures and Maintenance   5
    Maintenance: An Overview   5
    Upgraded Technology   6
    Good Maintenance Services   7
    Problems of the Plant Manager   8
    Situational Failures   9
    Change in Environment   9
    Factors Involved in Functioning of Maintenance   10
    Design of Maintenance Systems: The Challenge   10
    Automation and Maintenance   11
    Factors Affecting Maintenance Efficiency   12
    Maintenance Planning   13
    Benefits of Maintenance Planning   14

2. Maintenance Objectives and Functions 16–23

    Maintenance Objectives   16
    Maintenance Costs   17
    Down Time Costs   18

Responsibilities of Operators   19
Responsibilities of Maintenance Staff   20
Maintenance A–Z Functions   21

3. Maintenance Organisation 24–36

Maintenance Engineering   24
Organisational Prerequisites   25
Esprit De Corps—Team Spirit   26
Span of Control   27
Subordinates Development   27
Competence/Skills Utilisation   29
Hierarchical Levels   30
Conflict Management   30
Factors for Effectiveness of Maintenance   32
Objectives of Organisation Design   33
Types of Maintenance Organisation   34

4. Maintenance Systems 37–50

Classification of Maintenance Systems   37
Breakdown Maintenance   37
Routine Maintenance   38
Planned Maintenance   40
Preventive Maintenance   42
Reliability   43
Predictive Maintenance   45
Corrective Maintenance   45
Design Out Maintenance   46
Total Productive Maintenance   46
Japanese 5S   47
Contract Maintenance   48
Examples of Contract Maintenance   49
Cannibalisation   50

5. Design of Maintenance Systems 51–63

Influencing Factors on Systems   51
Criticality Determination   52
Downstream Effects   54
Age of Plant and Technology   56
Skill Availability   56
Bath Tub Curve   57
Maintenance Support Facilities   58
Operating Environment   58
Maintenance System Design Optimisation   61

Contents   vii

6. Condition Monitoring **64–72**

    Need of Condition Monitoring   *64*
    Plant Availability   *65*
    Predicted Trend   *67*
    Increased On-stream Plant Availability and Reliability   *67*
    Plant and Personnel Safety   *68*
    Optimal Maintenance Cost   *68*
    Methods of Condition Monitoring   *69*
    Choice of Equipment for Condition Monitoring   *69*
    Vibration Monitoring System   *70*
    Condition Based Maintenance   *71*

7. Non-destructive Testing (NDT) **73–81**

    NDT Concept   *73*
    Non-destructive Testing Methodologies   *74*
    Vibration Monitoring   *76*
    Soap and Fibre Optics   *76*
    Diagnostic Instruments A–Z   *77*
    Lightning Fast Material Test   *79*

8. Total Planned Maintenance (TPM) **82–100**

    Planning System Components   *82*
    Facility Register   *84*
    Equipment Record Card   *85*
    Maintenance Schedules   *87*
    Principles of Scheduling   *87*
    Scheduling Process   *89*
    Work Specifications   *89*
    Maintenance Records and Documentation   *91*
    History Record Card   *91*
    Defect Analysis and Down Time Records   *92*
    Maintenance Work Order   *94*
    Performance Report   *94*
    Information Control Analysis   *96*
    Control System for Planned Maintenance   *98*

9. Maintenance Turnaround **101–109**

    Outage Management   *101*
    Turnaround   *102*
    Turnaround Manager   *103*
    Opportunity Maintenance   *105*
    Network Analysis   *105*
    Basic Steps in a Project   *106*

SECTION II RELATED AUXILIARY FUNCTIONS

10. **Inspection and Lubrication** 113–121

    Why Inspection? *113*
    Frequency of Inspection *114*
    Planning for Inspection *115*
    Inspection and Testing Facilities *116*
    Lubrication *117*
    Lubrication Programme and Planning *118*
    Tribology *121*

11. **Calibration and Quality** 122–126

    Standards of Calibration *122*
    Calibration System *123*
    Calibration Documentation *124*
    Maintenance Quality *125*
    Innovation Fountainhead *125*
    Building up of Quality *126*

12. **Maintenance Training and HR** 127–137

    Need for Maintenance Training *127*
    Planning Background *128*
    Training Levels *128*
    Training Methodology *131*
    Maintenance Incentives *133*
    Incentives in Maintenance *134*
    Incentive Administration *136*

13. **Safety and Maintenance** 138–150

    Significance of Safety *138*
    Potential Hazards *139*
    Safety Principles *140*
    Hot Work Safety Permit *142*
    Fault Tree Analysis *143*
    Safety Symbols *145*
    First Aid Kit *146*
    Safety Culture A–Z Issues *146*
    In the Wake of a Quake *147*
    Unsafe Metros *148*
    Safety Norms Switched Off *149*
    Elevated Risks *149*
    Food Safety *150*

14. Computers and Maintenance 151–158
   Computer System   151
   Computer Application A–Z Areas   152
   Maintenance Costs   154
   Cost Benefit   155
   Management Information System (MIS)   156
   Maintenance Budget   157

15. Productivity and Industrial Engineering 159–170
   Productivity Concepts   159
   Work Study   160
   Method Study   161
   Recording Techniques   162
   Ergonomics   164
   Loads and Controls   168
   Work Study and Ergonomics   169
   Ergonomics and Maintenance   169

16. Activity Sampling for Work Measurement 171–180
   Snap Reading   171
   Normal Distribution   172
   Random Timings   173
   Areas under Normal Distribution   174
   Applications of Activity Sampling   176
   Advantages of Snap Reading   176
   Disadvantages of Snap Reading   177
   Work Measurement   177
   Time Study   179

17. Energy Saving by Maintenance 181–193
   Need for Energy Saving   181
   Issues of Maintenance and Energy   182
   Building Problems   183
   Boiler and Steam Distribution   184
   Boiler Maintenance   185
   Air Compressor System   186
   Heating Ventilation and Air Conditioning   187
   Conservation Guidelines   188
   Electrical System   189
   Lighting System   190
   Cyclic Replacement of Lamps   191
   Economising on Light   192
   Light Source Characteristics   193

18. Facilities Investment Decisions (FID) and Life Cycle Costing (LCC) 194–222

Asset Management   194
Decision Varieties and FID   194
Factors Influencing Facilities Investment Decisions (FID)   195
Selection/Replacement Problems   197
Economic Life Concept   197
Optimum Replacement Models   198
Time Value of Money   200
Capital Recovery Factor   201
Effect of Taxes and Depreciation   203
Cash Flow Illustration   203
Depreciation   204
Economic Analysis   206
Life Cycle Costing (LCC)   207
Aim of LCC   211
Application of LCC   212
Costing of Alternatives   214
A–Z of LCC   218

19. Evaluation of Maintenance Function 223–233

Need for Evaluation   223
A–Z Maintenance Challenges   224
Expectations: A–Z Parameters   225
Background of Maintenance   226
Equipment Discard Policy   226
MBO/SWOT and Managerial Styles   227
Evaluation Process of Plant Engineering   228
Evaluation A–Z Subjective Methods   229
A–Z Objective Criteria   230
Maintenance: Futuristic Scenario   232

SECTION III CORE SPARES ISSUES

20. Indian Spares Scenario 237–244

Capacity Utilisation   237
A–Z Financial Aspects of Spares   237
Spares Management A–Z Issues   240
Necessary Information   242
Initial Spares Provisioning   244

21. Spares Practices Survey 245–256

Research Methodology 245
Diagnostic Study 246
Organisational Aspects 246
Technology and Capital Equipment 247
Overhauling and Discard Policy 248
Obsolescence 249
Initial Provisioning 250
Categorisation and Codification 250
Standardisation 251
Lead Time Analysis 251
Requirement Planning 252
Selective Control and Stock Levels 252
Spares Bank 253
Import Substitution 254
Spares Information System 254
High Inventories: A–Z Causes 255
Research Survey Summary 256

22. Cost Reduction in Spares 257–265

Definition of Spare Parts 257
Spares A–Z Features 257
Categorisation of Spares 258
Financial Considerations 259
Choice of Equipments 259
Cost of Ordering 260
Inventory Charges 260
Cost of Stockout 261
A–Z Cost Reduction of Spares 261
Lead Time Reduction 262
Identification by Codification 263
Value Analysis 264
Variety Reduction by Standardisation 264
Spares Bank 265

23. Music-3D-Beyond Cost Criticality Method 266–273

Limitations of ABC Analysis 266
ABC Analysis 266
Vein-Ved Analysis 268
Spares Intelligence 268
MUSIC-3D Concept 269
Interpretation of MUS1C-3D 270
A–Z Advantages of MUSIC-3D 270

xii *Contents*

Other Selective Approaches 271
Service Level 272

24. Inventory Control of Spares 274–281

Objectives of Control 274
Stockout Cost 275
Ordering Charges 276
Service Level 277
Economic Order Quantity 278
Review Period 280

25. Maintenance Spares 282–287

Concept of Maintenance Spares 282
Poisson Pattern 283
Stock Levels 283
Service Level 284
Central Stocking 285
Illustration of Stock Levels 286

26. Simulation for Spares Control 288–293

Planning and Simulation 288
Simulation Method 288
Two-Bin System 289
Illustration of Simulation 289
Illustration on Initial Ordering 291
Use of Poisson Pattern 292

27. Insurance Spare Parts 294–299

Typical Insurance Parts 294
Stocking Policies 295
Financial Aspect 296
Spares Bank 297
Cost-Benefit Analysis of Stocks 298

28. Rotable Spares 300–307

Concept of Rotables 300
Financial Considerations 300
Queuing Approach 301
Application of Queuing Model 302
Solutions to Queuing Questions 302
Queueing and Rotables 303
Policy for Rotables 304
A–Z Issues in Queuing 306

29. Overhauling and PERT (Programme Evaluation and Review Technique) 308–315

Overhauling Spares Concept   308
Requirement of Overhauling Spares   308
PERT's Concept   310
Details of PERT   313
Procedure for PERT   313
Exercise on PERT   314

SECTION IV   RELATED ISSUES ON SPARE PARTS

30. Reliability and Quality 319–326

Need for Reliability   319
Concept of Reliability   319
Failure Analysis   320
Failure Parameters   321
Mean Time between Failure (MTBF)   322
Mean Time to Repair (MTTR) and Log Normal Distribution   323
Bath-Tub Curve   324
Inspection of Spares   325
Inspection Methods   325
Inspection Infrastructure   326

31. Procurement of Spares 327–334

Buying Relevance   327
Right Quantity   327
Right Time   328
Right Price   329
Right Quality   330
Right Source   330
Import Substitution   331
Right Contracts: A–Z Aspects   332
Delivery/Transportation   333
Right Systems   334

32. Logistics and Warehousing 335–342

Storage Objectives   335
Physical Storage: A–Z Ways   336
Preservatives: A–Z Ways   336
Receipts Management   338
Documentation of Outputs   338
Valuation and Verification   339

Multiple Warehouses   340
Logistics Management   340
Business Logistics   341

33. Pricing and Marketing of Spares 343–353

Pricing   343
Pricing: A–Z Factors   344
Pricing Strategies   345
Market Skimming Strategies   346
Forecasting and Planning   347
Order Processing   349
Distribution of Spares   350
Spare Parts Sales: A–Z Problems   350

34. ASS (After Sales Service) 354–362

Image of After Sales Service   354
Servicing Goals   355
Professionalising Customer Service   356
Organisation and Manuals   356
Cost of Servicing   358
Market Intelligence   358
Legal Aspects and Planning ASS   359
Spurious Spares and A–Z Problems   360

35. Multiechelon Distribution 363–371

Multiechelon Concepts   363
Multiechelon Characteristics   364
Classification of Multiechelon Problems   365
Inter-Echelon Interactions   365
Parameters in Multiechelons   366
Mathematical Approach   367
Mathematical Derivation   368
Optimum Stock Levels   370

36. Management of Obsolescence 372–379

Obsolete Spares Problems   372
SOS Items Concept   373
Obsolescence A–Z Issues   373
Solutions to Obsolescence   375
Movement Analysis   376
XYZ Analysis   377
Disposal of Obsolete Items   378

Contents xv

37. Reconditioning 380–386

Need for Reconditioning 380
Reconditioning Concept 380
Replacement vs. Repair 381
Reconditioning Time 382
Reconditioning A–Z Factors 383
Economics of Reconditioning 384
Infrastructure for Reconditioning 385
Technical Details of Reconditioning 386

38. Information System for Spares 387–395

Environmental Scanning 387
MIS Steps 388
Database 389
MIS for Selling: A–Z Aspects 389
Selling: A–Z Questions 390
Users: MIS A–Z 391
MIS Operation 392
Role of the Computer 393
MIS Reports: A–Z Aspects 394

39. Evaluation of Spares 396–405

Organisation Problems 396
Spares Planning Cell 396
Sales and Servicing 397
Role of Training 398
Need and Process of Evaluation 398
Evaluation of Purchasing: A–Z Aspects 399
Evaluation of Spares Sales: A–Z Aspects 400
Stores Evaluation A–Z Issues 402
A–Z Inventory Evaluation 403
Management Audit 404

SECTION V   CASELETS/SHORT CASE STUDIES

40. A Hazare Fertilisers Ltd. (AHFL) 409–415

Corporate Scenario and Structure 409
Spares Planning 410
Spares Control 411
Provisioning of Spares 411
Increased Spares Inventory 412
Tyre for 30 Tonne Trucks 412

xvi *Contents*

    Ancillaries for Import Spares   *413*
    Role of User and Finance   *413*
    Spares Bank   *414*

41. Vasanth—ASS Limited 416–423

    Company Details   *416*
    Failure of ASS   *416*
    After Sales Service Poor Image   *418*
    Supply Failure   *419*
    Price Differences   *420*
    Supply of Spares   *421*
    Self-Reliance Talk   *422*

42. Bhushan Refineries Limited (BRL) 424–431

    Company Scenario   *424*
    Organisational Aspects   *425*
    Business Logistics   *426*
    Reorder Levels   *429*
    Logistics Warehousing   *429*

43. Middle East Air Transport 432–436

    Company Background   *432*
    Suppliers Faults   *432*
    Forecasting Repair Task   *433*
    Conflicting Views   *434*
    International Pressures   *435*
    ABC-FSN Analysis   *435*
    Demand Analysis   *436*

44. Aruna Oil and Gas Limited (AOGL) 437–443

    Corporate Strategy of AOGL   *437*
    Organisation Structure of AOGL   *438*
    Assets and Equipments   *439*
    Spares Details   *439*
    Spares Provisioning   *441*
    Bid Evaluation Criteria   *442*
    Inventory Problems   *442*

45. Mahatma Gandhi Road Transport
    Corporation (MGRTC) 444–450

    Scenario Painting   *444*
    Depot Operations   *445*

Maintenance Policies   *445*
Overhauling Practices   *446*
Enter the Consultant   *447*
Gemini Tyre Retreading Machinery Details   *448*
Tyre Buffer   *448*
Tread Builder   *448*
Electric Curing Chamber   *449*
Mono Rail System   *450*
Inspection Spreader (Manual)   *450*
Envelope Expander   *450*

SECTION VI   INTEGRATED MANAGEMENT POLICY CASES

46. Ganapathy Ram Port Trust *453–464*

Enter the Consultant   *453*
Equipment Categories   *454*
Spare Parts Identification   *455*
Stock Levels   *455*
Procurement   *456*
Organisation of Spares   *457*
Supplier Profile   *457*
Procurement of Imported Spares   *458*
Maintenance Policies   *459*
Maintenance: A–Z Problems   *460*
High Spares Stock: A–Z Causes   *462*

47. Ganapathy Ram Steels Ltd... *465–476*

Corporate Background   *465*
Organisation Structure   *465*
Types of Audit   *466*
Products   *467*
Production Shops   *467*
HR and Personnel Policies   *467*
Management Development   *468*
Maintenance Policies   *469*
Overhauling Practices   *470*
Purchase Policies   *471*
Import Substitution: A–Z Problems   *472*
Inventory Management   *473*
Maintenance Problems: A–Z Aspects   *473*
Materials: A–Z Problems   *475*

xviii *Contents*

Discussion Questions on Maintenance and Spares 477–492

    Maintenance Perspectives   *477*
    Maintenance Organisation   *478*
    Maintenance Systems   *478*
    Maintenance Systems Design   *478*
    Condition Monitoring   *479*
    TPM and Documentation   *479*
    Maintenance Turnaround and Network Analysis   *479*
    Inspection Lubrication   *480*
    Human Relations in Maintenance   *480*
    Calibration and Quality   *481*
    Safety Engineering   *481*
    Computers in Maintenance   *482*
    Maintenance Costs/Budgets   *482*
    Productivity and Maintenance   *482*
    Activity Sampling and Work Measurement   *483*
    Energy-Steam-Boiler Systems   *483*
    FID and LCC   *484*
    Maintenance Function Evaluation   *484*
    Spare Parts Definition   *485*
    Spares Cost Reduction   *485*
    Spares Organisation   *486*
    Spares Control   *487*
    Inventory Policy   *487*
    Reliability and Quality Inspection   *488*
    Quality and Inspection   *488*
    Procurement of Spares   *489*
    Marketing and ASS   *490*
    Evaluation of Spares Function   *492*

Bibliography *493–495*

Index *497–500*

Preface

Several industrial engineers, mechanical engineers, skill managers, maintenance incharges and academicians had been writing to me to update this popular well-received title with complex field of maintenance and spare parts. The revision became difficult due to long illness of my coauthor Prof. A.K. Banerji and therefore the job of 'maintaining quality' was left to me.

With the help of academicians and practicing maintenance manager and based on my training and teaching experience in India and abroad, I attempted to present the prevailing state-of-the-art in a lucid manner. Wherever necessary, photos, exhibits, tables, illustrations, graphs and diagrams have been provided for the benefit of the readers.

This comprehensive research-based revised text analyses all aspects of concepts and applications of maintenance associated with spare parts principles and practices in six sections. It lucidly explains the nuances of maintenance fields, maintenance system, non-destructive tests quality and inspection, design out maintenance, discard policy, LCC, calibration, spares pricing, after sales service, AMC, marketing of spare parts, CPM, insurance, floats, financial aspects, policies, and evaluation of maintenance management functions.

Successful life cases and integrated policy cases on maintenance policy have been provided to enhance the analytical skill of the readers. It may be emphasised that there is no single solution or simple answer to the case, but forces the readers to look into actual problems in simulation.

We have sown the seeds of importance of maintenance and spare parts, but it is up to you to ensure that some of the seedlings become a sapling and ultimately turning into a big healthy banyan tree, bearing the fruits of knowledge and the process goes on to future generations. The material is class tested and the reactions are duly incorporated.

The second edition of the book is intended to serve as a textbook for undergraduate students of mechanical engineering and industrial and production engineering. Postgraduate students of management and engineering will also find this text useful. Practising mechanical engineers and marketing, ASS, AMC and inventory executives, will find this as a reference and application-oriented for updating their professional knowledge.

For easy comprehension and cohesive approach, the book has been divided into six sections. The first section deals with core maintenance while the second section discusses related auxiliary functions of the mechanical engineer. The third section deals with the categories of spares and the fourth section deals with related auxiliary spares functions. While the last two sections explore caselets and two integrated managerial policy cases. I hope this will make the reader a thorough professional.

I will be happy to receive any constructive comments, suggestions and criticism on the second edition, since improvement is a continuous process.

P. Gopalakrishnan

Preface to the First Edition

The relevance and the significance of maintenance and spare parts management cannot be overemphasised, particularly in a developing country like India. However, today this remains the most neglected area of management. Also, the distressing phenomenon of repeated breakdowns leading to low capacity utilisation in almost all sectors brings to focus the urgent need for formulating a well-organised spare parts management policy in various organisations.

Many plants remain shutdown for lack of spares and therefore huge working capital is blocked up in obsolete spare parts inventories.

Paradoxically enough, there is hardly any literature available on maintenance and spare parts management, particularly in the Indian context. We have strived to fill this gap through this text.

This book is the outcome of our long years of study, research and teaching experience as well our consultancy assignments in India and overseas. Besides, our close interactions with the senior executives attending various management courses in the Administrative Staff College of India and the feedback received from the students have redoubled our efforts and reinforced our knowledge. Finally, the various assignments we had in the Middle East sponsored by the United Nations gave further insight into the subject and crystallised our ideas.

The text consists of two parts: Part I dealing with Maintenance Management and Part II covering Spare Parts Management. Part I begins with an overview of perspectives, objectives, organisation and systems of maintenance. It then proceeds to discuss in detail such topics as design of maintenance systems, predictive maintenance, non-destructive testing, inspection and lubrication, calibration and quality, training, safety management, incentives, the use of computers in maintenance, and ergonomics. This part also provides an indepth analysis on activity sampling, energy saving facilities, investment decisions, life cycle costing and evaluation of maintenance.

Part II dealing with Spare Parts Management includes topics such as cost reduction approach, MUSIC-3D view of spares, and quality and procurement of spares. This part also gives a comprehensive coverage of spare parts marketing, after sales servicing, inventory control, different types of spares, simulation approach, MIS, and organisation and evaluation of spares.

The salient features of this book include two case studies on the Ganapathy Ram Port Trust and Ganapathy Ram Steel Mills, and the presentation of many topics in a capsule form—"a–z" aspects.

In writing this text, we have consulted our colleagues, friends, professionals and interacted with many students. We are indebted to them for their valuable suggestions and help. We are also grateful to the large number of organisations and to the many senior executives in India and abroad who shared their knowledge with us and who took pains to respond to our questionnaires. Our thanks are also due to Ms. Mahalakshmi Ramakrishna for typing the manuscript. Finally, we thank the publisher, PHI Learning, for meticulous processing of the manuscript and for bringing out this volume in a remarkably short period.

Any comments and suggestions on improving the content of this volume will be most welcome.

P. Gopalakrishnan
A.K. Banerji

Acknowledgements

I gratefully acknowledge the help of G.B. Iyer—former G.M. Maintenance Gwalior Regions Wagda, and presently a consultant in Bombay—for helping me in updating this title significantly as my coauthor A.K. Banerji had been severally ill since 2010.

I also thank the organisations who have supplied data for the case studies and Ganapathy Ram policy cases.

The publisher—PHI Learning has done a remarkable job in editing and bringing the title in a short time after meticulous corrections.

My wife—Indira Krishnan who withstood my idiosyncrasies for six decades after series of tragedies — needs to be thanked.

The patience and understanding she showed is remarkable. She helped me in maintaining my physical strength and mental attitude, with right inputs and right food, at the right time etc. by taking total productive maintenance actions, even though I am a super senior citizen.

P. Gopalakrishnan

SECTION I
Core Maintenance

1. Maintenance Management and Challenges
2. Maintenance Objectives and Functions
3. Maintenance Organisation
4. Maintenance Systems
5. Design of Maintenance Systems
6. Condition Monitoring
7. Non-destructive Testing (NDT)
8. Total Planned Maintenance (TPM)
9. Maintenance Turnaround

Chapter **1**

Maintenance Management and Challenges

RELEVANCE OF MAINTENANCE

The industrial age has given way to the information age. This transition calls for unmatched kinds of organisations filled with innovations as success in the past neither assures continued success nor a good reputation in future. Indian industries have invested crores of rupees in production assets. Indian industries facing global competition have to rapidly improve and innate for their survival. This depends on efficient use of the assets in the form of machineries and human assets, by building people and updating the productivity of assets by better maintenance of assets to ensure win-win situation for all stakeholders, employees, employer, customers and society at large. Companies must improve at a faster rate than their competitors, if they are to become leaders. For maintaining these challenges progressive firms are adapting best in class manufacturing practices and improvement processes, which include total productive maintenance (TPM). The progress of maintenance depends upon socio-political, macro- and micro-economic, technological, human and industrial factors which we will discuss throughout all chapters of this book. In this process, we will also focus on profitability, growth, security, safety, systems and procedures of maintenance management. Besides, other aspects of maintenance management, such as staff, equipment, materials and documentation will also be discussed. We will begin our discussion with improving management of assets ranging from batches to large single stream plants resulting incorporate profitability leading to India's progress in all sectors.

ASSET MANAGEMENT

Maintenance is one of the most crucial areas governing organisations. This is particularly true for an industry. Yet it is the most neglected aspect! In most industries in India, owing to a number of reasons, the capacity utilisation is less than 60 per cent. This lowering of capacity utilisation leads to higher inputs and lower outputs. This imbalance continues even when the *unit cost* charged to the customer is constantly being increased as an immediate remedy to this deep-rooted ailment. Maintenance, when neglected, leads to frequent breakdowns, resulting into costly repairs and faster deterioration of valuable equipment, besides causing incalculable loss of production.

The frequent power cuts throughout the country are, in most cases, caused by breakdowns of generating plants and their equipment. But the failure to deliver the goods could also be due to breakdowns taking place anywhere along the interconnected chain-links of all sectors of the economy leading to double digit inflation.

THE ROLE OF RAILWAYS, COAL AND POWER

While discussing links, the close link-up between thermal power stations, the railways (life line of India) and the collieries comes easily to mind, and one can readily visualise a situation where a thermal power station is unable to generate power for lack of its base raw material, i.e. coal. But where will the coal come from? For, there has been a major breakdown in the colliery, lack of power perhaps being a major contributing factor. However, eventually the appropriate personnel get down to their work and heap up coal at the coal pit-heads railway siding. However, at that very moment the railway rolling stock have multiple breakdowns, and since coal is not accorded the same priority as, say foodgrains, it has to wait. Then the railway maintenance staff, after considerable amount of work, get the ailing wagons back on to the rails. Again, there is no power to haul the wagons to the waiting coal. Finally, the solution is found and the coal does at last reach the power plant, but not surprisingly, a flaw develops in power generating plant, for it too is having a breakdown. These three sectors are major infrastructure sectors.

This is not an imaginary situation. In fact, there are numerous such instances on record: of trains being cancelled for want of coal, diesel, electric power, and of major industries like steel and power having to starve on meagre coal inventories as a consequence. Yet ironically, often it is also reported that coal is being stock-piled at the coal pit-heads, but cannot reach the crisis points due to shortage of railway wagons to move them.

It would not be out of place to say that considering the industrial climate prevailing in India, the Indian Railways are performing a laudable task, keeping in mind the sheer volume of human and goods traffic that it handles every day. We should also keep in mind that what may appear to be a failure on their part may not be so. For a shortfall at one point may have been caused by a sudden reallocation of rolling stock to meet the ever changing priorities that come up at the national or even the international levels.

However, while on the subject of maintenance, we have to concede that the turnaround time of railway wagons is a major contributing factor towards creating avoidable bottlenecks, for the wagons quite often take an unduly long lime lo come out of their workshops in a serviceable condition.

FAILURES AND MAINTENANCE

We have just observed that failures, big or small, anywhere along the chain of interconnected industrial activities have far-reaching detrimental consequences on production as a whole. The loss incurred by a particular sector due to its failures and shortcomings may not by themselves be very significant (in monetary terms). However, from the material point of view it may be quite significant because of the resulting delay in and even stoppage of supply of certain goods and services within the specified time when they are most needed. Thus a high state of maintenance efficiency is not only desirable, but absolutely obligatory for industrial well-being, at all levels and eventually at the national level.

At the very outset we have stated that failures and losses are caused by various factors, and that maintenance is one of the major contributing factors. Therefore, it is also true that healthy and efficient functioning must be contributing in equal proportion towards success. Though this appears to be a very simple and logical assumption, often the maintenance engineer/manager does not find this logic working in his professional life. Even though it can be stated without any undue exaggeration that maintenance is one of the most important ingredients for achieving higher industrial productivity and consequent higher returns on investment, yet the maintenance personnel are more often blamed than praised for their job. What really causes this to happen needs to be examined with care, for it has far-reaching consequences on the productivity and efficiency of the maintenance personnel.

MAINTENANCE: AN OVERVIEW

The aim of this book is to take an overview of maintenance function and to assess the problems and challenges that beset the maintenance departments and the maintenance man, and to find solutions and search for ways to face these challenges adequately. Being a low profile, repetitive, and low accolade winning job, maintenance needs to be handled with a great degree of sensitivity and perception, if a climate for its systematic development is to be cultivated. Therefore, it can be said with assurance that there can be no systematic development of this exacting profession, or any fruitful growth of this vital function, unless the top management in industry develops the requisite understanding and enthusiasm for the work done by the maintenance personnel.

It is a known fact that wherever the prestige and position of the plant engineers have been valued, assured, and become well established by the top management, the plants there have shown far better results on the whole. But such organisations are few and far between.

The need for a maintenance organisation is not being questioned by today's top management, but the due importance and the rightful place which it ought to occupy in the organisation are yet to be recognised by most of them. With the growing complexity of equipment and process, and the magnitude of losses suffered in production due to breakdowns, today's management can no longer look upon maintenance as only a subsidiary function to production, but as one of the main tools of planned productivity, which must be effectively used to obtain the highest availability of production equipment commensurate with maintenance cost.

As seen above, maintenance has an extended role to play and it is imperative that it is included for close and active consultation from the nascent design stages of an industry. Furthermore, it is essential that the maintenance personnel too be fully involved in the process/equipment selection, and thereafter, when the unit goes into production, they must ensure that they are in constant coordination with the operation of the plant, so that they can keep their own performance and that of operation at the most optimum level.

In today's industrial world, there is a great need to involve maintenance as an equal partner, in the entire gamut of industrial decision-making process. The span of accountability needs to increase and that too perhaps, not in just arithmetical proportions. Thus the maintenance department will have to be very alert.

The expanded role and its added responsibilities will make it imperative for the maintenance engineer/manager to mould and educate himself so as to be able to reach out beyond the narrow confines of his own particular specialisation, if he wishes to make a mark and function effectively. In his own field of activity he will have to accept the fact that the days of obsolete methodology and conventional tools are over long back.

UPGRADED TECHNOLOGY

As the technology is developing rapidly, well known principles and processes are constantly being stretched beyond their set boundaries by the more sophisticated tools and materials. Metallurgy with its advances takes a place of pride in the stretching of frontiers. The advanced technology also brings many challenges to the plant designer, such as more aggressive conditions of temperature, pressure and corrosion, but the same technology enables him to meet these challenges. However, the designer lives with the plant for only a short while, whereas the entire burden of meeting the challenges of upgraded technology, consistently, and over a long period of time, falls upon the shoulders of the plant managers and the plant engineers. Today, plants are bigger, more complex, automated and capable of sustained effort towards continuous outputs. To maintain them at high production levels, a whole range of services have to be provided.

These *services* have to be understood and mastered by the plant engineers of today, so that they can put them to use whenever and wherever required. At this stage we shall take a quick glance at the nature of the services that the maintenance man has to harness and use. However, considering the importance of these hi-tech tools in the sphere of maintenance work, it will be necessary at a later stage to deal with them at length.

GOOD MAINTENANCE SERVICES

The services that will be needed by the higher level maintenance workforce are the following:

(a) Non-destructive testing (NDT) facilities enabling the surveillance of plant.
(b) Data bank of information on plant problems and their solutions, preferably in a computerised form.
(c) Anti-corrosion treatment facilities.
(d) Welding and crack detection facilities, X-ray, dye-penetrant radiography, ultrasonics, magnetic particle testing, scanning electron microscope, interfero-metric halography, eddy current testing and so on.
(e) Rotating machinery (e.g. centrifugal compressors, turbines and bearings need special apparatus for dynamic balancing, aligning, vibration monitoring, and high quality bearing setting services and so on).
(f) Calibration facilities.
(g) Proper documentation, preferably computerised.
(h) Supporting services.

They would also require the following:

(a) Technical literature which includes—maintenance schedules, maintenance instructions, operating manuals, fault analysis charts/ diagrams, drawings, and specifications.

(b) An organisation to provide genuine high quality spare parts.
(c) An organisation to provide special tools, tackles and equipment.

Before going on to the basic problems that the Maintenance Manager faces, let us see what maintenance means,

Maintenance can be defined as those activities required to keep a facility in as-built condition, so that it continues to have its original productive capacity.

The responsibility of the maintenance function is to ensure that production plant and equipment is available for productive use at minimum cost, for the scheduled hours, operating at agreed standards with minimum waste.

The objective is the systematic and scientific upkeep of equipment for prolonging its life, assuring instant operational readiness and optimal availability for production at all times while making sure that the safety of man and machine is at no time jeopardised.

It seems very difficult to achieve the objective, and once we examine the problems the Plant Maintenance Manager faces, we will be able to assess the magnitude of his task.

PROBLEMS OF THE PLANT MANAGER

In providing the services, the Plant Manager faces a number of problems which are described below:

(a) The job is not glamorous. Since maintenance is seen as a means towards achieving an end, the person in-charge normally receives more blame than rewards. This very often demotivates the hard working maintenance personnel.

(b) The burden of maintaining a sub-optimum (in terms of maintenance) plant or machine falls on the shoulders of the Maintenance Department. Usually no concessions are made for this sub-optimal plant, even though they may not even have been consulted at the time of purchase, regarding its maintainability.

(c) The maintenance personnel become the prime targets of others in thrusting responsibility.

(d) They have to deal with poor and sub-standard spares that cause quicker breakdowns, and create more work.

(e) Production being the immediate end-goal, machines are worked beyond endurance, thereby making the job of maintenance personnel much more difficult and time consuming.

(f) The status of the maintenance manager rarely, if ever, is commensurate with the extent of his responsibility and accountability.

(g) Training of the maintenance force does not keep pace with the rapid changes and advances that are constantly taking place, partly due to the short-sightedness of top-management and partly due to apathy on the part of the maintenance personnel themselves.

Evidently, all the trouble that the plant manager is confronted with, stems out of these basic problems. However, it would not be out of place to elaborate and illustrate how some of these problems grow and become complicated.

SITUATIONAL FAILURES

Often, in the process industries there is a need for a plant to be available continuously. The lack of such availability is one of the major reasons for lower capacity utilisation in many areas of industry, services and utilities. And yet, we shall find that this is so despite the fact that costly manpower is not being discouraged by most of these organisations; for, large amounts are being disbursed by them regularly towards overtime to their personnel. Moreover, in most of these organisations, a backup of a comprehensive array of systems, manuals and handbooks have also been provided to serve as guidelines to their workforce. But the end results that they are left with is far from satisfactory, for the problem of breakdowns continues to exist, causing both loss and embarassment. The reasons could be many; among them one could cite the erratic utilisation of maintenance manpower, and the non-productive cycle, which is set in motion by breakdowns, leading to cost escalation due to enhanced downtime. This in turn invites adverse comments and criticism. The end result is demotivation which leads to inefficiency and further breakdowns.

CHANGE IN ENVIRONMENT

The environment for maintenance is changing rapidly and, therefore, there is a need to constantly absorb and upgrade technology if an organisation is to remain viable in the ever changing climate of today. In these changed circumstances, more indepth performance appraisal of maintenance will be needed, with each passing day, both internally and externally. "Internally" means to analyse whether or not the maintenance personnel are able to cope with the *roller-coaster* ride down this tortuous path of relentless change. (Any shortfall on their part delegates the burden of high cost of down time on the organisation as a whole.)

The external threat comes from the ever growing service sector which is becoming more and more easily available for 'contracted maintenance', and which is ready to take over at the very first sight of a stumble. Once contracting out begins to take hold of the maintenance men, soon they find themselves in an unenviable position. The contract maintenance keeps making deeper and deeper inroads into their domain of activities, leaving them more and more denuded each day of power, prestige and control.

FACTORS INVOLVED IN FUNCTIONING OF MAINTENANCE

(a) A mixed variety of plant, equipment, and machinery belonging to different countries of origin which have been imported and installed within a single factory;
(b) Ever changing technology and its impact on maintenance;
(c) Ageing plant in Indian industry;
(d) High replacement costs, for new acquisition as well as for existing older machines;
(e) Availability of external maintenance support facilities;
(f) Internal facilities relating to skill availability and its upgradation; technical documentation and its analysis; repairs and test facilities; facilities for calibration of equipment etc., and their high cost;
(g) User skills, and operating conditions, and its effect on the life of equipment and its performance.

The pulls and pressures of these factors which can impede the efficient functioning of maintenance can be well imagined.

DESIGN OF MAINTENANCE SYSTEMS: THE CHALLENGE

By and large, our maintenance system designs are as per original equipment manufacturer's (OEM) recommendations. The machinery needs, our own needs as per equipment criticality, the usage condition, the severity of utilisation, age of the equipment and the maintenance facilities and infrastructure, appear only to marginally affect our system designs. To remain effective and to achieve organisational productivity objectives, the maintenance systems will have to become more dynamic and responsive to the changing demands. Per force, we have to develop a *need-based cost-effective* approach to remain competitive.

To remain viable, organisations will have to take on the responsibility of the quality of the products they produce and the price at which they sell the same. To achieve these ends, the maintenance work being carried out has to be of the highest order. The quality of maintenance in an organisation can be easily gauged from the ability of its maintenance staff to reduce and prevent frequent and recurring breakdowns, the quantum of its rejects, requirement of rework, and of course, the overall quality of the service they provide. In most instances, maintenance organisations do not measure up to their tasks. If the objectives of an organisation have to be optimally met, then the management has to give serious thought towards improving the quality of maintenance performance.

The change-over to high technology will bring about an increase in importance to maintenance. More and higher skill levels will be needed for carrying it out and, therefore, maintenance skills will be at a premium. Both the organisation and the maintenance managers will have to prepare for this shift and the chief executives will have to devote more time and attention to maintenance functions. The maintenance/user cooperation and interdependence will assume greater importance for operational effectiveness and efficiency. The changes on the technological front pose tremendous challenges for maintenance personnel, more so than in any other sphere of activity. Maintenance function has a meaningful future—and the people have to prepare now for it.

AUTOMATION AND MAINTENANCE

As plants become more mechanised or automated, maintenance management have to become more competent and have to be better equipped with necessary resources of manpower, equipment and commensurate skill levels. In the initial stages of mechanisation, the organisation will be faced with a sharp rise in maintenance costs and problems. As maintenance personnel gain experience with the equipment, by being involved in its installation and debugging process, the costs and problems will gradually reduce. In India, we are at a stage of increasing *mechanisation* in all industrial sectors, and hence, there is bound to be a considerable degree of concern about the initial increases in maintenance costs, Thus, it is expected that this concern will usher in an era of the required organisational improvements. In large organisations, it would be worthwhile training small teams to specialise in maintenance of selected critical plant or equipment. By and large, more emphasis should be laid on the training of multi-skilled workers, who can identify the fault and rectify combined systems having mechanical, hydraulic and electronic components. Single-skill workers are no longer enough for a maintenance workforce.

It should also be noted that the amount of maintenance required is not directly proportional to the degree of automation in plants. It will depend on whether or not the new equipment is well tried and proven, whether it has been recently developed, and on the type of equipment and the degree of its complexity, precision, and sensitivity, and the extent to which it needs a controlled operating condition.

It is an accepted fact that there are extreme variations between companies, but whatever be the situation, die level of maintenance effectiveness in most instances in India is much lower than what it ought to be and can really be. However, it is well within the powers of the top management to enhance the effectiveness of maintenance function, provided it so desires. Unfortunately, many members in the top management team of companies are not quite aware of the real costs of maintenance. The present costs and the long-run costs, that equipment, machinery or facilities generate over a longer period of time, have to be understood fully and in their true perspective if a clearer picture is to emerge. For only with cognizance will they be able to cater to, and plan for maintenance correctly.

FACTORS AFFECTING MAINTENANCE EFFICIENCY

It is rather strange that more often than not the maintenance man is left holding the wrong end of the stick. Even where his diligence cannot be faulted, he seems to be perpetually facing a 'heads you win, tails I lose' situation. It would be well worth examining what goes into making such a demoralising situation. Some of the conspicuous factors that have a bearing on his effort and affect his efficiency are as follows:

(a) Inadequate research on reliability and maintainability and even where researched, their application not being made during plant design;
(b) Procurement on lowest lender basis, without regard to life-cycle maintenance costs;
(c) Procurement of equipment lacking proper specifications;
(d) Procurement of equipment without enforcement of adequate product support;
(e) Failure on the part of top management to appreciate the importance of maintenance;
(f) Failure to plan maintenance properly and to budget it correctly;
(g) Inadequate training for maintenance staff beyond the levels of basic skills;
(h) Insufficient general management training for Maintenance Managers.

There are a large number of factors that magnify the effects of the above mentioned items. These are now enumerated below:

(a) Better performances in other manufacturing departments produce unfavourable comparisons which at times could be unfair. For example, the economies in production, labour force, brought in with advancement are greater than in the maintenance sector; yet this is overlooked and conclusions are arrived at.
(b) Continual introduction of more sophisticated and systemised equipment without keeping in mind the fact that its maintenance will

need better trained, more skilled and experienced workers with more capable managers at their helm.
(c) Maintenance budgets seem to fall under the hatchet in the hands of top management in search of short-term economic considerations and gains.
(d) Maintenance by and large tends to be a labour intensive chore and, therefore, the cost of maintenance escalates rapidly along with the ever escalating labour costs.
(e) The rising costs of spare parts and materials unfairly get reflected as costs emanating from the maintenance department.
(f) A tendency to introduce built-in obsolescence by the plant designers.
(g) A tendency to introduce non-repairable sub-assemblies by industrially advanced nations.
(h) Increased output levels and, therefore, high down time costs on automated plants.
(i) Increasing ratio imbalance between maintainer and operator in highly automated plants.

MAINTENANCE PLANNING

Having taken a look at the factors that affect maintenance efficiency, it becomes obvious that a good deal of thought will have to be given to find solutions to these problems. However, prevention being better than cure, preventive measures will have to be sought and found. Therefore, it is essential to focus on maintenance planning, when a new project is being planned and formulated. The reasons for introducing a maintenance planning system and for including it in the preparation stage of die project are the following:

(a) protect the investment in machinery, plant and buildings through regular and adequate maintenance to ensure their long life;
(b) safeguard the return on investment by maximising plant utilisation with minimum down time;
(c) prevent waste of spares, tools and materials;
(d) control and direct the maintenance labour force;
(e) maximise the utilisation of labour and other resources;
(f) ensure adequate technical information for maintenance;
(g) record expenditure and estimate the costs of the work;
(h) control maintenance costs and establish records;
(i) assist in future budgeting;
(j) establish a safely system and
(k) to evaluate plant performance as a guide to future forecast.

Thinking ahead in any context is bound to show good results, but in a field, such as maintenance with its myriad problems and challenges, its value is tremendous. We shall now discuss some of the benefits accruing from maintenance planning.

BENEFITS OF MAINTENANCE PLANNING

(i) *Reduction in down time:* Even in the most unsophisticated kind of installations, a reduction in down time is brought about by the introduction of maintenance planning. This reduction becomes still more significant as the plant in question becomes more and more complex and sophisticated, where labour and material are both expensive and down time is the costliest factor of all.

Some major factors which bring about the reduction of down time are indicated below:

(a) Introduction of maintenance planning
(b) Rapid fault-finding systems and the use of diagnostic charts
(c) Performance monitoring on critical plant items, but installing monitoring facilities to obtain warning of failures
(d) Having efficient support organisations ensuring availability of genuine spares, tools and test equipment
(e) Rationalisation of plant and spares through standardization.

(ii) *Staff development:* This is achieved through the introduction of planned overhaul and maintenance, and by providing improved technical information systems. These reduce worker frustration and encourage job satisfaction which improves the working life of the personnel. Ultimately, all these result in higher productivity.

(iii) *Resource utilisation:* If overhaul is planned properly with the help of the 'critical path method' (CPM), then it ought to show the way to plan for all the necessary resources correctly, for men, materials, tools and tackles, accurately. Hence, planning must be done before taking up the job in hand. Moreover, it ought to also help assess accurately the amount of time required to complete the overhaul, and consequently, its return to efficient service. This time assessment can be progressively improved upon with the passage of each subsequent overhaul. Thus, it is evident that by resource smoothing and levelling of men and critical inputs, all the necessary inputs required for performing the task will be better utilised.

(iv) *Machine life:* This too will increase when regular planned maintenance is introduced. However, it is difficult to quantify the extent to which maintenance will lengthen the life-span of a machine, for that is dependent on not only maintenance but on the manner of use.

(v) *Cost of improvising:* Excessive down time is caused by the need to improvise because of the non-availability of spare parts and special tools. If the technicians are not able to do with what they have, the down time of course would have been still higher. With proper planning and a sharper perception on the part of the maintenance manager, not only could the down time have been reduced, the inventiveness and creativity that surfaces while improvising could also have been harnessed and put to still better use.

In whichever way one may look at the problems and challenges of maintenance management, one will find that the answers to these are to be found in proper maintenance organisation and planning.

Chapter 2

Maintenance Objectives and Functions

Maintenance is an integral part of an organisation in its entirety and, therefore, maintenance objectives should be established within the framework of the whole, so that overall organisational or corporate objectives and needs are adequately met.

MAINTENANCE OBJECTIVES

The maintenance objectives are as follows:

(a) To ensure maximum availability of plant, equipment and machinery for productive utilisation through planned maintenance in an excellent manner.
(b) To maintain plant equipment and facilities at an economic level of repairs at all times, to conserve these and increase their life-span.
(c) To provide the desired services to operating departments at optimum levels, through improved maintenance efficiency.
(d) To provide management with information on the cost and effectiveness of maintenance.
(e) To ensure the working capital at appropriate levels.
(f) To achieve all the above objectives as economically as possible.

The economic factor has to be kept in mind by all the contributing departments of an organisation because, whatever be the aims and objectives of that organisation, it cannot sustain for long without earning profit. It can be safely said that the primary objective of an enterprise is *profit*. As a corollary, the primary objective of the maintenance department is to extend help towards achieving this goal by creating capabilities within the enterprise to earn profit. Since the ultimate objective is profit, the production infrastructure and facilities have to be maintained at as minimum a cost as possible with maximum efficiency and operational availability.

The maximisation effort normally includes the following:

(i) *Preventive and planned maintenance.* This entails timely repairs, rectifications and adjustments, time-bound overhaul, inspection and lubrication of equipment. Such meticulous care is absolutely essential because defective equipment causes down time, which results in declining of output and production quality. Hence, down time reduction becomes an important area of concern.

(ii) *The health and safety of both workmen and machines.* These have to be carefully looked after. The orderliness and cleanliness of the work place has to be assured to improve total productivity and the conditions of the labour force, and to avoid accidents.

(iii) *Planning and scheduling.* These must be based on realistic work content, Similarly the time, labour and repair cost estimate should be made. These have to be time-bound because lapses in time-estimates will upset the schedules, priorities, and costing, and will totally retard the process of coordination and cooperation with other departments, particularly with production.

(iv) *Controls.* These have to be established so that checking can be done, i.e. whether or not plans are being adhered to, and progress is being made towards the attainment of assigned objectives.

(v) *Flexibility.* This too has to be built into the system, for resilience is a must in a job such as maintenance, where changes can be expected at any time due to sudden needs and rescheduling, in either the production or the maintenance department.

(vi) *Documentation.* This should be done with care in respect of all maintenance effort so that it can lend itself for analysis whenever required.

MAINTENANCE COSTS

It should be clear from the foregoing discussion that every effort of the maintenance department or section has to be subjected to strict monitoring and control, towards the achievement of its objectives,

The main objective of a properly run maintenance department is to have plant, equipment, and machinery available for productive utilisation during the scheduled hours, operating to agreed standards with minimum waste and minimum total cost. *The total cost* is the sum of maintenance labour costs and material cost, plus cost of loss in production. Exhibit 2.1 shows how the *lowest total cost* can be achieved. Its explanation is given below:

(a) If *maintenance cost* (M-curve) is at zero, then it indicates that no maintenance is being carried out at that point, and the cost of production loss (see P-curve) is at the *highest* or at its peak.

(b) As maintenance effort is gradually being introduced and increased (see M-curve), the production loss (see P-curve) slowly decreases.

(c) Following the total cost curve (T-curve) pattern, we notice that the effect mentioned in point (b) holds true till we reach the minimum combined cost level at point A (see T-curve). Thereafter, any additional maintenance effort being applied *increases* cost.

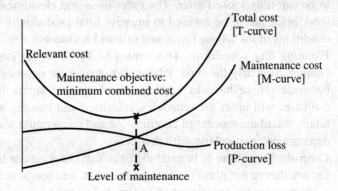

EXHIBIT 2.1 Down time control.

DOWN TIME COSTS

This clearly shows that 'maintenance optimising technique' indicates that point A on the T-curve is the *objective* for the maintenance to achieve; Why? It is because at that level we get the minimal combined cost. This observation is also equally true for the *service sector organisations*. Take, for example, the case of a material department where its cost is the sum of its own labour and material cost, plus the losses caused by inadequacy of its service, like stockout cost.

The relationship between down time and cost incurred in production loss can be easily understood. To understand this in the correct perspective, a few points must be made:

(a) Down time dots not always imply/cause a direct production loss. There are certain exceptions and conditions under which down time does not create a production loss. These include:

(i) The plant does not operate at full capacity utilisation level.
(ii) The finished goods inventory, after completion of production process, is not lifted for delivery to customers but is stored.

(b) Down time may be due to a number of reasons other than breakdown, such as the following:
   (i) Shortage or absence of operator
   (ii) Lack of proper tools, jigs, or fixtures
   (iii) Non-availability of specific raw material
   (iv) Improper planning, scheduling or machine loading
   (v) Power breakdown or load shedding
   (vi) Non-availability of calibrated test equipment.

(c) Down time should, therefore, be honestly recorded and charged to relevant and responsible departments, such as maintenance, production, material, inspection or tool room. Bar charts on down time of the machine should be analysed to know the reason for down time and increase efficiency.

However, it has to be understood that whatever may be the cause, if the machine remains idle and non-operational, it will invariably result in lowering of productivity. Furthermore, one of the major recognisable causes for lower productivity is high down time. Hence, every effort has to be made by the maintenance manager and others towards reducing, managing and controlling down time.

To lower and control down time, a *two-pronged attack* strategy is to be formulated. It should be aimed at making every possible effort jointly, both by the operator and the maintainer, to achieve effective control on down time.

RESPONSIBILITIES OF OPERATORS

Those who operate the machine should be made responsible for certain basic and simple tasks for maintaining it. They must take good care of the machine so that it gives him the maximum output. Some of the areas which require attention and action by the operator are described below:

(a) Carry out a visual check of the machine for cleanliness. Remove all metal scraps from machine and dispose them of safely.
(b) Check if safety devices are present and functional. If not, do not operate the machine.
(c) Operate the machine and check for any unusual noise, vibration, friction and obstruction.
(d) Check if lubrication cups are replenished to correct levels and grease nipples so that they are not dry; then operate the machine.

(e) Never keep any object particularly heavy objects (e.g. hammer, tools, jigs, fixtures, or materials like rods and bars) on the machine while working.
(f) Never operate the machine beyond laid down conditions of *speed* and *feed*.
(g) Always note down the number of tools brought by the operator and ensure that they are all returned, i.e. make sure that no tools are dropped inside the machine during the day's operation, which can cause a serious breakdown.
(h) Never fail to report a mistake committed by the operator due to his carelessness.
(i) Supervisors must *not* punish any operator once he admits that he has committed a mistake.

Simple *do's* and *dont's,* if followed as discussed above, can eliminate a large number of breakdowns. However, this requires concerted effort on the part of the management to educate and train the production staff and to give such effort as wide a publicity as possible.

RESPONSIBILITIES OF MAINTENANCE STAFF

The second part of the two-pronged attack is to be carried out by the maintenance staff. This can be planned on the following lines:

(a) Categorise machines according to the derived *criticality* criteria, and write out maintenance instructions and prepare schedules in respect of this criticality evaluation.
(b) Perform planned and predictive maintenance as per this schedule.
(c) Record all breakdowns with care and absolute accuracy, so that the parts that have failed may be segregated and analysed to find out the reasons for the failure. Once the cause has been diagnosed, corrective actions are to be planned and taken.
(d) Analyse repetitive failures to avoid costly disruptions time and again.
(e) Search for solutions and take (the following corrective actions:
  (i) Replace incorrectly selected items, and standardise.
  (ii) Redesign equipment, if required.
  (iii) Replace the faulty items or parts with better specification items.
  (iv) Identify and correct user mishandling and abusing.
  (v) Correct, where necessary, and improve all-round maintenance effort.
  (vi) Revise the maintenance instructions, keeping these factors in mind.

Down time management, when first introduced, may produce results contrary to expectations; it may even receive a setback, or the improvement may be insignificant or barely perceptible. There may even be occasions when the improvements are just temporary. Such results are bound to be discouraging.

However, there is always hope, and the Maintenance Manager in search of improvement must continue to follow this methodology conscientiously even when the initial feedback is far from encouraging.

The aim of any organisation is to produce quality goods acceptable to customers, at reasonable price, and to provide a margin of profit. As co-partners in such an effort, production and maintenance personnel share responsibilities, to run their enterprise on sound engineering practices and cost-effective principles.

MAINTENANCE A–Z FUNCTIONS

The specific functions of each sub-section of maintenance must be accurately assessed and spelt out so that the supervisory staff can achieve its own departmental objective. The basic maintenance functions involve the following steps:

(a) Develop maintenance policies, procedures and standards for company-wide incorporation.
(b) Design practicable and implementable schedules of all maintenance work, and spell out maintenance work specification or master process sheets.
(c) Ensure the availability of production plant and equipment to carry out planned/preventive maintenance.
(d) Carry out repairs and rectify or overhaul production equipment to ensure good operational status and availability.
(e) Ensure scheduled inspection and lubrication of machinery.
(f) Carry out calibration as per the calibration plan.
(g) Maintain and carry out repairs of buildings, utilities and allied equipment.
(h) Ensure and carry out faithful recording and documentation of all maintenance work.
(i) Periodic inspection of all assets to know conditions leading to stoppage of production.
(j) Review all recorded documents and specifications for procurement of new equipment, and be a member of facilities investment decision-making body, to ensure maintenance requirements are looked into and taken care of.
(k) Standardise equipment for replacement and purchase.

(l) Carry out frequent analysis of pertinent documents so that corrective actions can be taken.
(m) Initiate procurement action necessary for maintenance work.
(n) Prepare spare parts and material requirement lists; revise/scrutinise/review list of spare parts required, and monitor procurement of spare parts and material for maintenance.
(o) Ensure proper inventory control on the spare parts and their replacement.
(p) Initiate and carry out energy conservation programmes.
(q) Design and enforce safety standards as required.
(r) Ensure segregation of, and collection of combustible waste material, such as oil and cotton waste dipped in oil.
(s) Recruit and train personnel to carry out maintenance work, and provide adequate replacements for skilled personnel, who may have retired or moved away; update older workmen about newer skills.
(t) Plan and prepare maintenance budget.
(u) Ensure budgetary expenditures are evenly distributed and kept within planned budget.
(v) Develop and apply cost and budgetary controls.
(w) Continuously apply cost reduction techniques.
(x) Ensure cost effective maintenance.
(y) Develop and provide proper management information system to the management, particular attention to be paid for the top management.
(z) Upgrade management skills of supervisory and executive codes.

Every single person working in the maintenance department must know his or her exact role and responsibilities. At the same time he will invariably have to work together with his colleagues as a part of a team. It may be kept in mind that the objective of the maintenance department is to let it become the prime objective of each member of its team. This philosophy has to be inculcated as a habit in each of its members.

Advancement in technology necessitates the development of newer skills which tend to progressively require a higher level of education, understanding and training. Because of this trend, in future the ratio of maintenance engineers to workers will go on increasing.

Some of the new skills needed are:

(a) Maintenance of electronic equipment
(b) Maintenance of card, tape or disc controlled operations
(c) Maintenance of nuclear powered equipment or other similar hi-tech equipment, requiring specialised knowledge, experience, skill and training
(d) Maintenance of cell phones, space equipments and military hardware.

Hence, technical training will assume more and more significance as time progresses. Therefore, any organisation that wishes to keep pace with the times has to take training with all seriousness.

Nothing fails without symptoms, even the heart! Recognition of such failure—models, symptoms, cause of such failures, and severity of such failures and detectability, ability to decode the risk assessment of preventing abrupt failures—is important to predict and devise maintenance systems.

Chapter 3

Maintenance Organisation

To fulfil its role effectively, in the case of maintenance, just as in all other industrial activities, it is essential to have a balanced, rationalised, and healthy organisation to cater to and control its manifold activities. While developing an organisation to fit the maintenance needs of a company or industry, one must keep in mind the fact that there is no 'ideal organisation' as such, since no two plants, or no two concerns are exactly the same. Therefore, while assessing the basic parameters of the ingredients required for creating an organisation to carry out the maintenance functions efficiently, one has to remember the basic fact that modifications have to be worked out within the plant depending upon, say, specific plant operations, local conditions, the constraints and requirements prevailing, the state and level of technology, and so on.

MAINTENANCE ENGINEERING

Maintenance engineering and facilities management can be defined as the *effort* made to provide a service so that the physical attributes of a plant (e.g. equipment, machinery, building) and utilities can continue to operate on an optimum level.

In industry, the emphasis is on process plant, equipment and machinery; hence *plant engineering* represents the main function. On the other hand, in institutions such as colleges, schools, universities, hospitals, and municipalities the emphasis changes and is on classrooms, wards, operation theatres, laboratories, play grounds, gymnasia, swimming pools, and so on. The term used in this context is *facilities management.* The basic difference, as one can see, lies on the emphasis.

The terms *effort* and *utilities* and their functions need to be understood clearly, while dealing with the subject of maintenance organisation.

The term *effort* includes the following engineering functions:

(a) Designing
(b) Fabrication
(c) Erection
(d) Construction
(e) Plant ulilisation (which is aimed at production)
(f) Maintenance.

Designing is the function where new process or procedure or developments are incorporated and gives detailed account of complete fabrication drawings, with manufacturing tolerances and specifications so that facilities can be created for production.

The *fabrication, erection* and *construction* managers/engineers use these drawings and translate the designers vision on paper into physical realities. These three activities have to be meshed into one another in order to create a harmonious whole, that is, the factory, the workshop, the plant, the dams, the power stations, the bridges and whatever else, where production has to take place. However, that which sets apart the four functions of effort from production and maintenance, is the fact that having once fulfilled their particular tasks, each one of them bows out by turn, carrying with it, whatever praise or reward it might have received.

ORGANISATIONAL PREREQUISITES

The production man, making use of raw materials, manpower, machine utilities, sets the ball of manufacturing rolling. The end results of his efforts can be easily quantified as they are tangible and measurable. Thus, if he is taken to task for his failures, he is also given his share of praise.

The maintenance engineer stands like a rock of Gibralter behind the success or failure of the production department. He along with his team is not only concerned with the day-to-day problems of keeping the equipment and physical facilities in a state of good repair, and good operating conditions, but is also the caretaker of the life-line of any production unit, namely, its *utilities*. Utilities can be defined as that part of plant services, which provides and controls essentials such as steam, air, water gas, electricity, liquid nitrogen, and oxygen. The role of the maintenance man consists in doing a good job, and unless he is supported and encouraged by an enlightened and very perceptive leadership, he comes into the picture only when he shares, rather undeservedly/unequally, the failures of the Production Department.

While creating an organisational infrastructure, certain parameters have to be considered and kept in mind. These are discussed below in detail:

ESPRIT DE CORPS—TEAM SPIRIT

It is the *people* who make up an organisation. In order to organise a carefully selected group of people, one has to formally or informally bring them together to work closely and cohesively as a team. This can be achieved with the help of common guidelines like organisational charts, job specification/ descriptions, and clearly delineated lines of authority. Although, these are at best path-finders towards developing the right attitude towards teamwork, that which unites these people together is an abstract yet infinitely powerful force which is called *esprit de corps* or *team spirit*. This brings us to the other important factor in establishing and running of an organisation, namely, the *leadership* of the group or the team. For it is good leadership alone that can bring a group of people together and instill it with the necessary team spirit that will lead on to teamwork. Thus, the key to success for the maintenance workforce, just as in any other sphere, rests on the shoulders of an enlightened leadership. Treating trained human beings as valuable human assets and building up human workforce for relevant jobs is the leader's main job.

The plant engineer has to be a sincere, dedicated, and a professional manager. He should be technically competent, knowledgeable and cost-conscious in order to be an effective and efficient leader of the team, for he has to keep things going against all odds. As his efficiency is essentially measured by the extent of absence of down time, he has to plug all loopholes which lead lo wastage of time, talent and effort. This can be done by adopting a philosophy of better and improved planning and scheduling; a healthy and balanced programme of supervision; application of proper techniques; motivation of personnel; timely resolution of conflicts; and above all, providing the group with inspiring leadership. He has to be able to speedily and effectively solve both technical and social problems that come up within die sphere of his command.

The maintenance or plant engineering department in most industries does not have any management philosophy. Many feel that trained engineers do not need this type of direction and that it does not apply to their field of activity. However, this reflects a very narrow viewpoint. Every Plant Engineering or Maintenance Engineering Department must have a philosophy and policies that go with it. These must be reflected honestly and sincerely by deeds and actions of each and every person in the department.

Once a philosophy has been enunciated, the 'policies' of the organisation should emanate from the same. These policies should be easily understandable as guidelines to the people who fall within its framework.

These policies can be either formal or informal. The formal policies are written down and need reviewing and updating from time to time and can be used for much wider circulation and dissemination. It is the responsibility of the Plant Manager or Maintenance Manager to formulate these policies. They may include company rules bearing upon matters such as hours of work, attendance, leave, sickness and employee benefits, safety, security, overtime, the norms for forming associations and unions, grievance handling, suggestion schemes, transfer, promotion, and wage and salary administration, continuing education, maintenance of buildings, plant, machinery, facilities etc. The policy statement, needless to say, must fit the organisation.

SPAN OF CONTROL

While discussing the *span of control,* a large number of factors have to be taken into consideration. Broadly speaking, these factors should include the following:

(a) There should be a reasonable numerical proportion between the 'supervisor' and the 'supervised', for there is a limit to the number of people one can supervise effectively, just as there is a limit to the amount of monitoring a small group of people can tolerate.
(b) Nevertheless, given the nature of work and the levels of the people being supervised, a balance between the ratio of the number of people being supervised and supervisor is essential.
(c) If the ratio is higher than required, it may result in wasted employee effort and lack of monitoring effectiveness. On the other hand, if the ratio is less than the optimum, it may lead to wastage of supervising time and talent.

Though there is no hard and fast rule by which one can ascertain the amount of supervision a particular group needs, yet there are a number of very useful guidelines to assist one while taking a decision, which we shall now discuss.

SUBORDINATES DEVELOPMENT

The development and growth of subordinate staff is largely dependent on the personal skills of the superiors, and in particular on the skills of the *supervisor*.

The raw material, the implement, or the finished product that the supervisor works with and creates, is the *man* or the *men* he oversees.

Man is an egotistical, sensitive, emotional, unpredictable and sentimental creature with his pride and prejudices. His in-built inhibitions and anxieties, his likes and dislikes cannot, therefore, be handled like an inanimate bit of machinery, if the best is to be obtained from him. But, without any doubt, he is the most important cog in the wheel of the machinery of production and therefore needs to be handled with utmost sensitivity and care. This fact should be specially kept in mind in a low-profile job such as maintenance.

However, even while keeping in mind the fact that each individual is an entirely different entity we can still draw certain generalisations regarding both the degree of supervision required and the fulfilment of the need for developing one's subordinates. But, before we can do so we must keep in mind the following facts about the subordinate/the individual being supervised:

(a) The level of education
(b) The level of skill
(c) The level of intelligence
(d) The level of motivation.

Along with the individual, the nature and level of his job has also to be kept in mind. That is, whether it is (a) unskilled, (b) semi-skilled, (c) highly skilled, (d) interesting or repetitive and boring, (e) simple or complex and multifaceted, and (f) in need of periodic checks and controls.

In broader terms, one can state that the less educated and the less skilled the man is, the more supervision he will need, for largely supervision in tins case will amount to keeping the man at his job. Under these conditions, though constant supervision may be necessary, yet, because of the nature of the work, larger number of people may be supervised by a single individual. The more evolved, educated and skilled a person is, the more he will resent being watched over. And in such circumstances, the supervisor's own knowledge and skill will automatically come under close scrutiny. So, if he wants to get honest work done, he must ensure that he has that slight edge over the men he is watching over. Moreover, he must make sure that he does not demotivate the man by either being overinstructive or overbossy.

With the acquisition of education and of skills, a man tends to become touchy; thus the subtler the controls, the better.

When the more enlightened subordinates are largely left on their own, they tend to improve in their growth and development, whereas, in organisations with unduly close supervisory controls, subordinates do not have an opportunity to develop. And given the first opportunity, they may leave that organisation for another, where they can perceive the possibilities of growth.

Paradoxically, however subtle and distanced the supervising may be at these levels of expertise, it does not leave die supervisor the option to take on the supervision of workers in large numbers.

There are of course jobs that perforce need a good amount of supervision at regular steps. These, however, should not normally pose any problems because the subordinate is aware of the need for these checks. But in case a certain amount of sensitivity has been perceived on the part of the recipient, then the management can strengthen the hands of the supervisory staff by formalising and standardising these checks by giving clear-cut written instructions regarding the compulsory nature of these controls.

COMPETENCE/SKILLS UTILISATION

The organisation must be able to use to the best possible extent the unique capabilities of each of its individuals. A particular person may not perhaps be a good supervisor or a good manager yet he may well be an excellent engineer. The organisation must, therefore, be structured in such a way that it can recognise and nurture those traits which are of value in a person. Thus, the brilliant engineer who does not measure up to make either a good supervisor or a good manager, will be recognised by a good organisation and given his proper position, status, and monetary incentives, thereby conserving that particular quality or skill of his from dwindling away and getting wasted due to neglect. Such persons should be nurtured and developed so that they will not desert the organisation.

In large industrial cities like Bombay where there is a constant struggle going on for acquiring skilled manpower, demand being higher than supply, the price per force of the desired skills automatically gets hiked up. As such quite often an organisation cannot afford to employ people at the appropriate level of competence. Under these circumstances the organisation may have to restructure itself to handle the work with the available lower skill level of the workforce. But, when such conditions prevail, greater responsibility automatically develops on the shoulders of the supervisory level of the workforce, and a much closer monitoring of the work being done will have to be undertaken.

However, it is quite clear that while functioning within these constraints, no organisation can afford to compromise on the quality of its supervisory staff, if it wishes to maintain its position'.

Who has not heard that "too many cooks spoil the broth"? It is quite a normal tendency to have 'too many cooks' in industry as well, for one can very often see an assistant to an assistant to an assistant for a job which certainly does not need a triple chain of command. This obviously overdoing of things only results in the overlapping of duties, thereby causing confusion. This must be avoided by keeping a close watch on overlapping functions amongst the various levels of organisation for the better and healthier functioning of the organisation.

HIERARCHICAL LEVELS

Five fingers cannot be equal in any society and there are levels in any walk of life. Distorted loose definitions of *functions, responsibilities* and *authority* are major hindrances to team work. Each person must know the scope and limit of his position in clear terms. Hence, the clear enunciation of functions for each department, each section, and each person is necessary in order to avoid confusion and overlapping which leads to conflicts and resultant inefficiency.

The roles of *production* and *maintenance* in industry are as closely interlinked as that of the partners making up an alliance. So they share a complementary relationship and any disharmony prevailing between the two will undermine their progeny. Similarly, any lack of coordination between production and maintenance in an organisation can affect targets and the quality of products.

Both production and maintenance have their distinct roles to play, but normally it is the production that delivers the goods which is considered more important than maintenance.

Thus, in most instances, maintenance is a subordinate function to production and these results in major decisions being taken unilaterally which in turn results in the following:

(a) Acquisition of equipment and machinery without the consideration of maintenance problems.
(b) Production not releasing machine on time for maintenance, yet demanding them back after an unreasonably short passage of time, particularly during year end.
(c) The misuse of machines beyond their levels of endurance to meet production targets, thereby making the task of the maintenance team near impossible.

It is advisable, therefore, to have maintenance at the same level as the production head so that arbitrary decisions which lead to eventual industrial wastage are obviated.

CONFLICT MANAGEMENT

Conflicts are common between two fingers on the same hand, one claiming to be taller than other. Similarly, there are some peculiar problems/issues confronting the maintenance personnel. Far, apart from the raging 'caste war' that goes on perpetually between the Production and Maintenance Departments, there is a lot of discord between what we may term 'sub-castes' prevailing within the maintenance team itself, which is not beneficial.

(A) *Internal conflicts:* Internal friction such as conflicts arising between the departments say, between Electrical and Mechanical, or between Mechanical and Instrumentation, is not uncommon. At times there could be an undue degree of domination on the part of a particular section on other sub-groups, which could be deeply resented by them and this may result in disharmony and non-cooperation. One could easily visualise a situation where, for example, the electrical sub-group having opened up a sub-assembly has to wait endlessly for their mechanical counterpart to come and do their share of the work, or perhaps wait for the instrumentation sub-group to come and do their part in the desired sequential order. This shows poor coordination and inadequate integration 'within', which leads to frayed nerves, discord and meaningless delays. The Chief Mechanical Engineer must reconcile such conflicts in the interest of the firm.

(B) *External conflicts:* The repercussions of the conflicts which arise out of the inter-departmental friction and jealousies have far reaching detrimental effects. These can be examined as external conflict. For instance, where severe conflicts arise between the maintenance department and an array of opponents such as the production, materials, and finance departments or even the personnel department; a chronic inter-departmental conflict situation prevails because the entire onus of 'down time' is laid squarely upon the shoulders of the maintenance department. Critical comments are aired about the poor performance of equipment after the repairs, maintenance activities, etc., overlooking the other factors that contribute to the down time. In such a situation, the added criticism of the Line Manager (production) and of the top management brings about a breakdown of internal support or cohesion within the maintenance team and the sub-groups fall out and start exposing each other in order to save their own skin. Further, on such a situation, wrangling creates an all round lack of trust and each sub-group begins to feel demoralised. Being a service function, maintenance is subject to continuous criticism, with no avenues open for sharing in the limelight. The men who work in this sphere feel neglected and discriminated against, which lowers their morale, and hence their motivation. Thus, having become frustrated, they tend to have lower levels of initiative and drive and tend to stagnate. In the weekly/monthly coordinating meetings, each department points a finger at others, forgetting that other four fingers are pointing at you!

Maintenance executives being mostly on the receiving end, either become *introverts* or *aggressive*. The introvert among them avoids facing up to the problems that ail his work or his organisation and does not even attempt at finding solutions; nor does he try to reconcile with production. The aggressive type escalates the conflict further and generally forces issues instead of trying to resolve them. It is very obvious that both these attitudes and their fall-out is bad for the organisation as a whole. In the larger interests of the firm, like the hand controlling the fingers, the presiding officer must coordinate the conflicts and ensure healthy relationship.

FACTORS FOR EFFECTIVENESS OF MAINTENANCE

Apart from recognising the prerequisites of organising, one must also keep in mind several factors that will determine the effectiveness of a carefully structured maintenance organisation. To start with, one should have a clear view of the existing maintenance organisation, and its functioning both internally and in relation to the other departments with which it has to interact. At first, familiarising oneself with these factors may appear to be an easy task. But concretising such matters may not prove to be so simple. Keeping these basic factors in mind, the issues that need to be taken care of will emerge largely from the following set of questions. Finding the correct answers to these questions could be of immense help while structuring a new Organisation Design or modifying existing one.

(a) Is maintenance recognised as an important function? If so, how is its importance shown?
(b) Is it given the required authority which commensurates with its responsibilities, in terms of what is expected from it?
(c) Is it involved and consulted in the selection of equipment for replacement and new acquisition?
(d) Is it a part of the policy planning and formulation process of the company?
(e) Does it have the necessary specialisations required within the domain of the specialisation itself? And has it the requisite skill levels or the means and methods of attaining them? If so, have they been formally laid down?
(f) How good or healthy is the interpersonal relationship between the different specialisations, such as mechanical, electrical, instrumentation and so on within the body of maintenance?
(g) How good or bad is the level of morale and motivation of the maintenance executives and its workers?
(h) How is the interpersonal relationship between maintenance and other departments, such as maintenance vis-a-vis production, or maintenance against finance?
(i) What is the kind of industrial relations climate that exists in the entire organisation, and in particular within maintenance itself?

OBJECTIVES OF ORGANISATION DESIGN

What does one understand by maintenance organisation structure, and what is it supposed to do? It is obviously much more than just an organisational chart, which it is quite often understood to be. The connotation has a much wider scope, for basically it is a spectrum of 'interactions' and 'coordinations' that connect the process, the task and the people of the organisation to enable it to achieve its purpose or objective.

The structure has two major aims: the first objective aims at facilitating the flow of information within the organisation so as to help the decision-making process and to thereby reduce uncertainty. The design should be so made that it helps in the collection of desired information by managers.

In case the sources within the organisation are not adequately responsive in terms of *time* or *quantum* of information, then the structure of the organisation should be appropriate and flexible enough to help managers seeking alternative sources of information in doing so. Above all, it should be supportive and non-obstructive when engineers need to take recourse to new procedures, in order to overcome hurdles in the path of their completing assigned jobs. For example, the maintenance department may find itself in urgent need of some information regarding the supply of certain critically required spares, just before the annual shut-down or overhaul is due. If they were to wait for this information to come to them through normal 'material procurement' channels, then more likely than not, the information would reach them too late to be of any particular use. Under these circumstances, they would like to seek a direct answer from the regional spare parts sales office of the concerned vendor organisation without facing organisational road-blocks. As they have to react and act quickly, with ever changing needs, they cannot afford to wait for critical information to come to them through normal slow and meandering official channels.

The second objective of organisation design is to achieve *coordination and integration*. This is particularly so when departments are interdependent; organisation structure should help coordination and integrations. The state of *reciprocal interdependence* can easily be exemplified through the working relationship between "operations" and "maintenance" in die Indian Airlines. Reciprocity can clearly be seen at work here for the output of department A, i.e. operations, becomes the input of department B, i.e. maintenance, and the output of department B, becomes the input of department A.

In order to maintain the serviceable condition of the fleet of aircraft, there is a regular need for maintenance function to cater to the repair and overhaul needs. The aircraft under these circumstances is just like a sick child, needing nursing, care and medication, in order to restore it to good health.

After accomplishing the maintenance task the repaired and fully serviceable airplanes become the inputs to the operations to be assigned for utilisation on various commercial routes. In situations of *reciprocal interdependence,* a much more complex and comprehensive coordination of systems is required. The operations and maintenance departments in the airlines must communicate freely and frequently to know when the airplane will be coming to them so that dial they can plan and carry out their respective jobs effectively, efficiently and economically, without delay. This coordination can be achieved by having a continuous *feedback* system.

In the ultimate analysis, *organisation design* is the allocation of resources and manpower for a specific (maintenance) purpose, and the structuring of these resources in a correct manner in order to achieve this purpose. Basically, the organisation is designed to fit the environment and to provide the information and the coordination required in that milieu. While deciding on what kind of an organisational structure one should have and use, it is necessary to analyse, clearly and specifically, the special characteristics of the environment dial one is operating on, and in particular, recognising the demands that the environment will make on the organisation in terms of information and coordination for its successful functioning. Once the environment has been recognised and understood, then the designing of the organisation structure no longer remains a difficult task.

TYPES OF MAINTENANCE ORGANISATION

There are basically three types of maintenance organisations;

(a) Decentralised
(b) Centralised
(c) Partially decentralised.

If the plant/factory is large like a steel plant/petrochemical complex, and its units are located in far flung areas, making inter-unit communication difficult, then it would be advisable to have a *decentralised* organisation in such circumstances. That would imply having a separate maintenance set-up for each one of its units. The maintenance organisation in each such unit would function under the direct administrative control of its production chief. In a case like this it is advisable to have an interchangeable cadre for the production and the maintenance executives. The production chief could then be selected either from the production or the maintenance streams, depending upon the seniority and suitability amongst the available personnel in either streams of specialisation. This would ensure a uniform and all-round generation of motivation in both cadres.

If the factory is small, compact and accessible and the inter-unit or interdepartmental communication speedy and easy, it is then that a centralised maintenance organisation is recommended. This *centralised* maintenance organisation is placed under the Chief Maintenance Manager who is of the same rank or hierarchical level as the Production Manager and both of them report to the General Manager, who is responsible for the overall functioning of the factory. However, the production line managers in this system often feel that, since they have no control on maintenance personnel, their work would often tend to suffer. Obviously, under such circumstances, accountability has to be strictly maintained, and the responsibility for shortfall should be fixed by the management if healthy, time-bound operations are to take place.

The third type of organisation is known as a *partially decentralised* organisation. This is a modified form of the 'totally decentralised' organisation, which is suitable for large plants with far-flung units.

In this partially decentralised format, the day-to-day maintenance of equipment (line maintenance or first line maintenance) is carried out by a group of maintenance workers who are attached to, and are responsible to, the Production Manager of that unit. But important maintenance functions, like planning and scheduling of maintenance work, drawing up of schedules, master-process sheets, work specifications, documentation, maintenance costing, major overhauls, procurement of spares, are all kept directly under the Chief Maintenance Manager. Such an organisation adequately serves the needs of the Production Manager, while enabling a centralised maintenance policy to be adopted and followed at the macro level. While the maintenance crew attached to the units look after the day-to-day maintenance needs, as indicated by the Production Manager according to his *priority* and to his complete satisfaction the broader maintenance needs are decided by the Chief Maintenance Manager. Both the Chief Maintenance Manager and the Production Chief will have to be of the same level of authority in such an organisation. But, under these circumstances the maintenance personnel feel the stress of dual pressures—on the one hand from the line executives and on the other from their functional maintenance incharge.

It would, however, be judicious to keep in mind the fact that these three types of recommended maintenance organisations are not housed in water-tight compartments, and the 'fitting' organisation for maintenance in each case would eventually depend upon the individual needs of that particular plant or factory. The goal of maintenance professional is to "somehow" manage and keep the organisation interest above the conflicts so that, there will be a long term recognition of the maintenance profession at a par with other departmental heads.

It is regretting to note that many organisations, particularly in the infrastructure project construction sector operating with cranes and concrete mixers, maintenance is virtually absent with machine operators managing the show even today!

Contract maintenance—where the maintenance work is carried out by outsiders is slowly gaining acceptance in small–medium organisation to avoid labour problems and also to reduce the maintenance cost. This aspect will be dwelt in the next chapter.

Like the proverbial cave man/our great grand fathers, not consulting a doctor and managing somehow, there are, even today, a few organisations where maintenance is not even mentioned, but the operator samehow manages the equipments. This is close to breakdown maintenance system.

Chapter 4

Maintenance Systems

Any organisation which is involved in machinery, plant, equipment and facilities must have a clear-cut maintenance policy in order to ensure its well-being. However, each individual organisation, big or small, simple or complex, public or private, using highly advanced or simple technology, must choose the maintenance system which best meets its individual needs in implementing the policies. Therefore, it becomes necessary for the policy makers to get acquainted with various kinds of maintenance systems that need to be evaluated before a judicious choice can be made.

CLASSIFICATION OF MAINTENANCE SYSTEMS

Usually the maintenance systems can be classified under the following heads:

(a) Breakdown
(b) Routine
(c) Planned
(d) Preventive
(e) Predictive
(f) Corrective
(g) Design Out Maintenance (DOM)
(h) Total Productive Maintenance
(i) Contracted Out Maintenance
(j) Cannibalisation

BREAKDOWN MAINTENANCE

Breakdown maintenance is also at times referred to as *repair maintenance*. But, for all practical purposes, this is not a system at all. The basic concept behind breakdown maintenance is not to do anything until and unless the machine ceases to function. Hence, no servicing is carried out excepting for a little bit of cleaning and lubrication, which is done by the operator himself. The only attention the machine receives is at the time of failure. In most cases in a set-up such as this, there is no maintenance man available and there are no spares kept even for immediate foreseeable needs, nor are there any maintenance manuals or handbooks kept at hand to be referred to by the personnel. On a superficial analysis, this may appear to be an economical system, because in such instances management does not take into account how expensive this little bit of economising can prove to be. However, if it were to compute the cost of running the plant and of time lost due to breakdowns, and the cost involved in repairs, then the management would certainly opt for a genuine maintenance system geared to their needs.

Such a system, if at all it can be called one, can continue for some time, but unfortunately once breakdowns begin to take place, they have a tendency to recur at an ever increasing frequency. Hence, once the cycle of breakdowns starts, they are prone to occur over and over again at an alarming frequency. Thus, every breakdown begins to have a sense of emergency. When the plant breakdown takes place, everybody gets involved in somehow patching up process, money flows, and of course while the defective machine lies idle, no production takes place. To tide over the immediate crisis, ad hoc repairs are somehow done but at a tremendous cost. Idle capacity is utilised with a sigh of relief until the same problem is repeated. The reprieve is so short lived, because nobody is concerned about or prepared to face the chronic state of disrepair which creates such an emergency. This apathy continues because of the lack of information on (1) the reasons for the breakdowns, since no records have been maintained; (2) how to carry out the repairs; (3) how to acquire the spares; (4) who is qualified enough to undertake the repair; and (5) the exact fund from which payment has to be made for these works.

ROUTINE MAINTENANCE

Routine can be defined as a "procedure followed regularly" or "as a cyclic operation recurring periodically". Let us, for example, examine the following routine checks belonging to a fictitious organisation, to understand the nature of this cyclic operation recurring periodically:

(a) Check all compressors on Mondays.
(b) Lubricate and inspect two machines daily. This cycle has to be repealed after all the machines are completed.
(c) Service all machines after 1000 hours of operation.

(d) Give your attention to: Monday—lathes; Tuesday—presses; Wednesday—grinders; Thursday—millers; Friday—machine tools; Saturday—electrical machinery.

(e) All electrical motors to be inspected in the first week of every month and all starters in the second week.

In all these examples, routines are being established by defining the time available for each task and the number of units to be attended to, within that time-frame. This means that a frequency is being established which, therefore, sets a pattern. Thus, by estimating the time needed to do a particular job, the period required for one cycle is known. Hence, the frequency that a particular unit will attain in a year, in half a year, or in a month is known. Or the frequency of service gets defined. Alternatively, we can define the frequency of service which we insist upon. Then, the total number of man-hours can be calculated to carry out the job. The advantages of this system are that it:

1. is simple to follow and establish;
2. does not require any clerical work;
3. achieves a high degree of prevention by intercepting developing faults;
4. can give somewhat good results;
5. prolongs life of machinery, equipment and facilities.

But, this system needs a supporting crew for repairs.

Compared to the breakdown system approach, each machine under routine surveillance will at least be serviced once a year, if not more often. Then the defects that are slowly developing are likely to be discovered and looked into, and eliminated before they can cause any major stoppage. Also, a very high number of failures which would otherwise have occurred due to total lack of attention would be avoided. When a routine is defined, its ruling, say, a decision to service a particular set of machinery every fifteenth day, has to suffice and hold good for the maintenance staff, ignoring all past records, data, and actual needs. While establishing a *routine system,* we do not take into consideration the details and the needs of the machine. We establish an average service interval that can be implemented, taking into account our total manpower and die machines which have to be taken care of, for example, whether the tail rotor bearing of a helicopter needs greasing and oiling after every 10 hours of flying time, and the main gear box after every 25 hours of flying; if this difference is ignored, both will be serviced only after 15 calendar days. Hence, blanket orders to service, lubricate, rectify, adjust are given and no checklists are prepared. It may be noted that this has to be supported by a separate crew who would only be repairing the defects. The basic difference is that the routine servicing does not cater to the needs or the expected requirements of the machine. This is far better than the breakdown maintenance system. However, as we can see, it is not a fine-tuned method for fulfilling differential levels of maintenance needs.

PLANNED MAINTENANCE

Planned maintenance is the maintenance organised and carried out with fore-thought, control, and records, to a predetermined plan. In the planned maintenance system, the emphasis is on the machine's needs, and the expected requirements from the machine. It has to be centred around the original recommendations made and prescribed by the original equipment manufacturer (OEM). The Maintenance Manager has to use all his experience and expertise to superimpose refinements and improvements on the manufacturer's recommendations. However, in doing this, the following factors have to be kept in mind:

(i) *The extent of utilisation:* This means that whether this unit or machine is working for one, two or three shifts a day, every day, that is to say 365 days in a year, or it is being used sparingly?

(ii) *Severity of utilisation:* Whether the machine is working under normal load conditions or not, and are the other conditions of operations being adhered to, which have been prescribed and visualised by the manufacturer?

(iii) *Operating conditions:* Depending upon the conditions under which the operations are taking place, such as the degree of corrosiveness, humidity, temperature, dust, snow, or heavy rains, which the machine has to contend with, the question arises: Is more attention being paid to counter-balance the pernicious effects of the prevailing environment? This question has to be answered with due seriousness, because each one of these accentuating conditions will have a direct bearing on the type of attention each machine will need to maintain it in a serviceable condition.

(iv) Are there any other specific factors that may affect the equipment, for example, the particular kind of wear and tear on large fans being continuously subjected to dry fly-ash in a thermal plant?

In the planned maintenance system, instructions are more detailed and thorough. It is also possible that it may be found necessary to have differently timed servicing on the same unit, and this type of system is capable of accommodating this variance. For example, in a blast furnace, the automatic mechanism may need a weekly check, while the burners may require daily checks, and the electronic instrumentation only a monthly calibration check. It may, therefore, be noted that once the frequency of checks for all the various items in a particular unit has been established, only then the annual servicing dates can be established. This is a major difference between the routine maintenance system and the planned maintenance system. The planned maintenance system, therefore, requires the work to be planned in advance. The planning could be on the basis of three month's roll-on plan, in which the first month will be a fixed or a confirmed plan and the second and third months will only be tentative plans. The first month's confirmed plan will then be broken down into fortnightly plans, then weekly plans, and thereafter daily plans. This involves a total planning effort and a faithful implementation of the plan and its recording is a *must,* if it is to yield satisfactory results.

This system has detailed instructions which have to be followed strictly. These include directions for inspection, repairs, rectification, replacement of parts, and so on. There has to be enough flexibility in this plan so that changes in the plan can be accommodated to ensure reduced chances of failure during the intervening period extending up to the next scheduled servicing date. According to the judgement and assessment of the maintenance engineer, this system provides for apportioning as much attention as an equipment or its part needs. Obviously, to start with, the shorter the intervals between inspections, and the more detailed the servicing schedules, the better will be the insurance against breakdowns.

As experience is gained, the frequency between the checks/servicing can be progressively reduced, and certain elements of work found unnecessary and eliminated. Initially, the checklists and the detailed schedules of inspection and servicing are elaborate, as the maintenance personnel are not yet quite familiar at this stage with the newer equipment. As time goes by, and the maintenance personnel get experienced with the equipment and its servicing, reviews have to be carried out, and the elapsed time between cycles increased; many items of checks/work earlier considered essential are eliminated and some new ones are introduced. The concerned departments will be informed about work stoppages.

This is a continuing process and to achieve the maximum advantage from this style, planning has to be the focal point. It must be thorough and comprehensive and the recording must be accurate and faithful. Recording of data is futile if regular analysis is not carried out to improve efficiency. This will help in producing improved maintenance manpower schedules, production manpower schedules and utilisation and production plans by indicating as correctly as possible the service timings and frequencies. The analysis of failure data will clearly bring out the reasons for failure, which may be caused by inferior parts or poor material, bad workmanship, both by production or the maintenance workers, faulty design, and faulty operation (indicating that the operator needs training). In the planned maintenance system, one important area which has to be taken care of is *calibration*. This function is so important that it has to be catered to separately and specifically because, without calibration checks, one is not sure of the output results. Hence, the need for calibration with properly trained manpower becomes imperative.

PREVENTIVE MAINTENANCE

Preventive maintenance is a must for every machine and must be carried out in a stipulated time. It refers to those critical systems which have to reduce the likelihood of failures to the absolute minimum, for example, operation of an aircraft which flies from one place to another, or critical installations within power plant, and the forced ventilation system in underground mines. To prevent breakdowns, preventive planned servicing is carried out with the specific objective of detecting/locating weak areas and ensuring perfect functioning by even replacing certain critical parts which could still be used, but for the assurance or reliability we demand. Thus, after each servicing, the equipment is considered to be as good as new and is expected to give a high level of reliable performance. It may be noted that in addition to this servicing, planned inspection and servicing will continue to be carried out in the interval between two planned preventive servicing which are also aimed at achieving elimination of possible breakdowns. Thus at an added cost, a very high level of reliability of performance is assured or obtained. By proper scheduling and planning, it is possible to substantially increase the equipment life and increase profitability.

Planning and implementation of a preventive maintenance system with the added cost of replacement of certain parts/components before they actually do fail, makes this an expensive system, and hence ought to be applied with caution and care to really critical equipment only. Statistical methods are usually employed to determine the 'life expectancy' of components/parts/spares to fix the replacement periods accurately. Preventive maintenance increases the life of equipment as a stitch in time saves nine.

In large complex plants, therefore, *preventive engineering* has been established and is becoming an important need and function. This basically involves a detailed analysis of breakdown—failures and incidents. An *incident* is defined as a happening which might lead to, or is a prelude to a serious accident or failure. The analysis indicates the areas of concern where real effort is needed, and where one can reduce chances of breakdown by redesigning, substitution, changes in specification, material or design improvement, and modifications. Preventive maintenance reduces the word urgency as all activities are known in advance and decreases down time.

It is the duty of the top-management to appreciate that such an analysis requires expertise, experience, devotion and intelligent direction. Since this analysis is an activity akin to research work, the approach necessarily becomes different and, therefore, cannot be expected to be performed by the Maintenance Manager, who is normally fully occupied and busy with his day-to-day work, and is neither trained nor equipped to undertake such a task.

These findings have been achieved by making a detailed and thorough analysis of the documents given below, and by assessing the responses to interviews held with concerned persons. The documents to be persued are:

(a) Defect reports and investigation report
(b) Accident and incident reports
(c) Premature withdrawal of equipment reports
(d) Modifications suggested by OEM with details, drawings and their classification into safety modifications, operational modifications, and normal modifications
(e) Environmental testing laboratories reports or accelerated climatic trial reports
(f) Preventive maintenance charts identifying the better material, time etc.
(g) Cost reduction techniques employed
(h) Value engineering reports. It is similar to overhaul for where parts are required in advance.

RELIABILITY

In a preventive maintenance system, we seek high reliability of operations at an accepted added cost. *Reliability* is, therefore, an assurance that the equipment will function for the duration it is intended to, and will *give, failure-free* performance. It is the probability of successful operation of the component/equipment for a specified time or usage in a given environment. Successful performance may mean failure-free operation or at operations with not more than a specified number of failures during that specified period.

By definition, an equipment which never fails is an *ideal equipment*. Unfortunately, this is not practically possible as the present state of reliability engineering has not yet reached that far. It is, therefore, prudent to accept some 'unreliability' and as small a chance or probability of failure as the technology and cost permit, and then make efforts to minimise the effects and duration of these failures when they do occur.

Apart from accepting as low a chance of failure as possible, the user department always wants a rapid return of the failed equipment to service. This needs a *repair system* with a quick response time. The equipment design and the system of repair and its infrastructure ought to be so geared that the response time is minimal. This is the *maintainability* feature of the equipment.

Availability of any equipment for operational use and exploitation is a function of reliability and maintainability. Both these factors deeply affect availability. A highly reliable equipment, which will obviously be very expensive to design and produce, may have breakdown very rarely, but if it takes too long to repair and to be restored, then the availability will be poor. Conversely, even if it takes little time to repair and is brought back to service quickly, the availability would still be poor, if its frequency of failure or breakdown is too high, which means its reliability is poor.

When more and more reliable equipment is needed/demanded the cost is bound to go up. This cost includes *developmental cost,* including reliability trials and testing, expensive quality assurance programmes, quality controls, and checks during production, trials, testings and selection of high grade and high quality materials/components. Reliability is quantifiable and is thus a design parameter and function. Numerous problems may arose while doing preventive maintenance, as the operators normally do not carry out instructions properly. Reliability is also discussed in Chapter 30.

The user, on the other hand, demands better performance, high accuracy, and sustained and uninterrupted operation from existing plant, machinery and equipment. This naturally results in designing of increasingly complex equipment. In this process, perhaps, in the initial trial phase, higher than acceptable failure rates are indicated. This necessitates design changes to reduce the number of parts, often by combining the characteristics of several parts. But the composite parts bring in complications in manufacture, its attendant difficulties and cost. Thereafter, the producer brings in changes for ease of manufacture or uses mass production methods. Thus, a reliable but sophisticated and complex equipment is delivered to the user. However, this highly reliable equipment when it fails, creates a series of problems—being a complex piece of machinery, many parts are inaccessible. Moreover, one cannot search inside directly to diagnose the problem as it is inapproachable. Thus, it needs special tools and methods to locate and reach the problem area, and requires specialised processes to be adopted by a trained and skilled person in that field. This means that the total response time to put it back to *operation* will necessarily be too long. Therefore, all the benefits of high reliability achieved at very high cost is negated. Thus an acceptable level of availability can be assured, sacrificing expensive reliability. A multiple trade-off in levels of *reliability, maintainability, availability* and *cost* has to be maintained. Bearings, clutch plate and lock screws are purchased in advance to preventive maintenance. Preventive maintenance assures full capacity of utilization as the machine will not be stopped.

Maintenance prevention is practised in throw away items like ball pens, where the parts cost and labour of maintenance will be more than the cost of the item itself. Hence the item/machine is thrown away and replaced by a new item. This is similar to throw away machines and replace by new ones in other systems.

PREDICTIVE MAINTENANCE

Predictive maintenance can be defined as "methods of surveillance used to indicate as to how *well* the machine is, while performing its intended tasks".

In chemical, fertiliser, and petroleum refining industries, the plants are subjected to severe conditions and are put to continuous operations. To be able to get the maximum number of on-stream days of operation, the system of maintenance operation should be such that it will be able to reduce down time to the absolute minimum. Hence, continuous plant monitoring, and diagnosing the actual condition of the equipment by means of *on-stream non-destructive testing* methods are being increasingly used. The objective being the ability to predict an impending failure well in time, thus avoiding failures which could cause heavy penalty costs and even create health and safety hazards. Therefore, an ability to forecast equipment behaviour by *condition monitoring* is a prerequisite for predictive maintenance. Condition monitoring is a method of extracting information from a plant/machinery and enables us to indicate its condition or 'health' in quantitative terms. Hence, it is a very important diagnostic tool to the maintenance engineer. It provides him increased on-stream availability, while optimising on maintenance costs and ensuring plant and personnel safety at the same time.

CORRECTIVE MAINTENANCE

Corrective maintenance is defined as *maintenance carried out to restore (including adjustment and repairs on item) machinery which has ceased to meet an acceptable condition.* For example, you may have looked after your car fairly well, and managed to keep it in a good condition, yet a time will come when however well-tended the vehicle may have been, age will catch up with it, for after the car has done a certain amount of mileage, normal wear and tear will begin to take its toll, and the car will need to be restored in order to meet an acceptable condition. Perhaps the car will be in need of a new clutch plate as the clutch would have begun slipping, or new brake pads would have to replace the worn-out ones. Or for that matter, an entire engine overhaul may become necessary. All these jobs—big as well as small—carried out upon the car are examples of corrective maintenance.

DESIGN OUT MAINTENANCE

Design out maintenance (DOM) is a system that strives to eliminate, and if that is not possible, then to minimise the need for maintenance to the lowest possible level. It is, therefore, also known as *eliminative maintenance*. Hence, it has to be thought of, and applied to the product at the design stage itself, so that machinery, plant, and equipment are so designed as to need the least possible amount of attention or maintenance during their economical lifespan. Some examples of this concept are visible in the present-day motor vehicles. Things like permanent bearings, bearings using solid lubricants and permanently sealed units have become commonplace.

In today's Indian car, the Maruti, there are plenty of examples of components which, although do not eliminate maintenance completely, yet definitely aim at reducing the maintenance effort. For example, the radiator of this car does not require a weekly replenishment of water, nor does its battery need to be checked and filled with distilled water every now and then. These systems have been so designed that replenishment occurs automatically in the period intervening between scheduled checks. Only on the occasion of scheduled checks which have been prescribed by the manufacturer is the water or distilled water replenished in their respective containers. The design in this case is based on the *modular concept,* of module by module replacement, which makes the maintenance easier.

This concept has been applied to great advantage by the manufacturer, for by providing user convenience, safety, and reduction/elimination of maintenance, they have increased sales and have thereby increased their own profits. Some examples which come easily to mind are disposable razors, cigarette lighters, ball-point pens, cheap quartz wrist watches, torches, disposable syringes, etc., all while utilising the system of maintenance by elimination have also eliminated the hazard of contracting diseases through improperly sterlised hypodermic needles, disposable diapers and so on. In these instances, the concept goes a step further and becomes *maintenance by disposal.*

TOTAL PRODUCTIVE MAINTENANCE

Total productive maintenance (TPM) is known in Japanese Industry as *productive maintenance* (PM), with all employees participating through small groups. The first organisation to introduce TPM was Nippondenso Co., a large manufacturer of automobile electrical parts, belonging to the Toyota group of companies, which has a name for high quality standards and productivity. Preventive maintenance was introduced in Japan in 1951, from the USA. This led to the introduction of productive maintenance by Nippondenso in

1960, which meant operation personnel was responsible for production work, and maintenance personnel remained responsible for maintenance work, i.e. each group performing their designated roles. However, it was only in 1969 that Nippondenso introduced TPM. They had totally automated their manufacture and assembly of parts by then. This made it impossible for the maintenance crew to maintain the very large number of automated facilities. Therefore, it was decided that the operators of automated equipment were to be responsible for routine maintenance themselves. Based on this concept and their experience of quality control (QC) circle, it was decided to evolve PM with all employees participating in this small group.

The basic concept of TPM is to change the attitude and improve the skill of all personnel by using quality equipment. Earlier, as usual, the job of the operator and that of the maintainer were separate; each was divided into groups of operators and maintainers. The operators had thus no interest in the maintenance. Changing this attitude on the part of all employees was the first major task of TPM, i.e. the operators maintain the equipment they use (self maintenance). Therefore, training the operators in maintenance skills and knowledge was the next step. Improving the human beings is possible when both *willingness* and *ability* are present. This is achieved by driving home the need for eliminating losses in all forms and achieving optimal overall equipment effectiveness. All these activities improve the worker and his equipment. Thus the organisation prospers.

JAPANESE 5S

According to the Japanese methodology, five S's improve the methodology of the worker.

The five S's are formed by the first letter of the romanised version of the Japanese words, *Seiri* (orderliness) *Seiton* (tidiness), *Seiso* (purity), *Seiketsu* (cleanliness), and *Shifsuke* (discipline). And TPM carries out these five S's thoroughly.

To maximise the effectiveness of man-machine systems, TPM aims to eliminate three types of losses:

(a) *Down time losses* caused by unexpected breakdowns or set-up and other adjustments.
(b) *Speed losses* due to idling and minor stoppages which arise from temporary blockage or stoppages in chutes; and those due to difference between designed speed and actual working speed of equipment.
(c) *Defect losses* arising from rejects, defects and rework. Also, reduced yield due to losses which result from the time gap between the *start* of production and *stabilised* production.

Total productive maintenance strives to produce overall equipment effectiveness, through a combination of availability, performance efficiency, and rate of quality products, the concept being that the overall equipment effectiveness = availability × performance efficiency × rate of quality products. TPM is the latest and many progressive organisations are adopting this.

CONTRACT MAINTENANCE

Today, there is a shortage of experienced maintenance personnel who are capable of servicing complex and sophisticated plant and equipment. Even if they are available, they can be too much expensive to be retained on a full-lime basis by most organisations. This has forced many of them to look around for an alternative.

Contract maintenance can provide the answer to this kind of a dilemma. This is a system where for an agreed fee the contractor supplies trained manpower along with the requisite supervisory staff and the necessary tools and equipment required for the job. Contracting out maintenance chores also has certain other advantages which are briefly discussed below:

(i) *Down time reduction:* Even organisations which have an adequate maintenance workforce to keep their work going at an even keel may find themselves unable to cope, when faced with a sudden breakdown at a crucial juncture of the production cycle. At such a point of time, it becomes imperative for the management to find ways and means to reduce the down time to the shortest possible period. If a maintenance manpower contracting agency were available then they could have immediately inducted competent manpower in adequate numbers without any formalities and delays. They could have worked, if necessary, round the clock in order to help tide over this crisis situation. Once crisis is over, the organisation could go back to its normal maintenance back-up once again. Needless to add, such a shot in the arm could indeed be a life saver.

(ii) *No liability on a long-term basis:* The carrying cost of trained manpower which may be required occasionally can be well imagined even by the least initiated. For along with the needlessly enlarged pay-roll, the company would also have to bear all the other attended responsibilities and expenditures which go with keeping full time employees. Provident fund contributions, bonus, retirement benefits, medical care, leave travel concessions, and innumerable other liabilities would have to be borne, apart of course from the headache of finding work for idle minds and idle hands before they can turn into the 'devil's workshop'.

(iii) *Elimination of acquisition cost:* This involves the need to do away with very expensive or very bulky machinery, which may not be required often to warrant the expenditure on purchase or the expenditure on expensive storage and inventory carrying, and the need to eliminate very expensive machinery which is required only occasionally, like the stringing equipment (costing approximately ₹1 crore) required for the laying of transmission lines, or costly and bulky construction and earth moving equipment. In today's overcrowded environment, space has become one of the most expensive items. Hence, saving storage space by taking machinery and equipment on contract wherever possible becomes a sensible option.

When an organisation uses the contract maintenance system for an agreed fees, it does not have to burden itself with a variety of trained manpower for all the specialised services it makes use of from time to time. However, the major problem lies in identifying and selecting good and reliable contractors. For the contractor should be reliable and capable of delivering the goods as and when required. Obviously, the best method would be to associate oneself with a contractor who can be commended by his own past records, provided such a contractor is easily available.

EXAMPLES OF CONTRACT MAINTENANCE

There is a wide range of items which are suitable for contract maintenance. Starting from the most sophisticated area of computers, it lends itself equally well to mundane tasks, like the maintenance of typewriters. Some of the areas in India where it is in vogue are the following:

(a) Maintenance of computers
(b) Maintenance of telephones
(c) Replacement and cleaning of lamps, chokes, etc. in factories and the caring of the perimeter lighting and fencing
(d) Maintenance of sub-stations
(e) Fire detecting systems and fire-fighting equipment
(f) Maintenance of typewriters and duplicators, and of xeroxing and photocopying machines
(g) Airconditioners and airconditioning plants
(h) Maintenance of vehicles (In many cases, organisations have done away with having vehicles of their own; a blanket contract meets all their vehicular needs, eliminating thereby capital acquisition and garaging costs and the permanent responsibility of maintaining the fleet.)

(i) Maintenance of the water-supply system and the pumping stations at industrial townships
(j) Maintenance of cinema projectors, slide projectors, and overhead projectors
(k) Maintenance of earth-work, buildings and roads
(l) Works contract for painting and allied jobs; for sewage, drainage and conservancy
(m) Maintenance of gardens, playgrounds and house-plants.

Today we can visualise a tremendous growth in the infrastructure/facilities being created by trained technocrats who are taking on a wide range of services on contract. The time is not far off when they will become big competitors and a major challenge to the conventional maintenance man in industry.

CANNIBALISATION

This is not a system at all and must not be practised in any industry, but unfortunately it is practised in industries. If you have 10 identical machines, the part/subassembly/assembly from one machine is removed and fixed in the broken machine. Slowly one-by-one the parts are removed which disappears from the machine. This process, if continued permanently 10 machines become 9, and then after some time, the number is reduced to 8 and soon thereafter the fixed capital is eroded completely! Very few organisations do practice this and instead of ordering 10 machines they order two more for cannibalisation, particularly if the manufacturer may not be able to supply spare parts.

Chapter **5**

Design of Maintenance Systems

INFLUENCING FACTORS ON SYSTEMS

Designing the maintenance systems for an organisation is a difficult and complex task, as it has a number of factors which interact with and affect it. This is the main challenge and task of the Maintenance Manager, which he has to perform with extreme care and caution. As a scientific approach is imperative for the success of designing a maintenance system, the first step is to carry out the *criticality* analysis of plant and machinery which can guide about the extent and the kind of maintenance effort needed for a particular type of machine or facility. Usually this is known as vein analysis for equipments and VED analysis for parts discussed in Chapter 23. Based on this, a judicious blend of breakdown, planned/preventive and predictive maintenance systems can be designed.

The other factors that need to be carefully considered are the internal facilities that need to be created to support the designed systems. Also, their cost, and the possibility that they have to depend on external sources, like OEM (Original Equipment Manufacturer) and the facilities offered by them, its location, proximity to own factory location, lead time needed to obtain their help and also the facilities available in the geographical location for contracted maintenance and their availability and quality. In certain situations, due to the lack of facilities being available externally or due to certain internal factors, the organisation may be forced to opt for creating its own support system and resort to depend on internal maintenance capabilities alone. Thus, on an aggregate, the maintenance planning effort will depend largely on the size of the organisation, unit location, and connected logistic problems, local availability of maintenance infrastructure facilities level and nature of technology involved, secrecy needed (for Defence or atomic energy plants) and skill levels required. Due consideration has also to be given to designing of the organisational structure and the extent of centralisation and decentralisation that will be required to meet the needs of the system so designed.

CRITICALITY DETERMINATION

Criticality determination is the starting point for planning of maintenance effort. It is quite obvious that more critical the equipment, die more maintenance effort it will warrant to keep it in a serviceable condition. It is apparent, therefore, that a concerted effort will have to be made at the earliest stages of the endeavour, if a meaningful plan of action is to emerge for implementation. For it is this criticality which will also determine as to which equipment, machinery or plant will require the greatest or the least amount of care and attention, and the amount of attention each grade of criticality will require.

If in a critical plant, critical item is not available, the whole plant will be shut. The effort made towards categorisation will show that some of the equipment and machinery will require different types of maintenance:

(a) Predictive maintenance (which could also be called condition monitoring) or preventive maintenance (As both predictive and preventive maintenance efforts are very expensive, it becomes imperative that careful sifting be done so as to not allow those equipment, machinery and plant which do not require so much maintenance and care to be mixed up with this group erroneously.)
(b) Normal planned maintenance.
(c) Breakdown maintenance (which has to be carried out only as and when the need arises; this kind of *ad hoc* maintenance work should only be recommended for the least critical categories of equipment).

In order to be able to achieve maximum plant availability—most economically, the able planner will have to device a system of maintenance, which will be a judicious mix of the different types of systems for specific equipment as per criticality determination.

Criticality determination is the process wherein one attempts to scientifically and systematically pin-point and list the most important steps/organs of a process, plant or factory by determining how important is their healthy functioning for the well-being of the whole system so that they can be given correct weightage in terms of care and maintenance.

However, it would be worthwhile to keep in mind the fact that this is an extremely delicate and difficult task. In order to arrive at a judicious and cost effective assessment of the task ahead, all the skill, experience and expertise of the maintenance managers will have to be pooled together. Even so, criticality cannot be determined by a one-time effort, for solutions to the problems will emerge from the evolution of thought process based on facts and figures over a period of time. The correct precedence of critically can be determined only after a considerable amount of trial and error.

Criticality determination is an exercise that has to be performed collectively through a series of *brain storming sessions* by the concerned persons of both the maintenance and production departments. A concerted and coordinated effort on the part of the managers of both these streams of specialisation (maintenance and production) belonging to the plant or factory for which the maintenance design is being planned and formulated is absolutely imperative. Only in conjunction they can get a clear picture of the entire plant's operations, its commitments, its weaknesses and its strengths. The condition and age of machinery and equipment, the use and the pressures that are imposed on them, and their value and importance determine the criticality in terms of their contribution towards the successful fulfilment and meeting of operations tasks, and production goals.

Some of the important areas to be looked into while determining criticality would be, the cost of down time, severity of utilisation, standby availability, downstream effects, hazards involved in the event of failure, age of plant, skill availability, maintenance support facilities and infrastructure, user skill, operating environment, condition of usage, and severity of utilisation.

Basically, criticality may be gauged by posing a few simple questions. For instance, what would happen, if a particular machine or equipment were to breakdown? Would production come to a sudden halt? Or, will it be possible to perform the work with another machine? Or, perhaps by switching on to another machine during the interim period without hampering production? If alternatives are easily available, then obviously that particular machine is not critical.

But maintenance planning in order to be worthwhile cannot be based on such an oversimplistic and unidimensional assertion of criticality. The factors which govern criticality have already been mentioned. They are multifaceted and have blurred boundaries, for the factors overlap, interact and shape one another. Therefore, it is essential that all the parameters be examined closely and assigned their correct weightage before deciding about the degree of criticality of each item.

As already discussed, criticality determination is an exercise that has to be carried out collectively through a series of brain storming sessions by the concerned persons of both the maintenance and production departments. Each one of the factors has to be placed in turn before the group concerned and questions asked so that answers regarding criticality, which should as far as possible, be jointly agreed to by the two groups, can be elicited.

In this section, we shall discuss each one of the important factors and see how they influence criticality.

The first question that comes to one's mind when a breakdown occurs is: How expensive is the breakdown? For all assessments eventually are cost related; hence let us begin with this important question.

To know the extent of breakdown expenditure, one has to accurately work out the exact cost, costs accruing from a breakdown could be immediate as well as far-reaching. Cost of lost production, labour cost, cost of replacements and spare parts come immediately to one's mind. However, the other aspect, namely, 'impact' cost could well be even more and needs to be taken carefully into account. For example, along with the cost of loss in production one must calculate cost in terms of the idle time of the production labour, and of all those others who have been rendered idle by the same breakdown downstream. Then, there could be the cost to be paid as penalty for delayed delivery of an order, or worse still, the cost of an order being cancelled. Further costs may be incurred with an urgent need to air freight a particular indispensable spare part or some necessary raw material to get on with the work of rectifying the breakdown.

The actual cost of down time can be calculated only by adding the time taken by the machine to be repaired and to be put back into operation and the operators idle time.

Once all these parameters have been calculated, then an evaluation will have to be made by top management as to what they would consider as very high cost and what would be high cost, and so on, in a continuing order. Having considered the problem from all the angles and with the table of cost of down time before them, the top management will then take a decision about the cutoff point, where the cost of down time will place a particular piece of machinery in the *critical* or *vital* category. Practically, you may not get answers in a single shot.

The *demand* made on a machine for a particular service that it provides to its users, and the extent of its loading or utilisation per day, can be termed as *severity of utilisation*. Its critical relationship with the *product* or *production output* must be ascertained and gauged correctly.

For very critical machinery, it is always desirable to have a parallel system, which can be put to use in case of equipment failure. This is called standby availability, and is an important factor to be considered while designing maintenance systems. For assemblies/parts, it is called 'insurance' items.

DOWNSTREAM EFFECTS

Downstream effects are the important criteria as they can have serious long-term implications. Let us take the example of a chemical plant where, say, a slurry has been prepared, which needs to be continually stirred and allowed to chemically react for six hours at a fixed temperature. If a breakdown were to take place upstream in a critical machinery at any time during this predetermined phase, then not only would the entire slurry go waste, but it could also lead to shut down of the chemical reactor itself, for as a result, it would in all probability be needing a thorough chemically processed cleaning or worse still a relining of the vessels and vats where the slurry had gone waste. Thus, we can see how an upstream breakdown has far-reaching repercussions downstream. Far apart from the product cooking in the vat going waste, the repairs and refurbishing bills, the losses to be borne due to loss in production time and all its other attendant cumulative expenses can lead on to a total shut down. The financial loss due to non-delivery of goods and so on, can prove to be a rather costly for any industry. It is obvious, therefore, that a critical area such as this needs to be given due consideration, and takes a long time to evaluate this cost.

In the event of failure, breakdown or malfunctioning, certain machinery and plants can bring terrible hazards to man, his property and even his environment, for the fall-out of such failures can be very dangerous, damaging and destructive. The hazard could be in the form of poisonous, toxic or inflammable gases, or in the form of torrents of engulfing water unleashed upon the unwary, by the malfunctioning of a simple thing like a flood-gate. However, while talking of hazards which may occur in plants and factories, one must concede that in case of a failure it will be the plant personnel themselves who will be affected first. Only when the magnitude of the breakdown becomes larger, will its pernicious effects be felt externally and in ever enlarging circles. In those circumstances where failure can lead to hazardous ends, the criticality of the *safety services* will be self-evident.

A thoroughly efficient and well-maintained safety service must at all times be available at such sites, which can prevent, and combat if necessary, those threats which may emerge in case of failure. The danger arising from the emission of inflammable, explosive, toxic, poisonous or in any other way, harmful substance should be examined and evaluated with care; for not only can it become the perpetuator of untold human misery, but it can also completely ruin a company. For the company may have to pay a very heavy price in terms of not only money, but reputation as well, and the possibility of a total closure could also be the outcome of a failure with disastrous consequences. The Fukushima radiation in Japan, the Bhopal Gas Tragedy or the Chernobyl Disaster are exemplifying cases in point. Therefore, all plant which handle hazardous materials would be having an array of machinery and equipment which should be placed under the *critical* or *vital* category, for breakdown in any one of these spheres could lead to terrible consequences.

Hence, such machinery would warrant a maximum of maintenance effort and attention, to avoid breakdowns. In fact, this effort must not be slackened even when a *standby* equipment is kept on the ready and available. Moreover, even the standby to the vital piece of equipment will have to be given the care and status of *critical*. Some firms have a standby for each standby.

AGE OF PLANT AND TECHNOLOGY

An important criterion of maintenance planning to be kept in mind is the *age* of the machinery in use. As ageing occurs, the need for maintenance increases quantum-wise. This is a factor that needs particular attention in India, for here one invariably continues to use machinery and plant way beyond their optimum economic life span. For example, the average age of plants in India is 22 years, whereas it is just 7.5 years in Japan. With such extensive use, it is obvious that the probability of failure increases. This trend needs to be carefully evaluated and watched, because it per force lowers the level of *reliability*. Because of this, certain machines from the merely essential category may be included in the 'vital'/'critical' category. With ageing and consequent vulnerability, these machines need to be continually monitored and evaluated, so that actions may be initiated accordingly without bringing on a crisis through breakdown. This can result in an increase of *condition monitoring* effort and perhaps also need preventive maintenance checks. An increase in the frequency of inspection and the cycle of planned maintenance will also perhaps have to be taken into account. A closely linked problem to this will be the non-availability of spare parts in this extended life period. No wonder Indian maintenance engineers become so good at improvisation and substitution. The marketing force introduces some change in the equipment once in 5 years, sighting as technological upgradation.

SKILL AVAILABILITY

There could be certain items of equipment in a complex, complicated and technically sophisticated modern plant which is important in the chain of production requirements. Though its welfare is vital to the well-being of the operation, it is possible that the maintenance organisation may not yet be able to acquire the necessary skills to repair or rectify these components, for that matter they may not even be quite confident about their diagnostic abilities to pin-point a fault as and when it arises.

With the rapid advancement in technology, skills required to cope with the state of the art technology often lags behind, the effects of which are felt by all areas and functions. However, in the case of production, the benefits positively outweigh the negative ones. Initial training is a must for the production personnel before they start working on any new machinery. However, soon they get a lot of praise for the results and outputs consequent to harnessing new technology. Fragile machines with highly flexible controls and stupendous levels of productivity may be easy to handle and control, but can prove to be very difficult when in need of repairs and rectification. Normally with brand new machines in hand, little or no attention is paid to the need that may arise some time or the other. More often than not, the maintenance man begins his *on-the-job training* with a crisis already at hand. Thus, the more advanced the technology, the more difficult does skill acquisition become for the maintenance personnel. The different phases in a machine's life can be easily illustrated and understood from Figure 5.1. And from it one can easily deduce the different types of skills that the maintenance department will have to acquire in order to keep the machine operating at an even pace.

BATH TUB CURVE

From the curve shown in Figure 5.1 we see that the initial phase can be termed as the *infant mortality stage*, where numerous teething problems arise which need to be identified and rooted out with the help of the collaborators/technical assistance teams, who should be normally available under the *warranty*.

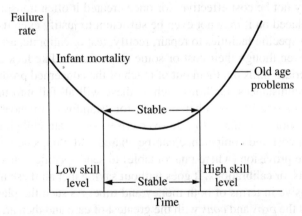

Figure 5.1 Bath tub curve.

Then there is a *stable* period when normal day-to-day upkeep is all that the machine will require. Thereafter, comes the phase of 'old age' problems, where special skills are needed to keep the 'old horse' going apace. Obviously, maintenance has a serious responsibility at hand from this stage onwards, because it is up to them to acquire, develop and refine skills to keep production going at economically viable levels. Hence, skill upgradation must run parallel to machinery acquisition, and this upgradation can only be acquired through the adoption of a judicious programme of training, which must begin well before the acquisition of the plant. We will discuss more about this in Chapter 30.

However, well planned the training programme is, there will be a period in the early stages where the confidence levels will still be required. Till such time as adequate progress is made and key personnel have acquired necessary skill levels, the equipment will have to be classified under the critical/vital category. The answer to all these initial problems lies in having proper technical collaboration/assistance, tie-ups with the original equipment manufacturer (OEM) or annual maintenance contract (AMC). Their trained personnel should be available in the country of installation in the initial phase of equipment's utilisation. The added advantage accruing from it would be *on the job* training possibilities that this situation would provide. In view of its importance we will again meet this in Chapter 12.

MAINTENANCE SUPPORT FACILITIES

The creation of support facilities and an infrastructure to cater to the demands of the entire range of maintenance activities is not only very expensive but also time consuming. It is also quite likely that the creation of an expensive facility may not be cost effective, for once created it often transpires that the demands placed on it may not even be sufficient to justify even its existence. Yet certain specific facilities to repair, rectify, test or calibrate, are absolutely essential, even though their cost or some other reason, like lack of adequate and appropriate skills put them out of reach of the concerned production unit. In such circumstances, facilities such as these will all fall into the 'critical' category. For one has to either provide for 'standbys', or have a sizeable number of 'float' available to tide over crises, or create clubbing facilities, where both cost and competence can be shared and alongside with this one should have provision (where unavoidable) to send specific items abroad for repairs, tests, or calibrations. It goes without saying that all these methods are very expensive in terms of both money and time. As such the planners must evaluate all the *pros* and *cons* with the greatest of care and then arrive at what should be the most cost-effective solution.

OPERATING ENVIRONMENT

The user skill, operating environment, conditions of usage and the severity of utilisation are the factors that either individually or collectively have a direct impact on the life of equipment and machinery and its performance. Any one of these aspects could create a *criticality situation* demanding special or critical attention from maintenance. The factors that can lead to this state of criticality need to be looked into with care. They are, operating speed, temperature, load, vibration, pressures, and factors such as dust and corrosion. It is a well-known fact that if load is doubled, then wear and tear increases ten times. And intermittent load, dust and vibration can cut life expectancy down to one-third of what it ought to be. It is possible that overloading cannot be eliminated fully. Therefore, it is essential that the operators be trained properly and be made fully aware of the detrimental effects of 'overburdening' a machine so that they endeavour to keep it to the minimum. The brunt of all these pernicious factors has to be borne by maintenance and, therefore, care has to be taken in maintenance planning, of a number of things like schedules, the frequency at which maintenance work will be undertaken, the quality of the spares to be used, the methods and periodicity of inspection, condition monitoring, a watch on environmental assailants like dust and moisture, and warning systems to alert one of the highly injurious overload situations.

So far, we have been discussing various criteria that can be used for criticality determination. Let us now examine how criticality can be quantified in terms of maintenance costs, so that maintenance system designs can be charted, keeping in mind this significant and vital aspect clearly and without any ambiguity.

If, in an organisation, documentation is maintained accurately, regularly and in detail, then it can be put to very good use, for if the job cards and the history record cards are properly maintained, one can choose machine or equipment-wise information about how much has been the annual maintenance cost on each one of them. In large and progressive organisations the need to go for such data should not be there, for normally, in such organisations most of the information is fed into a computer, from where information can be retrieved in a few seconds.

Let us take, for example, a list of 100 items in a very large organisation, which have been scrutinised closely and looked into with care, in terms of cost. And the scrutiny reveals that there are just a few items which have taken major share out of the entire annual maintenance costs. Now coming back to these 100 hypothetical items, one can begin with tabulating them, numbering them serially, then identifying them by their machine identification or machine code numbers, thereafter, the annual maintenance cost in Rupees which have been incurred on each of these heads should be placed alongside, and then converted into percentage cost in terms of annual maintenance; cumulative cost per cent; and percentage of cumulative number of items. In Table 5.1, the items have been tabulated in the descending order of the annual maintenance cost.

Table 5.1 Annual Maintenance Cost

Item	Machine Identification Code No.	Annual Maintenance Cost (₹) (per cent)	Cumulative Cost (per cent)	Cumulative Number of Items (per cent)
1	009	20	20	001
2	001	17	37	2
3	003	16	53	3
4	005	14	67	4
5	019	8	75	5
6	022	5	80	6
7	099	1	81	7
8	145	2	83	8
9	347	3	86	9
.	.	.		
.	.	.		
.	.	.		
100				
	Total	100	100	100

It will be seen that only six out of the entire 100 items account for 80 per cent of the annual maintenance cost which is indeed very high. Therefore, if cost were to be the only criterion, then these six items would logically deserve the maximum of maintenance attention and care. If maintenance information is computerised, and failing that, well documented, then this kind of information retrieval should be simple and quick and would be of immense help in criticality determination.

However, we are well aware that there are many other factors that affect criticality, which are specific to type of the problems of a particular plant, industry, or installation. There is a need to separate the essentials from the non-essentials among the varieties of machines, items, spares and raw materials that are being dealt with, by the maintenance department. A simple way to do so would be to code each item by criticality, and this coding could be added to the material or spare part code to help in easy identification and sorting out by a computer:

CODE α: *Most critical* production equipment, that is, equipment which has no standby

CODE β: *Critical* production equipment with standby

CODE γ: *Non-critical* equipment

CODE A: *Vital*—critical equipment, which on breakdown will cause loss in production

CODE B: *Essential*—semi-critical equipment, which on breakdown may result in loss of production, but where invariably production loss can be recovered or made up

CODE C: *Normal*—non-critical equipment which on breakdown does not affect production.

Or

CODE X: *Purchase value*—an equipment bought at a cost of ₹10 lakhs should obviously be given greater maintenance attention than an equipment bought for only ₹10,000.

CODE Y: *Functional value*—if vital equipment essential for production becomes unserviceable, then it will result in direct loss of production; hence such equipment must receive greater maintenance attention.

CODE Z: *Attention value*—equipment that frequently breakdowns or is prone to breakdowns should receive adequate maintenance attention.

MAINTENANCE SYSTEM DESIGN OPTIMISATION

While carrying out a detailed criticality analysis, the following factors should be taken into account:

(a) Down time cost
(b) Relevance to direct production efforts
(c) Standby availability
(d) Intensity of demand of that service, i.e. severity
(e) Downstream effects
(f) Dangers to safety to life and property
(g) Skill availability
(h) Spares availability.

It may be remembered that maintenance system design is not dependent upon criticality alone, but on certain other factors as well. These factors are:

(i) Age of plants
(j) More variety of equipment (e.g. different countries of origin)
(k) Replacement costs
(l) External and internal maintenance support facilities and infrastructure.

Maintenance system design optimisation is not a static concept but a dynamic one and it changes as time goes by. A method that can be suggested is to design a matrix, assign weightage and thus get criticality indices for various equipment in the organisation. By superimposing the OEM's recommendations on these, one can add a reliability analysis on to it as well. Thereafter a validity test will also be required to be done to complete the exercise. After having done this, the entire plant, equipment and machinery can be divided into four types or categories:

(A) Vital
(B) Essential
(C) Important
(D) Normal.

(In Chapter 23 we will discuss this in more detail.)

This is known as the VEIN analysis. Having done the analysis, a curve of criticality can be plotted on the Y-axis, and the type of maintenance on the X-axis (Figure 5.2). Basically one has to distribute the equipment into a mix of breakdown and planned maintenance (PM). The planned maintenance can be further sub-divided into planned, preventive and predictive maintenance systems depending upon the equipment, its criticality, complexity and its need and cost.

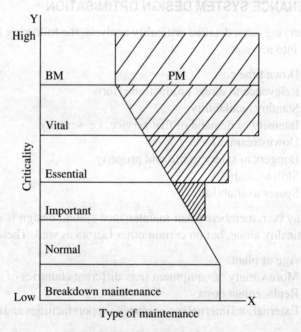

Figure 5.2   VEIN analysis.

By looking at the allocated VEIN analysis chart, it will be noted that the vital equipment will have a small percentage of breakdown maintenance (BM), as breakdowns can never be eliminated, but a very high share of PM. Then the 'essential' category of equipment will have slightly larger share of BM and a fairly large percentage of PM. Finally, the normal equipment will perhaps suffice by only having breakdown maintenance. This kind of a mix will have to be decided upon because the direct cost of PM is high, but the consequential cost of BM is also high. A picture will emerge which will provide the organisation with a proper mix of breakdown maintenance and planned maintenance which is again a mix of planned, preventive and predictive maintenance systems, as required by respective plant, machine and equipment. This will be providing savings in PM costs where not necessary and give the plant high serviceability and availability. This has to be continuously updated with the feedback received along with the defect analysis and the down time analysis reports.

The work force requirement for breakdown and preventive maintenance is given in Figure 5.3.

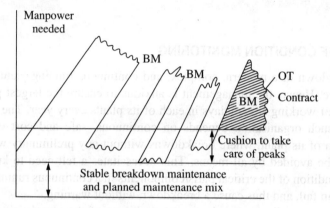

Figure 5.3 Workforce requirement for breakdown and preventive maintenance.

The basic manpower needed for operating on a stable equipment is shown at a stable BM and PM mix (Figure 5.3). Added to this will be the cushion to take care of the peaks. This is the level at which manpower needs have to be fixed. Any additional breakdown has to be taken care of, with operating overtime or contracted maintenance. Thus an optimum workforce required can be calculated for utilisation.

The moot point to be noted is that maintenance effort should be apportioned according to criticality. And criticality has to be determined by proper maintenance engineering analysis, taking a variety of factors into consideration, as discussed earlier, in this chapter we will meet criticality and its implications with other parameters in Chapter 23.

Chapter 6

Condition Monitoring

NEED OF CONDITION MONITORING

The shutdown of a modern, complex, and continuous running plant is very expensive. Hence, the management is anxious to ensure the largest possible number of working stream days in each of its plants every year. The success of any such organisation depends on continuous, safe and cost effective operation of its plant, where breakdowns without any preliminary warnings have to be avoided by all means. This necessitates a felt-need to know the exact condition of the critical parts/components in a continuous running plant which can fail, and thus cause a breakdown without warning.

This means diagnosing the exact conditions of the plant/equipment by means of on-stream, non-destructive testing to avoid unnecessary shutdowns. Today, a large number of methods have been developed for accurately measuring the factors and characteristics of major parts, without affecting these parts in any way either physically or functionally. These actual measurements can be compared with predetermined acceptable standards or against those set as given by OEM, to enable the Maintenance Manager to arrive at a decision as to whether the characteristics depicted by these measurements are within the acceptable limits or not. Non-destructive testing facilities and methods have been of tremendous help in condition monitoring of equipment while in use.

Condition monitoring of a machine or equipment can be easily compared to the health monitoring of an individual by diagnostic tests/checks/instruments. For example, the annual medical check-ups which executives undergo each year is in reality a boon for them because they are told by the doctors/specialists if they are perfectly fit medically, or certain symptoms are indicative of a malfunction and there is need for certain corrective surgery or medication to control these. Normally, persons around 45 years of age do undergo a series of check-ups under company directive, which include, for example taking an electrocardiogram (ECG), indicating the condition of their heart, a blood pressure check-up MRI scan and a blood sugar level test on fasting and after breakfast to monitor the blood sugar levels. All these tests are in fact health condition monitoring checks. If the blood pressure is high, then the doctor may advise reduction in weight and recommend regular exercises, and lay down certain dietary restriction such as reduction in the intake of salt, oil, junk food, fatty items, etc. If not controlled with these cautionary measures, then certain medicines would be prescribed to bring down blood pressure or maintain it within acceptable limits. If these checks are not done at regular intervals, then a person may continue to have high blood pressure and be ignorant about it. If it continues to go unchecked, it may unfortunately lead to a heart attack or a stroke, that is, a breakdown without warning, leaving the patient sick, and in need to be hospitalised and treated for a long time. An unwarned breakdown causes both a loss of health and happiness and imposes a heavy penalty in costs for treatment, serious danger to safety, and to life itself. Thus, it will be seen that condition monitoring checks can indicate where problems are causing trouble and provide precise information on the health conditions and thus enable one to arrive at certain decisions well in advance of the actual breakdown. Condition monitoring involves surveillance, that is, obtaining of data specific to the particular equipment, and coming to certain conclusions known through 'diagnosis' by analysing the data.

PLANT AVAILABILITY

It is an accepted fact that preventive maintenance practices do undoubtedly increase the on-stream availability of plant and machinery. However, the main difficulty remains in making the correct choice of preventive maintenance intervals, which are initially devised on the recommendations of the OEM and our own experiences. But, despite making the choice of intervals based on these two factors the problem which still continues to bother the Maintenance Manager is whether he is overmaintaining or undermaintaining the equipment, for both are undesirable. Then, the question still remains as to what should be the optimal maintenance interval and how can it be decided upon and arrived at? These choices are important, because:

(a) Overmaintenance or too frequent maintenance increases expensive down time resulting in added cost due to expenditure incurred on maintenance men and material and waste of precious time.
(b) Too much handling may result in familiarity, breeding contempt which can lead to carelessness and hence to malfunctions caused due

to human errors, in wrong assembly and the replacement of original spare parts with spurious ones, or due to following an *incorrect technical practice*. These may cause a malfunction to occur in a perfectly healthy operating machine.

(c) Undermaintenance or too long an interval between two successive checks may also result in high incidence of failures, which is *not* acceptable.

It may be better, therefore, to follow such a system, which will not be *calendar based* but *condition based*. This would mean that we should have a continuous knowledge of the machine's condition based on certain critical pre-decided parameters. Also, we should know the changes taking place on these parameters as time goes by. In a preventive maintenance system, a machine is taken on preventive maintenance checks, depending on a predetermined cycle, irrespective of its actual condition, whereas in the case of predictive maintenance, a machine is taken for maintenance only when its status or condition so demands. This is also known as *condition-based* maintenance.

If trained personnel are employed for the analysis of surveillance data, then they can allow a plant to be safely kept in operation as long as the predetermined monitored parameters remain within the acceptable limits. A look at Figure 6.1 will make this clear.

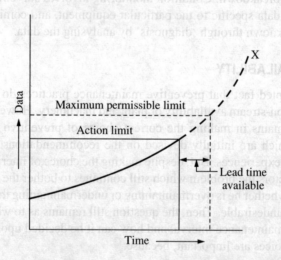

Figure 6.1 Analysis of surveillance data.

In the majority of cases that fall under the category of *mechanical failure*, the actual failure occurs gradually over a period of time. It is the culmination of a slow deterioration process that takes place regularly. Here, an appropriate parameter could be continuously monitored, so that one can detect the commencement of this process and allow the plant to continue to operate till the maximum permissible limit is reached. The important thing is to monitor the condition of the machine continuously and to know what constitutes the critical conditions of the monitored parameters, to decide on allowable tolerance levels.

PREDICTED TREND

'Action limit' curve (Figure 6.1) shows the prescribed limit within which certain monitored parameters must remain for normal operation. This curve when extrapolated can indicate (the *maximum* safe permissible limits of operation, that is, the point up to which the tolerances permit safe operation. Thus, the analysts of monitored data can permit the operation of the plant till the maximum permissible limits are reached, thereby providing the Maintenance Manager a warning, and making available enough lead time between the action limit, and the maximum permissible limit time, to accurately plan for the maintenance.

The objectives of preventive maintenance and condition monitoring are to provide (a) increased on-stream plant availability and reliability, (b) increased plant and personnel safety, and (c) optimal maintenance costs. These are now discussed.

INCREASED ON-STREAM PLANT AVAILABILITY AND RELIABILITY

As already explained, in the *predicted trend*, a given machine can continue to be kept in operation as long as the monitored parameters continue to remain within the laid down limits. Between the action limit and the maximum permissible limits, there is enough time available with the maintenance personnel to make adequate advance preparations. Thus, the total down time can be cut down as all the preparations would have been made in advance. Particularly so, as it is also known as to which particular part of the machine will need replacement and which would need special attention. It may also be noted that if it is a critical machine on which the entire production function depends, then even a marginal increase in the running of the plant, (provided it is possible to do so) will more than offset the cost of operating and installing the condition monitoring system. Also, a sense of confidence is created amongst the production personnel, when they can see for themselves what the monitored data and its analysis and its predictions are, and based on these, verify that the machine can actually be run safely till it reaches the maximum permissible limits.

PLANT AND PERSONNEL SAFETY

Advanced technology plants of today operate at very high speeds, high temperatures and pressures. If any one of these factors were to go out of control, the consequences can be well imagined and in certain plants, the process fluids can be toxic, inflammable, corrosive or poisonous or a combination of more than one of these injurious substances. Hence, it is to be accepted that sudden failure of such equipment without any warning can, and often does cause serious damages to plant, personnel and at times even to property. Hence, a properly designed condition monitoring system could with ease help avert such disasters by giving sufficient warnings of an impending failure.

One of the major contributions of the use of condition monitoring devices is to increase safety. The incorporation of condition monitoring system which gives warning of impending dangers to operating personnel in terms of audio or visual signals is a must in all high-tech plants, as soon as the machine or plant approaches the maximum limit of safe operating parameters, it gives out these, often, life-saving signals. A variety of such devices are fortunately being put to more and more use in the industry today.

OPTIMAL MAINTENANCE COST

The cost of maintenance consist of (a) maintenance man-hour costs, (b) spare part costs, (c) maintenance consumable costs, (d) tools, test-equipment and other utilities, and infrastructure facilities cost.

Condition monitoring tells us in advance the following information:

(a) When a machine is to be taken on maintenance, based on its condition, and so it is only taken on when the condition so warrants it. Thus, one has enough lead time to plan accurately and in detail for this chore, thereby reducing the down time and ensuring the optimal utilisation of maintenance manpower.

(b) Whether any particular part is likely to fracture, and cause a breakdown. However, as we do not use the plant, thanks to this foreknowledge, till the time where an actual breakdown occurs, we can minimise the cost which would have been increased in making its replacement by attending to it in time. Thus spare part costs can be reduced, and the stocking of spare parts can also be better controlled and managed, depending upon the lead time available for obtaining spares.

(c) The requirements of spare parts, the consumables, and of requisite tools, can therefore be planned and obtained well in time, thus reducing the total down time. It is a well-known fact that quite often the expensive down time of a machine is not effectively utilised and the down time gets unnecessarily extended due to lack of anticipation on the part of management of the possible or future requirement of spares, and the non-availability of tools and test equipments, thus resulting in the inefficient use of the costly infrastructure facilities which have been created.

METHODS OF CONDITION MONITORING

Condition monitoring can be carried out when the equipment is in operation, which is known as *on-line*, or when it is *off-line*, which means when it is *down* and not in operation. While on-line, the critical parameters that are possible to be monitored are speed, temperature, vibration and pressure. These may be continuously monitored or may be done periodically (time-interval). Off-line monitoring is carried out when the machine is 'down' for whatever reason, and the monitoring in such a case would include crack detection, a thorough check of the alignment, and the state of balancing and the search for tell-tale signs of corrosion and pitting and so on.

A large variety and choice of monitoring equipment are available in a wide ranging spectrum of cost, complexity, sophistication and versatility. Electronic measuring and recording devices are also available for those situations where monitoring of a large number of points have to be done simultaneously.

CHOICE OF EQUIPMENT FOR CONDITION MONITORING

The big question that arises is as to which of the machinery and plant in a concern have to be chosen for condition monitoring; a blanket coverage would neither be necessary nor meaningful in the context of incurred cost. One has to decide which of the equipment or plant is not essential for production. The decision will depend upon whether their breakdown will in any way hamper production; if they do not, then such machines can straightway be eliminated. Hence, to arrive at a reasonable solution, the entire plant, equipment and machinery will need to be divided into two broad categories:

(a) Critical
(b) Non-critical

These have been discussed in many chapters including "Design and Maintenance" and "Beyond Cost-criticality".

Some of the major criteria for this decision-making process will have to be thought out with care and clarity. These are briefly outlined:

(i) *Criticality of the machine/equipment with regard to the process or production:* One must ask whether the production work can continue if this particular equipment/machine is unserviceable, or this equipment can be short-circuited and the process still continued, or its task can be switched over to another machine, that is to say, is a standby available? If the answers to these questions is *Yes*, then the machine is *not* critical. On the other hand, if the answer is *No*, then it is critical.

(ii) *Availability of standby:* If the original equipment is critical, then the 'standby' will also be a critical equipment and would need the same attention/care, and will always be kept in a fully serviceable state. But the standby will not require condition monitoring when not in use.

(iii) *Hazards involved in the case of failure:* Whatever be the implications in terms of production, one must find out how dangerous a particular piece of equipment can be in case of failure and then decide its criticality. If it endangers safety, property, or life, then it is a critical item and needs condition monitoring and the incorporation of a warning system to help avert a failure.

(iv) *Downstream effects:* What will be the effect of a breakdown, down the stream? Will it affect the ultimate production? Will it create a loss of any process material? If so, how expensive will it be?

(v) *Cost of down time:* How high is it? If it is disproportionately high, then condition monitoring is needed.

(vi) *Cost of condition monitoring equipment:* This is to be weighed against the costs that would be incurred in case condition monitoring is not required.

After having categorised the machine/equipment into the above-mentioned three categories, we will find that the critical equipment will invariably be the most complex and expensive of all, and all its parameters would need monitoring continuously and would be in need of having a standby. The semi-critical machines would need monitoring but to a lesser extent. And the non-critical equipment will require only periodic checks of certain very important parameters, or may even need no monitoring at all.

Let us take the example of the most important and commonly used monitoring equipment, viz. the *vibration monitoring* device used for diagnosing vibration related problems in all types of rotating machinery.

VIBRATION MONITORING SYSTEM

Vibration monitoring techniques can be chosen depending upon time, effort, level of sophistication and cost. A simple portable vibration-meter with a properly matched accelero-meter can be made use of to get the root mean square (RMS) value of the vibration velocity, displacement or acceleration. The RMS value of these signals gives very accurate information about the condition of the machine, and is believed by experts to be the single most important information, representative of the condition of a machine.

The vibration signals can be fed into a spectrum analyser for a frequency analysis. Alternatively, the signals can be recorded on a magnetic tape and analysed later on a large variety of frequency analysers which are available in the market, starting from the most inexpensive portable ones to large, complex and expensive laboratory models. If the machine is very critical, then, an on-line continuous monitoring system can be used, which will monitor the vibration signals received from pre-selected points on the machine and continuously keep analysing these signals as long as the machine is operating. As a further sophistication, this entire process can be controlled by a computer.

Monitoring is a vital decision, which is based on criticality, likely damages that can occur in case of neglect, and the machine's history of failures, and its analysis. Inspections are normally carried out fortnightly, depending of course, upon the criticality.

After recording, the vibration readings are compared with the standard vibration as recorded in the initial stages of the use of any machine or that given by OEM. Having collected this data over a period of some months, a trend can be plotted. Thus, an 'alarm' limit can be set which, if exceeded, would mean that the vibration level of a machine is nearing the 'dangerous' level. The fixing of a 'maximum allowable' level and an 'alarm' level of vibration are a difficult task, for they are based on a number of unquantifiable parameters such as past experiences, operating parameters, environmental conditions and personal experience and expertise.

CONDITION BASED MAINTENANCE

In the final analysis, predictive maintenance or condition based maintenance (CBM) is very cost effective in those areas where the cost of unplanned breakdown is very high. Predictive maintenance provides the ways and means by which impending failures can be detected in time, for with advance information made available, remedial actions can be planned for, thus improving overall efficiency. Even the useful life of certain components can be increased beyond their recommended life span (calendar life or running hours). However, it should be remembered that effectiveness of condition monitoring depends upon the availability of specialists and highly skilled back-up staff trained in the analysis and diagnosis of monitored parameters.

This method is being accepted more and more in today's sophisticated and high-technology plants, because of the (a) capital intensive and continuous nature of operations, (b) high cost of breakdown and shutdowns, (c) severe operating conditions, (d) specialised, sophisticated technology, (e) high safely hazards and consequent penalties.

An extremely important area of advantage for CBM is the conservation of energy. A lot of energy waste takes place due to the improper running of machines. Excess power consumption is due to poor condition of 'bearings', heavy friction losses due to poor lubrication, overheating, sparking/short circuiting due to incorrect insulation/connections, over loading, etc.

A healthy alliance between *predictive maintenance practices* and *modern energy management techniques* can lead to an effective *energy conservation programme*. Vibration monitoring has a wider area of application, and is regularly being used for monitoring compressors, fans, turbines, motors, pumps, generators and other allied components in the industry. According to an estimate, if proper vibration monitoring systems are applied to such equipment which need this type of monitoring, then a possible saving of approximate by 30 per cent of maintenance cost can be attained along with a bonus of about 15–18 per cent savings in energy cost. This kind of a saving ultimately improves the total productivity of an organisation, and thereby improves its profitability as well.

Chapter 7

Non-destructive Testing (NDT)

NDT CONCEPT

Non-destructive testing methodologies have been designed for the accurate measurement of characteristics of parts of a machine, without in any way affecting that part *physically* or *functionally*. These testing facilities are developing very rapidly all over the world, and are progressively being made available indigenously as well. They are frequently being used and relied upon, for application and use in *predictive maintenance practices and systems*. The operating parameters which can define the health or condition of an operating machinery need to be monitored thoroughly.

In order to monitor the health of a machine, there are several parameters; these parameters are now discussed.

(i) *Process variables:* These include pressure, temperature, flow of process fluids, lubricating oil, seal oil, and cooling water circulation.

(ii) *Mechanical running condition variables:* These variables are:
   (a) Temperature of vulnerable machine components
   (b) Mechanical vibrations
   (c) Machine RPM
   (d) The relative motion of moving elements with respect to stationary elements of a machine
   (e) Crack detection.

(iii) *Rotating machinery malfunctions.* These include imbalance, misalignments, foundation problems, fracture and warpage of casing due to a variety of loads such as thermal loads, and bearing failures.

NON-DESTRUCTIVE TESTING METHODOLOGIES

In order to monitor the above parameters, a variety of non-destructive testing (NDT) methodologies have been developed; some of these are described below in brief:

(a) *Boroscope:* This is a tool that can be inserted into inaccessible places through small openings, and with its aid one can see a highly magnified image of the part under examination, through an eyepiece, a prism, or a reflector which has lighting arrangements.

(b) *Flexiscope:* This is an instrument that is used to see contoured surfaces and U-bends, which are otherwise not within easy reach or view.

(c) *Liquid dye penetrant:* This dye can seep into minute surface openings by capillary action to detect surface cracks, porosity and lamination.

(d) *Magnetic particle detection:* This technique is used for locating surface and sub-surface discontinuities in ferromagnetic materials. When a test piece is magnetised and finely divided, the ferromagnetic particles sprinkled over the surface forms an outline of the discontinuity formed by the magnetically held collection of particles, indicating the size, location, shape and extent of the problem of discontinuity.

(e) *Eddy current testing:* This method is employed to measure electrical conductivity, magnetic permeability, grain size, heat treatment condition and hardness. The eddy current detects seams, laps, cracks, void and inclusions and sorts out dissimilar metal compositions.

(f) *Ultrasonic devices:* These devices use beams of high frequency waves to detect surface/sub-surface flaws. The waves travel through the material with attendant loss of energy and are reflected at the inner faces. The reflected beam is analysed to define the presence/ location of flaws, cracks, laminations, shrinkage, cavities, pores, inclusions, etc.

(g) *Radiography:* Due to the variation in the density and thickness, the radiographed item absorbs different amounts of radiation energy, with the unabsorbed radiation passing through the part. The radiation can be recorded on film, or on to photo sensitive paper and viewed through a radiographic viewer to locate defects.

(h) *Hardness testers:* A variety of hardness testers, such as the Rockwell hardness tester, the rebound hardness tester and other allied tools, are available today. They are all portable and can indicate resistance to penetration and wear.

(i) *Creep tester:* Creep is a phenomenon that results in permanent deformation and damage when an item has been operating at high temperatures and stresses and measures the changes that have come about in the dimensions. This has to be done at periodic intervals.

(j) *Spark testing:* In this test, a visual examination of the spark pattern that emerges when an alloy is held against a grinding wheel is used as an indicator for classification of ferrous alloys according to their chemical composition.

(k) *Leak testing:* The assembly is filled up with air up to a specific pressure, and then a testing is done with a soap solution, halogen, helium or freon, to locate the leak and this is called *leak testing*.

(l) *Thermal testing:* The measurement of temperature can indicate abnormal or overhot conditions in thermal testing the measurements can be done by making use of contact thermometers, heat sensitive paint, heat sensitive stickers, infra-red thermometers or optical pyrometers.

(m) *Acoustic emission testing:* Acoustic emission is defined as a *high frequency stress wave* generated by the rapid release of strain energy that occurs within a material during crack development or plastic deformation. This method is capable of detecting the minutest of increasing flaws. There is no other NDT method which can match its capability.

(n) *Holography:* Holography is a three-dimensional image of a diffusely reflecting object having an arbitrary shape. Both the amplitude and the phase of any type of wave motion emanating from the object are recorded by encoding this information in a suitable medium. This reading is a *holograph*. It can be obtained by using visible light waves and is known as *optical holography* or ultrasonic waves called *acoustical holography*. It is used for detecting/locating debonds within honeycomb core sandwich structures, unbonded regions within pneumatic tyres, cracks in hydraulic fittings, stress of all kinds, corrosion, cracking in metals, fatigue in turbine blades, etc. Acoustic holography has been used commercially for inspecting various types and sizes of welds.

(o) *In situ metallography:* The particular part to be examined is polished and etched in its location and the surface structure is transferred to a thin film using suitable chemicals. Then it is examined under a microscope to review the metallurgical changes that have taken place; such as graphitisation, carburisation, intergranular cracking, and grain-growth. This technique is called *In situ metallography*.

VIBRATION MONITORING

Strain monitoring gauges are used to monitor the condition of parts subject to strains due to variations in their operating conditions. This is used extensively in aircraft testing during design stages and after overhauls during test flights. Vibration in rotating machines is caused due to factors such as imbalance, misalignment, looseness, cavitation, turbulence, wear, bearing damage, coupling damage, and rubbing. As the defects grow in magnitude, the vibration levels increase. By trend monitoring, the faults can be identified.

In vibration monitoring, portable vibration analysers, digital vector fillers, spectrum analysers and oscilloscopes, and the data generated can pin-point the cause of vibration (see Figure 7.1).

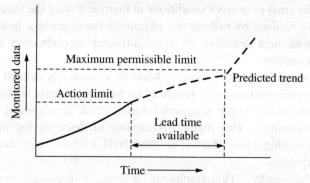

Figure 7.1 Vibration analysis.

SOAP AND FIBRE OPTICS

The major characteristics of vibration analyses are *displacement, velocity* and *acceleration* (which show the presence of vibration); *frequency* (which shows what causes vibration); and *phase* (which helps in identifying the source from where vibration is coming, and is used in balancing); *low frequency spectra* (which gives information about the imbalance and misalignment in bearing); and *high frequency spectra* (which are used to detect faults like bearing damages, gear damage, etc.).

*Spectrometric Oil Analysis Procedure (*SOAP) is a technique used to monitor the condition of machines by analysing the concentration of metal elements present within the used oil samples taken and analysed at regular intervals with the help of a spectrometer. If the lubricating oil from a machine shows any unusual concentration of any metal, then 'SOAP' analysts can show with certainty from which part of (the machine the concentration comes from. For, by comparing the various patterns of wear elements, one can diagnose the problem with certainty without fully dismantling the machine. Such precise determination of the problem helps maintenance engineers to take decisions and undertake planned corrective maintenance at the right time. It also reduces the costs borne on lubricants, by advising on extension of periods between oil drains wherever possible. Thus, 'SOAP' not only minimises premature failures and helps bring about economy in oil costs, but also assists in planning overhaul properly, thereby increasing equipment availability by averting unnecessary maintenance work.

As early as in 1989, a consortium of 30 companies completed laying down of a 13,000 km long fibre optic cable across the Pacific Ocean. This cable of the size of a garden hose could carry 40,000 telephone conversations simultaneously, as compared to 6000 calls handled by copper cables and satellite. Optical fibres can transmit millions of bits of information every second, which means cost savings and high quality transmission, as there are no signal delay or echo on fibre optic calls, which are immune to electromagnetic interferences. Today fibre optic cables have wide applications.

As fibres are so thin, lightweight and sensitive, they are exceptionally suitable as sensing devices. They can be used to measure acoustic signals, magnetic fields, acceleration, temperature and pressure. Sensors of the size of a grain of sand can be placed on the up of fibre and slipped into the human body to test blood pressure, monitor the heart, and measure the temperature inside a tumour which is treated with radiation.

Fibres can also be employed to find oil and gas fields underground. Today, various aircraft designers are developing 'smart' aircraft panels, composed of fibre optic sensors that can sense when a plane is being tracked by radar, or if there is too much ice on the wings. Developments are on, to apply use of fibre optic to defence applications, like improved hyrophones for detecting a submarine.

DIAGNOSTIC INSTRUMENTS A–Z

More and more advanced portable diagnostic instruments are being developed day by day. There are a host of them; some of these are now described, giving their particular functional details:

(a) *Pocket-sized thermistor thermometer* is shaped like a pocket watch with battery and probes, and gives a temperature reading within a few minutes.

(b) *Ultrasonic hardness tester* is an instrument which is used to read surface hardness in *Rochvell C*, of bearing, races, shafts, etc. A light weight probe, which is held against the surface, makes a reading. The answer is delivered in 2–3 seconds.

(c) *Ultrasonic carona detector* is a device employed to hear the 'carona' in the voids in cables or in the splice insulation slung across insulators before the carona can damage the insulation.

(d) *Laser beam source and detector readout* permits the alignment of shafts, fixtures, or structures to be made even when the items to be aligned are hundreds of feet apart. This is done to a precision of 0.001 inch.

(e) *Pistol-grip static meter* measures the electrostatic charge on any surface at which it is aimed from a foot away.

(f) *Portable sonic resonance tester* measures the thickness and soundness of concrete or wood, and can check the uniformity of fire bricks or metal.

(g) *Eddy current tester* has a pointed probe which spots tiny discontinuities on or below the metal surface without touching the object being scanned.

(h) *Pencil-probe leak detector.* A neon light within the transparent probe flashes whenever the point of the probe gets near a freon leak.

(i) *Tension checker* pencil sized for V-belts.

(j) *Thermopile heat flow sensor* can be connected to any vacuum-tube voltmeter, and then it can be calibrated to read the extent of heat loss due to insulation (B.T.U. per square ft. per hour) or to check the efficiency of the different areas on a heat transfer surface.

(k) *Buried cable fault locator* allows even unskilled crew to trace the course of buried pipes or cables to determine their correct depth.

(l) *Phase angle meter* measures the phase angle difference up to 360° between two about to be connected circuits.

(m) *Safety checker for the portable electric tools* can check wrong connections and ground leakage within seconds. This safety device can also check whether or not adequate grounding of each wire has been done.

(n) *Temperature sensitive crayons, stickers, or paint* change colours permanently or melt to indicate when the rated temperature of the items has been reached.

(o) *Stethoscopes* can pick up mechanical problems like bearing problems, which develop within, to overcome external noise and disturbances. There are electronic stethoscopes as well as those commonly used by doctors everywhere.

(p) *Smoke bombs* are used to ascertain the wind direction.

(q) *Thermistor thermometer* is a probe, unlike the pocket version, attached to a flexible lead on portable instruments. It reads the temperature directly and accurately.

(r) *Vacuum-tube voltmeter* is characterised by its ability to read voltage across points in an electronic system without drawing current from the circuit.

(s) *Fibre optic inspection probe* enables one to make an examination of the internal mechanism of a closed or inaccessible gear case or housing. Being a very small probe, it is a very useful tool for the inspection of gear teeth, or for locating broken/lost parts within a rotating machine.

(t) *Stroboscope* is a highly accurate speed indicator. The reading appears on the dial and along with it, it can show vibration, misalignments, belt slippages and hunting. It is a very useful tool for maintenance personnel, but needs adequate experience on the part of the user, if it is to yield good results.

(u) *Circuit breaker tester* enables to check the condition of relays safely by routine testing. This tester is used to ensure that tripping takes place at predetermined values.

(v) *Ultrasonic listening device* can hear frequencies up to 45,000 cycles which are far above the capability of the human ear by translating them electronically into frequencies we can hear. Operators wearing headphones can walk through a noisy plant and can easily locate leaking valves or unlubricated bearing by scanning with a directional microphone.

(w) *Torque meter* checks the output of predetermined tension loading.

(x) *Infra-red thermometer* can measure the average temperature of an area at which it is aimed as in the case of steam-trap or heat radiating surfaces. This can be done without going close to the area, from several feet away.

(y) *Light meter.* This is a portable simple device which measures the level of illumination or quantity of light which reaches the working surface, in units of lux lumens per square metre.

(z) *Stop-watch* is a time measuring device used in work measurement.

LIGHTNING FAST MATERIAL TEST

Ultrasound has been a valuable tool in non-destructive materials testing for years. But the demands of modern production conditions are increasing all the time. A more reliable process now delivers testing results at a rate that is up to a hundredfold higher (Figure 7.2).

Scientists have generated 3D images with the aid of an innovative software. Also, they've increased the testing rate of hundredfold.

Expectant mothers are familiar with a procedure that—when the physician examines them with an ultrasound apparatus that displays lifelike images of the fetus on the monitor. The application of this technology has been customary in medicine for years; in materials testing though, it has been used only in relatively rudimentary form to date.

Researchers at the Fraunhofer Institute for Non-destructive Testing IZFP in Saarbrücken have adapted the conventional sonar procedure—a simple ultrasound method—and have succeeded in generating 3D images with the aid of innovative software. At the same time, they have increased the testing rate a hundredfold.

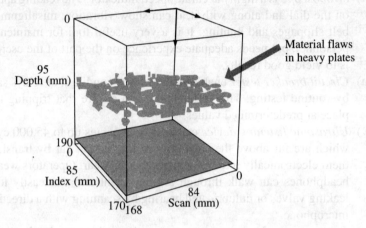

Figure 7.2 Ultrasound.

Many areas of quality assurance or production for the construction industry call for reliable testing methods—be it pipelines, railway wheels, components for power plants, bridge piers or items mass-produced by the thousands, there is a need to ensure that deep down the items produced are free from tiny fissures or imperfections.

For many years, ultrasound has proven a valuable tool in non-destructive materials testing. An ultrasonic transducer radiates sound waves into the workpiece, and the time the signals require to travel and be reflected back indicates where material defects are located. Scanning workpieces in this way is relatively time-consuming, since, an inspection tact can only register a single beam angle. Thus, many measurements must be performed to assemble the composite image suitable for evaluation of inspection results.

However, this approach is too slow if ultrasound testing is to be integrated in ongoing production or applied to large components. That is why Dr. Ing. Andrey Bulvinov and his team at IZFP have developed a method that works at up to 100 times the speed.

We no longer use the sonar method that emits a sound field in just one particular direction. Instead, we use the probe—which experts refer to as a 'phased array'—to generate a defocused, non-directional wave that penetrates the material. What we get back are signals coming from all directions, and the computer uses these signals to reconstruct the composite image.

In a manner similar to subterranean seismic testing, it analyses physical changes the wave encounters in the material—diffraction and heterodyning—and uses this information to determine the conditions within the material itself. He says, "We follow the sound field and calculate the workpiece characteristics on the basis of that. Similar to computer tomography in medicine, in the end we receive three-dimensional images of the examined object—where any imperfections are easy to identify. The startling thing about this approach is that with it, a fissure is now visible even if the ultrasound was not specifically directed at it."

I-Deal Technologies, an IZFP spinoff, markets testing systems based on this principle. "The method is suitable for virtually all materials used in the aerospace as well as the automobile industry, particularly for lightweight materials." Managing Director Bulavinov emphasizes. "Our method is even suited for use with austenitic steel—a type of steel that currently can be tested with traditional ultrasound methods only to a very limited degree," he continues.

Source: www.industry20.com

Chapter 8

Total Planned Maintenance (TPM)

Total maintenance planning includes all those activities that *plan, record* and *control* all work done to keep a plant, an installation or factory at acceptable maintenance levels. This includes *long range planning, short range planning* and *day-to-day maintenance work planning*. This is also sometimes referred to as total productive system.

PLANNING SYSTEM COMPONENTS

Long range maintenance planning entails extensive consideration and precise assessment of maintenance requirements for the proposed production plans. *Short range planning* makes arrangements for taking care of the overhaul of newer equipment, the installation of testing facilities, the creation of facilities matching the needs of the newer NDT and surveillance techniques, and the training of manpower to meet the demands placed on them to man these services. All these are planning functions and must be carried out effectively so as to achieve good results. The *critical path method* (CPM) is an excellent planning tool, and its use can result in making effective time and cost estimates, and bring about savings in time and cost by improving the control mechanism. Maintenance planning is one of the most important and a useful concept which is being put to use in modern-day maintenance management practice to make its working cost effective and efficient.

TPM is the new management tool, about which the Japanese are talking of more earnestly in today's competitive business horizon as the equipment plays an important role on which the total quality of product depends. Total elimination of machine breakdown failures affecting the plants operation is a major concern. TPM mainly focuses on zero breakdowns, zero defects, zero accident, zero wastage, and high morale. TPM with TQM and quality articles. Kaizen + teamwork increases productivity. These result in greater effectiveness work, defect free work, production department attending to repairs, and reduce manufacturing losses.

Total maintenance planning includes the following:

(a) Inspection
(b) Lubrication
(c) Planned maintenance
(d) Preventive maintenance
(e) Predictive maintenance
(f) Corrective maintenance
(g) Modifications
(h) Retrofits
(i) Refurbishing
(j) Overhaul
(k) Replacement of equipment
(l) Discarding of equipment
(m) Standardisation
(n) Material requirement planning
(o) Spares planning
(p) Documentation
(q) Spare parts manufacture
(r) Exercising of spare parts and sub-assemblies held under long-term storage. (Particular attention has to be given to rubberised components.
(s) Training of personnel to fulfil all these roles.

Let us now enumerate some of the necessary elements that go into a planned maintenance system.

(i) *Maintenance plan* helps to prepare a Master Plan for the year, and then it should be broken up into monthly, weekly and daily plans.
(ii) *Maintenance schedules* have to be made for planned/predictive/ corrective and preventive maintenance, for inspection, lubrication, overhaul, and so on.
(iii) *Maintenance control system* is necessary to indicate and initiate activities according to the Master Plan at scheduled periods and to record and assess its progress and control it as per the periodic plans.
(iv) *Maintenance work specification* involves preparation of documents defining the exact work to be done for each type of requirement.
(v) *Resource schedule* provides statement of requirements and their distribution for optimised man-power allocation and other critical resources.

(vi) *Liaison with production* by an accepted and agreed upon principle and methodology has to be laid down, whereby production will accept and permit the time-frame within which maintenance will be carried out.

(vii) *Maintenance records* and documentation of each and every maintenance activity and its analysis has to be done starting with the Facility Register, which lists all items that have to be maintained.

(viii) *Maintenance spares provisioning cell* has to constantly interact with maintenance and revise, update and indent all materials required to meet maintenance tasks.

(ix) *Maintenance support organisation:* This will provide support in respect of technical information, drawings and complete range of technical literature, and make arrangements for tools, tackles and test equipment and calibration facilities.

(x) *Cost control:* This involves a system to correctly estimate maintenance costs, and to control cost.

(xi) *Management information system:* This entails the creation of a reporting system to management.

(xii) *Training:* This is intended to make the personnel attain the required standards for operating the management information system.

FACILITY REGISTER

A Facility Register is basically a complete list of all the machinery, plant, equipment and buildings which have to be maintained. This is the starting point as the Maintenance Manager must know what are the items held in the organisation, which need to be maintained. Unfortunately, many organisations do not have any means of exactly knowing what items they are holding at any given moment of time. In the case of a large organisation, ways and means should be devised to subdivide the entire information for easy reference. This can be done either by usage, by technical practices, or by the maintenance methods which are being used.

Usage classification is done by dividing a plant into sections for maintenance needs. As the maintenance needs of each section is different, the shut down can be done section-wise for maintenance, without affecting the rhythm of overall production.

Classification based on technical practices separates items of different technological streams like electronic, electrical, instrumentation, mechanical, hydraulic or civil. Finally, there is classification done by the maintenance methods used for specific items. Only some items are maintained when an annual shutdown takes place (for whatever reason), some are maintained when failure occurs (as standby is available), some for preventive checks, some for predictive checks, while some are considered as 'rotables' which are placed on the *modular replacement* system from where they are returned to stores, and later sent over to the workshops for overhaul, to be finally returned to the stores as a new and fully serviceable item ready for re-issue. All the information regarding this classification is best stored in a computer. But, medium/small organisations can do this by using a *cardex system*. Care has to be taken to ensure that each card identifies the items it refers to, giving the particulars of the manufacturer, its location, cost etc.

EQUIPMENT RECORD CARD

A typical Plant Record Card or Equipment Record Card needed for TPM is illustrated in Table 8.1.

Table 8.1 Plant/Equipment/History Card

EQUIPMENT CODE: 246789729		ELECTRICAL OVEN				CARD NO. 173/88
		LOCATION: FORGE SHOP				DATE: 14.3.1987
DETAILS OF PLANT: OVEN ELECTRICAL DIMENSIONS 2MX/MX/N TEMP. 200°C						SERIAL NO. PR 312749
						TYPE: EF/2397/87
ANCILLARY EQUIPMENT 1. 2. 3. 4. 5.						PURCHASE ORDER NO: M 793/86 DATED 23.1.1986
MANUFACTURER: WELSE LEY CORPORATION MACORMACK STREET CHICAGO, U.S.A.		AGENT: DUTT & DUTT 6, DALHOUSIE SQUARE CALCUTTA.				PRICE—$5975
						INSTALLATION COST ₹
SERVICES NEEDED						ACQUISITION COST ₹
ELECTRICITY	GAS	WATER	STEAM	AIR	OTHERS	
SPARES SUPPLIED ORIGINALLY						TOTAL COST ₹
S. NO.	NOMENCLATURE	OEM-REF/NO.	STORE CODE	QUANTITY		QUOTED LIFE
LUBRICATION						ESTABLISHED LIFE
NO.	LOCATION	LUBRICANT	METHOD	FREQUENCY		HISTORY RECORD CARD NO.
DESIGN AND OPERATIONAL SPECIFICATIONS						REFERENCES ERECTION MANUAL OPERATING MANUAL MAINTENANCE MANUAL

This record is very useful when a repeat order has to be placed for procurement. This information is extremely useful when seven to ten years have elapsed and the necessary information is not available from any other source. This record card carries fairly detailed information, which proves very useful for consideration by the maintenance and materials executives. The information becomes more explicit when viewed in conjunction with the information contained in the History Record Card. For easy cross referencing, a reference to the History Record Card is made on the Equipment Record Card at the time of its being made.

A sample of an Equipment History Card is given in Table 8.2. Sometimes this item is also known as Log Card.

Table 8.2 Equipment History Card

EQUIPMENT CODE: 246789729	ELECTRIC OVEN	CARD NO.		
PLANT IDENTIFICATION NO.	LOCATION: FORGE SHOP	DATE OPENED		
		EQUIPMENT RECORD CARD NO.		
DATE	DETAILS OF WORK DONE, MODIFICATIONS CARRIED OUT	PARTS REPLACED	MAN-HOURS SPENT	DOWN TIME	SIGNATURE

One should always keep in mind that whenever an equipment is moved from one unit to another, its Log Card or History Record Card *must* move along with it. This ought to be made a mandatory management practice, whenever the transfer of any equipment is being made from one location to another. This card should faithfully record all corrections, replacements, repairs, and other modifications that have been carried out on that piece of equipment from the day it was inducted to the day it is scrapped, or disposed of. The analysis of this recorded data brings out certain important information, like the total down time, the frequency of occurrence of specific faults, and which parts or spares are frequently being replaced. This information must be used effectively so that ways and means can be devised to improve the performance of that equipment. The same information should be used for making standardisation and replacement decisions as it provides documented details with regard to each parameter, which can be used for comparing between different makes or different manufacturers of an equipment.

MAINTENANCE SCHEDULES

A maintenance schedule indicates, what is the work to be done, how often it is to be done, by whom it is to be done, and the estimated time required to complete the work. Separate schedules have to be prepared for each type of maintenance activity which has to be carried out on each of the items as per the Facility Register. This is a Herculean task and needs a man endowed with expertise, skill and experience on the one hand, and enormous patience on the other.

The preparation of the schedules should be planned in a phased manner, and to be of value, it should be jointly done by the Maintenance and Production departments, and this must be adhered to, more specifically when dealing with direct production equipment. A simple maintenance schedule is indicated in Table 8.3.

Table 8.3 Maintenance Schedule

Tradesman	Frequency	Details of Work	Plant Identification No.
			Estimated Time
Electrical	12 weeks	Check insulation and earth bonding	2 hr
Mechanical Fitter	4 weeks	Check slippage of clutch plate and rectify	3 hr
Instrument Fitter	6 months	Remove pressure and electrical instruments, check for damages and rectify	3.5 hr

The guideline for the maintenance engineer should be laid down by the manufacturers. But one cannot depend exclusively on these instructions. In order to have a workable system, the actual conditions of operation, the severity of use, and the skill level of operators, etc. will have to be kept in mind. Then one can arrive at a decision, as to when the maintenance activities can be performed, and whether these can be performed while the machine is in operation or it needs to be stopped.

PRINCIPLES OF SCHEDULING

Maintenance scheduling is the sequential arrangement by which maintenance is done. The decision on a sequence is based on the priority, the availability of spares, material and specific tradesmen. Scheduling can be effective only if there is confidence, mutual cooperation and understanding between the production and the maintenance departments.

To decide on a priority, we have to fall back on the *criticality determination* criterion. Most critical equipment have to receive prompt attention in maintenance caring and should be the first item to be scheduled. The other priorities could be emergency breakdown, preventive maintenance, predictive maintenance and/or other maintenance systems prevalent in the organisation.

A practical and acceptable *schedule* can be developed by following a few simple points which are briefly discussed below:

(a) *Schedules* must be made in two parts: the *long-term* schedule to be made 8 to 12 weeks in advance, and the *short-term* schedule, just a week or two in advance. A long-term schedule will help in preparing and planning in advance for material, spares, tools, and test equipment; it will also help in obtaining consent from the Operations Department to release the equipment for taking them for maintenance. The short-term schedule will be finally broken up into daily schedules for day-to-day implementation and for assuring that close control would be exercised.

(b) The *schedule* should be based on preplanned methods, activities, and, if possible, on scientifically determined time standards.

(c) The total system ought to be clear and understandable and at the same time simple to operate and to implement.

(d) *Revisions in schedules* are expected, and hence, they must be kept flexible enough to accommodate any change. This can be achieved by keeping aside approximately 20 per cent of the capacity for non-priority tasks. Such cushioning can take care of the emergency or priority jobs that would be available during this period. These can be accommodated without causing any major disruptions of work by switching over this capacity towards more critical needs.

(e) Visual aids like wall charts or scheduling boards, should be made use of, which show job status, the date of issue of work orders, the work progress, long pending jobs, etc. This makes it easy for all concerned to obtain the latest information/status of work,

(f) The maintenance requirements have to be carefully balanced against the available capacity to accomplish the tasks.

(g) Important information like backlog report, manpower availability report, and materials and spares availability report are of great use. *Backlog report* is a list of unattended, or if attended, then incomplete jobs, with indication of estimated man-hours required for each. This is used for feeding on to the schedule of the future weeks/days. The spares/materials report indicates the arrival of newly indented items so that pending jobs can be scheduled and attended to.

(h) *Production plan* is issued by Operations Department and includes machine loading and utilisation plans. It helps maintenance staff to schedule machine repairs to a convenient time and opportune moment.

SCHEDULING PROCESS

Everyday in the morning the Supervisor assigns work individually to each worker and keeps him informed as to what work is expected of him the next day. This allows the worker enough time to know what he has to do, and having/had advance intimation, he can make timely arrangements for the spares, the materials and the tools which he will be needing the next day, and thereby get ready for the job to be done in advance. However, while this arrangement may look very easy, it is rather difficult to put it into practice.

For success in scheduling process we should have full information so that the schedules can be of optimal value. The information needed for this are: manpower availability report, latest status reports, materials and the production plans, backlog reports, and maintenance request received, etc. Thereafter, the Maintenance Manager has to depend upon his experience and judgement to draw up a viable schedule, of course, keeping in mind the infrastructure backup available and the capabilities of his men to help him draw up plans which will give the best results.

All these aspects must be discussed with the supervisory staff, and their opinions and comments respected and given heed to, and changes where necessary incorporated in the schedule, based on their suggestions. Normally, this tentative schedule ought to be discussed with die concerned personnel, including the maintenance and production supervisors, die planning personnel and the Maintenance Manager at least a week in advance. Wherever changes are needed, decisions should be taken and implemented. As this becomes an agreed schedule, including the consent of production, it has far better chances of succeeding. This schedule then becomes the firm schedule for the next week and enables one to know about the advanced action to be taken.

It is essential that scheduling meetings are held a week in advance to discuss the next week's tentative plan to acquaint the production personnel of the maintenance plans; obtain their consent to work on production equipment to be released by them; and assure a reasonable guarantee of success for the plan.

WORK SPECIFICATIONS

The exact details of the work to be performed on each item on schedule varies according to the complexity of the equipment, and the type of maintenance being done. In its simplest form, the Maintenance Job Card (Table 8.4) itself makes reference to the volume and the chapter of the OEM manual to be referred to. This kind of work is only possible when highly efficient and educated workers are employed by the organisation, who can be relied upon and trusted to follow the steps indicated in the concerned manuals correctly.

Table 8.4 Maintenance Job Card

EQUIPMENT CODE NO:		EQUIPMENT DESCRIPTION		LOCATION	JOB CARD NO. DATE
MANUAL	VOLUME	SECTION	CHAPTER	SPARES AND MATERIALS	COST

SERVICES NEEDED				
TRADESMEN	NAME	TOTAL HOURS WORKED	SUPERVISOR CHECK AND SIGNATURE	MAN-HOURS COST MATERIAL COST TOTAL COST

Detailed work methods and specifications are necessary, when proper technical control is desired. This is possible when the correct method to work has been laid down by method study and standard times have been generated by work measurement, with a view to improving efficiency and productivity. But even otherwise, the Maintenance Manager has to lay down the estimated time for a job comprising of each of its elemental activities, to help in manpower scheduling and in exercising control.

A sample of a detailed work specification (overhauling a compressor) is shown in Table 8.5.

Table 8.5 Detailed Work Specification

EQUIPMENT CODE: 734567891	EQUIPMENT DESCRIPTION	SHEETS 1–3	SPECIFICATION NUMBER
	COMPRESSOR		
JOB DESCRIPTION	SPECIAL REQUIREMENTS	SAFETY NEEDS	
OVERHAUL	1. OVERHEAD CRANE 2. FIRE TENDER IN ATTENDANCE	HOT WORK PERMIT—YES	
WORK DETAILS	ESTIMATED TIME	MATERIAL SPARES	TRADESMEN NEEDED
REMOVE COUPLING, HUB, BEARING AND SEALS			
LIFT UPPER CASING			
REMOVE ROTOR			
CLEAN INTERNALS AND CHECKS			
CHECK ROTOR FOR CRACKS BY DYE PENETRANT TEST			
PLACE ROTOR, INSTAL UPPER CASING			
FIX SEAL BEARING			
FILL OIL AND CIRCULATE			
ALIGN AND TEST RUN			

Such samples can be prepared and presented in many ways, photocopies can be issued to workers, or they can be put inside plasticised envelopes and issued. A Master Card for each specification is always held in the Maintenance Control Centre.

MAINTENANCE RECORDS AND DOCUMENTATION

Recording maintenance information involves money to an organisation, and it is wasted, if the recorded information is not analysed to aid the decision-making process to improve overall efficiency. Therefore, it has to be decided as to what kind or type of information is needed and the kind of use it will be put to. Only then the depth and details to which recording needs to be done can be decided.

A large variety of forms in different formats are available, from the simplest to the most comprehensive ones for use in the organisation. The management has to take a major decision on the various documents that are to be used and how the actual recording will be done on them.

HISTORY RECORD CARD

A sample History Record Card has been illustrated earlier. This is one of the most useful and essential records which must be maintained in the Maintenance Control Centre. A periodic analysis of this document will help the maintenance in a variety of ways:

(a) The frequently repeating faults, finding out their causes and deciding about the corrective actions to be taken.
(b) Finding out the parts or spares that need frequent replacement and their causes. (This will determine the exact cause and help take corrective action.)
(c) If breakdowns occur soon after the maintenance team has worked on the equipment, then it may indicate a weakness in the maintenance quality or even on the inadequacy of skill level.
(d) Certain repeating failures which occur despite spares being replaced may either indicate the use of spurious parts, or that the spares are in need of improvement from the design or material content point or are in need of being produced under the guidance of a better quality assurance programme. For taking corrective action a dialogue with the OEM through their after-sales service engineers is required.
(e) Certain faults and breakdowns which may occur due to mishandling by the operating staff would need to be set right by imparting training for die proper handling of the equipment. This is possible only if the Production and Maintenance Managers are willing to do the analysis jointly.

(f) Decisions regarding equipment to be selected for standardisation and for replacement becomes much easier when the complete background is available for comparison between the different makes of the same equipment from the points of view of ownership cost, maintenance cost, down time cost and availability.

DEFECT ANALYSIS AND DOWN TIME RECORDS

This is another type of record, where the focus is on *defect analysis*. Appropriate columns are provided as shown in Table 8.6 to record the date

Table 8.6 Defect Analysis Record

EQUIPMENT CODE	PLANT IDENTIFICATION NO.		DESCRIPTION	LOCATION		CARD NO.
DATE	COMPONENT	DEFECT NOTICED	CAUSE IDENTIFIED	ACTION TAKEN	MAINTENANCE MAN-HOURS SPENT	TOTAL DOWN TIME	MAINTENANCE MANAGER'S SIGNATURE

when the defect occurred, the component under question, the defect noticed, the cause of the defect, corrective actions taken, the maintenance time spent on the defect, the total down time of the machine, the signature of the Maintenance Manager.

Here, the Maintenance Manager's supervision is needed or is desirable, because a defect analysis and its proper investigation is a research type of activity which requires a special type of attitude, acumen and involvement.

The analysis of down time, as shown in the format (Table 8.7), if done in close coordination with defect analysis and investigation, can lead to ascertaining the cause and its contribution to the total down time. It can thus focus the attention to specific areas of investigation, to improve maintenance and availability.

We now give an example which shows the down time (cause-wise) for two months.

The two months indicated are June and July 1989. The total down time has increased from 40 hours in June to 50 hours in July 1989, thus reducing the percentage availability of the Boeing 737 from 83.3 per cent in June to 79.1 per cent in July 1989. The most prominent cause is electrical malfunctioning

Total Planned Maintenance (TPM) 93

Table 8.7 Cause-wise Down Time Analysis

EQUIPMENT CODE: 123456789	PLANT IDENTIFICATION: VUADX NO.	DESCRIPTION: BOEING-737		LOCATION: HYDERABAD				CARD NO. 27/1/1989	
MONTH	PERCENTAGE OF TOTAL AVAILABLE HOURS THE EQUIPMENT WAS AVAILABLE FOR USE	TOTAL AVAILABLE HOURS	TOTAL DOWN TIME HOURS	CAUSE-WISE DOWN TIME IN HOURS					
				ENGINE	AIRFRAME	ELECTRICAL	INSTRUMENT	AVIONICS	HYDRAULIC
JUNE 1989	1 2 3 4 5 6 7 8 9 10	240	40	5	6	12	7	2	8
JULY 1989		240	50	6	5	20	9	3	7

which can be seen to have gone up from 12 hours in June to 20 hours in July 1989. Such an analysis can indicate the areas requiring attention and the order of priority in which they must be attended to. This also narrows down the areas in need of detailed investigations. Yet another extremely important outcome of this exercise is that it can show the existence of a trend in frequently occurring faults.

MAINTENANCE WORK ORDER

A maintenance work order is an authorisation to an individual to attend to a specific job. It has complete details of the exact work to be done and records the costs incurred for its apportionment to specific work centre or profit centre. It also lists the facilities, spares tools and tackles required for the job, and the details/time scheduled for the commencement and completion of the task. The priorities are also indicated, along with the type of maintenance to be carried out. Further, the order indicates whether it is to be planned, with occurrence of breakdown or need for overhaul stated clearly. The sample work order card given in Table 8.8 is from the thermal power sector, and hence has an additional column indicating whether there is a permit to work (PTW). This is a major safety need. It is an elaborate form, but the design of the form must be catered specifically to the need of the concerned organisation, and can be simplified.

PERFORMANCE REPORT

The top management is busy and not inclined to have voluminous data being presented to them. Hence, the performance report should be short and simple, so that it can compare the actual performance to the agreed plan. It should enable one to compare the present figures to those of the past performance and draw conclusions from them.

The performance report given in Table 8.9 makes use of two important ratios, to measure the maintenance performance effectiveness. The first is the percentage of total available time that a particular critical machine needed for maintenance efforts. The second is the ratio of man-hours spent on scheduled maintenance jobs to the total man-hours spent on maintenance. If the maintenance performance is good, then the first ratio should be going down and the second ratio going up progressively.

This format can be used for only one equipment and monthly ratios are to be indicated; or under the column 2, the list of a few critical equipment or machines can be included, and month-wise performance ratios can also be indicated. This may need a clear-cut instruction from the top management, indicating for which critical equipment they need information.

Total Planned Maintenance (TPM) 95

Table 8.8 Sample Work Order Card

WORK ORDER CARD				STATION EQUIPMENT NO.	
TYPE PM/ BKD/OH	SYSTEM	EQUIPMENT NAME		LOCATION	WORK ORDER NO.
PRIORITY	FAULT M/E/1/MGR	PTW REQUIRED—YES/ NO TYPE—MECH/ELECT.		DATE PLANNED	
DETAILS OF DEFECT/PLANNED WORK				IMPACT OF BKD (A) SHUT DOWN (B) GENERATION CUT (C) RISK (D) REDUCED EFFICIENCY (E) STAND BY REDUCED	
ORIGINATOR OR PLANNER	SIGNATURE DESIGNATION	DEPT.	DATE	TIME HOURS	MINUTES
DESCRIPTION OF WORK/JOB TO BE ENTERED BY MAINTENANCE ENGINEER				MANPOWER REQUIRED E/M/E/1/C	
				ESTIMATED DURATION	
				START OF MAINTENANCE DATE:	TIME
				COMPLETION OF MAINTENANCE DATE:	TIME
				READY FOR OPERATIONS DATE:	TIME
SPARES/CONSUMABLE REQUIRED/USED				SPECIAL TOOLS AND TACKLES REQUIRED	
CODE	DESCRIPTION		QTY.		
SPECIAL SERVICES REQUIRED				JOB ASSIGNMENT EMPLOYEE NAME	NUMBER
REMARKS: AREA MAINTENANCE ENGINEER				SIGNATURE	DATE/TIME

Table 8.9 Monthly Performance Report

From: Chief Maintenance Manager	To: General Manager		Monthly Performance Report			
Month/ Year	Equipment Name Code No.	Downtime Hours Planned Actual	Available Hours in This month = Working Days × hr/day	Maintenance Performance Ratio = Col 3b/4 × 100	Chief Engineer's Remarks	General Manager's Remarks
June '89						
July '89						
August '89						
Man-hours Spent on Scheduled Work Total Man-hours Spent on Maintenance			= July 1989			= August 1989

INFORMATION CONTROL ANALYSIS

Only a few important and essential representative formats for documents have been indicated. These formats will need to be tailor-made to meet the need for which the document is being prepared. The most important purpose of the document is the information it lends towards analysis, which leads to positive actions to improve performance. The other is the information taken from the records, which may include the following:

(a) Total down time hours, and the trend
(b) Shut-down hours for maintenance during this reference month to the same month previous year or the preceding months
(c) Percentage achievement of monthly planned maintenance work
(d) Ratio between planned work and emergency work
(e) Cost trends
(f) Total maintenance cost comparison for the same items bought from different manufacturers (e.g. compressors)
(g) Trends and guidance on standardisation
(h) Trends and guidance on plant replacement decisions
(i) Spare parts consumption pattern and rate
(j) Identifying plant items requiring high down time or maintenance costs needing detailed study.

A control system determines and controls the actions taken to channelise the activities to be carried out, based on a well-thought out plan, such as a maintenance plan or a production plan.

In the maintenance effort, such a plan includes the following:

(i) *Work order release system:* As explained earlier, the work order is a system of authorisation given to a person to carry out a task as per plans. Hence, the very issue of a work order is the beginning of control, as it is an order to start work as previously planned. This can be done as follows:

(a) The supervisor himself issues the work order to the workers so that he can clarify their doubts and give them clear instructions. But, this system can cause delays, as the supervisor could be busy in an emergency and the whole workforce is left idling and wasting their time while waiting for the verbal instructions to come,

(b) Instead, a board can be used at a central location (MCC) where each worker has a pigeon-hole or a flap marked with his name. The supervisor could place work order as per the plan for each worker in his flap the previous evening, which can be collected by the worker as soon as he reports in the morning for work.

Subsequently, as per the expected time to complete the previous job, fresh work orders can again be placed by the supervisor so that the worker on completion of the first task can come and collect the next work order without any waste of time.

Many organisations now provide a locker for each worker, in which the work orders are placed in advance.

(ii) *Technical literature:* The literature required for carrying out task such as work specifications or the Master Process Sheet (where required) is also attached to the work order.

(iii) *Material and spares requirement:* These are usually arranged in advance and are delivered by another person. If not, then the material requisition slip is also attached to the work order, so that it can be collected on the way before the start of actual work. But this will mean wasting a lot of the workers' time.

(iv) *Tools and tackles:* The procedure is similar to that for materials and spares requirement.

(v) *Quality of work:* Control is possible if there is a continuous check by the supervising staff, or by the shop quality assurance personnel or even by the production supervisory staff. Surprise checks by the Maintenance Manager will go a long way in getting an output of high quality work, and so will appreciation and rewarding of good work.

(vi) *Cost control on labour performance:* This is possible by comparing the actual figures with the planned ones. However, performance can be improved by proper motivation, encouragement, appreciation, impartial treatments, proper training and humane treatment of the workforce. There is no better example than the example set by oneself and the executive must, by his knowledge, skill and managerial capabilities, be able to lead his men properly and set an example in terms of performance for the others to follow. It can be best achieved by the application of standard time to maintenance work. However, it is expensive to instal and usually needs considerable amount of persuasion to get the labour unions to agree to it.

(vii) *Progress of work:* This can be easily seen by looking at the scheduling board, which must be kept updated. Completed jobs must be clearly identifiable; separate colour coding can be used. Whenever there is a shortfall detected in the achievement trends at the periodic review meetings, action plans have to be formulated and implemented jointly by the maintenance and the production departments to bring back the performance, as far as possible, to the level expected, by the original schedule.

(viii) *Feedback of information:* Exercising of control depends essentially on getting a correct feedback. Hence, efforts have to be made to institute a proper management information system for maintenance, which will enable the Maintenance Manager to obtain quick, reliable and relevant information as and when required.

A new system in vogue is known as the cost target programme. The aim of this programme is to encourage, try and achieve a reduction in the maintenance labour and material costs by making available attainable maintenance costs to them for comparison with their actual costs, the main objective being to reduce maintenance costs to the minimum possible level without sacrificing acceptable quality and quantity of production, and achievement of satisfactory standards of output of all production facilities. This can only be achieved by a cooperative and coordinated efforts from both production and maintenance. The main problem is to set the correct target which, though not easy, would yet be attainable with planned and concerted efforts. Thus, targets must be realistic and achievable; otherwise demoralisation would set in.

The most practical and effective way to use the analysed data is to list out (in descending order) those machines that have caused the greatest number of problems and incurred heavy expenses during each preceding month, by concentrating maintenance effort on those machines that have caused the greatest amount of down time, and those which have broken down most frequently and have thus been most expensive to maintain. Instead of tackling imaginary problems, when one has facts and figures at hand, one can certainly cut down failures, cost and down time, and exercise effective control on those machines that are the most expensive to maintain and work towards their continued improvement.

CONTROL SYSTEM FOR PLANNED MAINTENANCE

A basic model of a control system for planned maintenance is given in Figure 8.1. The nerve centre is the Maintenance Control Centre which has all the important and valuable master-records, such as:

(a) Maintenance schedules
(b) Facilities records
(c) History records
(d) Work orders/Job cards
(e) Work specification
(f) Records for defect analysis
(g) Monthly performance reports etc.

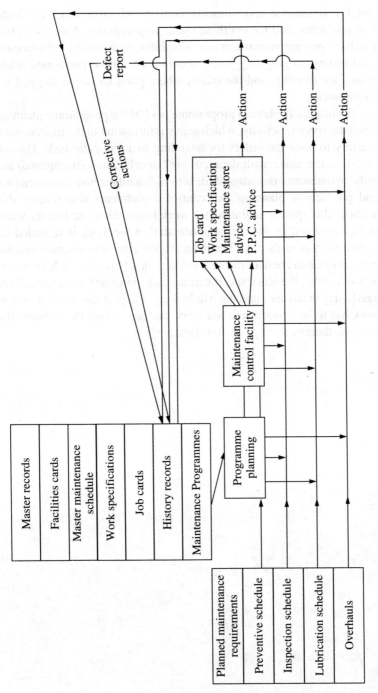

Figure 8.1 Control system for planned maintenance.

Planned maintenance requirements consists of inspection schedules, lubrication schedules, and the overhaul cyclic requirements. Apart from that we have to have preventive maintenance schedules, with checks, adjustments, repairs and testing. Segregating some actions that can be performed while the machines are running, and the others where plant has to be stopped for maintenance work.

As per the originally planned programme per PM the programme planning cell initiates the specific activity, which sends information to the maintenance control facility to issue the orders for attending to a particular task. The job card or work order, along with detailed work specification (if required) are issued to the maintenance department for action. Simultaneously, maintenance stores and production, planning and control departments also receive this situation about this specific maintenance work being taken on. Finally when the work is completed by the maintenance staff, a feedback is recorded on the relevant history records card. The defect report after investigation and the corrective action taken are documented and kept in a Master records room. At times the work order, the work specifications and spares and tools needed are all arranged a day in advance and the whole thing is kept at the location, where actual work has to be performed. And work can begin from the moment the workers arrive, thereby saving precious time.

Chapter 9

Maintenance Turnaround

OUTAGE MANAGEMENT

Maintenance *turnaround* is defined as the planned outage to completely overhaul the plant or machine, so that it becomes almost new and has a full operating life. With the advent of advance technology, and the induction of sophisticated and complex equipment into our industry, the turnarounds have become more difficult and complicated than before and often, therefore, take a long time to be completed. The total cost of turnaround tends to be exorbitant, due to the cost of labour and material involved in carrying out the overhaul, and the cost of production loss due to idle plant. Hence, the need to keep the outage as short as possible without in anyway compromising with the quality of the overhaul, is of utmost importance. This is possible, provided a total advance planning has been done, and the work to be executed has been identified, known and planned for. Properly trained manpower, equipped with proper tools and tackles with all. required/anticipated spare parts is made available along with duly calibrated test facilities, ready to carry out the final checks and tests. Hence, the requirement for advance planning and the need for quality outage management becomes self-evident from the complexity of the problem of coordination. Condition monitoring and diagnostic instruments are increasingly being used in industry today and play an important role in providing us with information on the physical health or condition of the plant. By analysing continuously monitored data, the exact condition of the equipment can be gauged to decide as to what type and quantum of work need to be done and which parts or components need replacement.

This kind of advance information helps us in planning and scheduling the work and manpower needs, to allocate and arrange resources and spare part requirements, and to keep in readiness the specific work specification, the master process sheets, etc. as far as possible. These should be done in advance to avoid any hold-ups during the actual overhaul, thus avoiding any last minute surprises to keep the outage from being under control.

Since the turnaround has to be achieved in as short a time as possible and involves the working of a large number of specialist tradesmen for the job, it creates a serious scheduling and sequencing problem. Progress of work is closely linked to the availability of spares and other materials like sub-assemblies. Hence, the whole planning is to be done and a Master Network prepared. From this network, sub-networks can be made for easier monitoring and control. Only then the completion can be achieved as planned, and an effort made to reduce the outage on subsequent occasions, due to the earlier planning.

TURNAROUND

Let us take the example of a complex, continuous running plant like a petrochemical complex, a refinery, or a thermal power station. In this case, the following areas would need special attention while preparing for a turnaround:

(i) *Planning:* The number of days planned for the turnaround is an important criterion, as it will ultimately decide the availability of on-stream days. Invariably the top management demands a reduction in the outage, and hence there is a shortage of time available as compared to the planned number of days decided upon earlier. This imposes a serious planning constraint and burden on the Chief Maintenance Manager. Such a situation lends itself eminently to planning by CPM which helps in making better time estimates in creating improved controls, and making for savings both in time and cost. To a certain extent, CPM also provides the tools to reduce the total outage time if *crashing* (buying time) is resorted to thereby possibly meeting the top management's time constraints at a cost, provided it is technically feasible and permissible.

Thus, planning will begin about 12 months in advance, to accomplish the general interrelationship and sequence and time estimates, material requirement planning, and its procurement, and if need be the personnel should be trained to required levels of competence during the same period.

(ii) *Material requirement planning:* All materials needed are predicted on the basis of give-off-lists or 10 off-lists, either supplied by the OEM or by the organisation itself. If the organisation has been maintaining accurate consumption details and has made an inventory list of the material and spares consumed in, say, five overhauls, then it is possible to predict the spares requirement list by *trending*. These items need procurement lead times to reach the stores (see Chapter 35). As they start arriving, they should be segregated, listed, lagged, and put in a bonded store. None of the items should be issued for routine operations and maintenance needs, until and unless an emergency warrants it, and is authorised by top management, and a replacement order placed immediately and monitored regularly till its arrival, and re-stocking takes place in the stores at *top most* priority. The MRP must include items needed for major modifications to be incorporated or retrofits needed during the turnaround.

(iii) *Indigenous manufacture:* This relates to the components that would be needed, but are not available through spare parts vendors or OEM.

(iv) *Pre-shutdown phase:* A detailed briefing is to be carried out among all concerned personnel, particularly among the engineers, about the work, the schedules, the resource allocation, specific responsibilities, and target dates as per the set targets. These may include actual briefing on some specific areas of concern, like safety, and it may be necessary to conduct a series of such briefings. It must be ensured that no question posed by the personnel goes unanswered. The material in bonded stores should be shifted to the sub-stores located at *site* for effective control.

(v) *Shut-down:* Actual shutting down of the plant has to be done following the correct laid down procedure, and all necessary actions should be taken so as to permit work to be commenced directly. This may include washing, purging, emptying, certain sub-systems and safety systems actuation.

(vi) *Turnaround:* This is the phase during which actual work is performed as per plans.

(vii) *Start-up:* The time when maintenance completes its task and operations personnel take over is known as the start-up time. During this phase a number of activities are performed prior to the actual start-up.

(viii) *Post-turnaround phase:* All aspects of the turnaround performance are critically examined and analysed, and improvements to be made in future noted down for the next turnaround,

TURNAROUND MANAGER

Process oriented organisations even follow the practice of nominating a Turnaround Manager whose job is to complete the turnaround smoothly, correctly and in time. In fact, he becomes the focal point of all actions and activities pertaining to the turnaround. He liaises with all concerned departments, like materials, production, safety, operations, etc., to ensure effective coordination.

The Turnaround Manager has a variety of tasks to perform and responsibilities to fulfil. Some of the typical actions and steps in planning which he has to take are indicated below:

(a) With the help of reports and discussions held with the staff from different departments or functions such as condition monitoring, NDT, safely, planning, maintenance and operations, he has to finalise the exact list of work which has to be carried out. These lists of major tasks are then to be reviewed, rationalised and finalised. The need for modifications to be incorporated is to be taken into account and provisions made, if certain modifications are to be carried out to the plant as decided by the Modification Committee.

(b) In cooperation with the concerned people, the Manager has to prepare the schedule of work and resource allocation and lay down procedures and systems which are to be followed.

(c) He has to prepare the Master Network and, thereafter, break it up into manageable sub-networks preferably department-wise segments for facility in control.

(d) He has to decide on progress review cycles and the periodicities of progress review meetings and the persons who should attend, the venue of meeting and time. He is the convenor of these meetings.

(e) Particular checks have to be kept by the Manager on activities on the critical path and their progress. A similar exercise has to be performed for the sub-critical activities as well.

(f) He has to install the management information system (MIS) and provide actual information to the top management.

(g) In cooperation with the material requirement planning (MRP) group, the Manager must ensure the availability of all materials which are stocked in the bonded stores before the scheduled turnaround.

(h) He has to decide on how losses in time can be avoided and decide on what should be the method for a quick decision-making process.

(i) He must ensure the availability of trained technicians.

(j) He has to lay down procedures in collaboration with the inspection staff as to how each item will be inspected during overhaul and only those within laid down tolerances will be cleared for reuse by inspection; the remainder will have to be mutilated and scrapped.

(k) He has to devise verification method for correct installation.

(l) He has to do supervisory job and check for start-up procedure.
(m) He is responsible for test run and the recording of important parameters and, finally, the verification and declaration of plant as *operational*.

OPPORTUNITY MAINTENANCE

A plant is shut down for turnaround, depending upon planning and convenience. At times, a major breakdown or failure forces the plant to be shut down at a time before the scheduled turnaround date. In such a situation, the turnaround should be advanced to take full advantage of the down time. In any case, the defect has to be repaired. It is possible to take advantage of such a situation only if the organisation is prepared for it. Quite frequently, certain organisations follow the practice of turnaround due to major failures. To convert a forced outage (due to major breakdown) into a planned overhaul is quite an art and needs proper anticipation, planning and considerable managerial talent.

Turnaround is a vital area of maintenance activity and, therefore, needs the special attention of the management. It must be completed within the scheduled and planned number of days, and every effort must be constantly made to reduce the number of these days. The condition monitoring techniques and their application combined with analysis of the monitored data makes it possible to tune the turnaround optimally. The critical path analysis helps in planning accurately for both time and cost, and its cost-effective monitoring and control. By using modern management concepts and tools, managers and workers can be trained and moulded into a motivated, dedicated and cohesive group of highly trained people who are fully committed to their tasks.

NETWORK ANALYSIS

Today, there is a growing realisation about the importance of the time element in industry, commerce, public services, developmental works and defence. Failure to complete any project on schedule means financial loss and lower rates of return on investment. Hence, there is a need to maximise utilisation of expensive plant, equipment and machinery and other resources. It is essential, therefore, that planning and control of projects is precise and accurate. Warning should be given sufficiently in advance about difficulties that may arise. This will enable the management to initiate corrective actions to execute the projects on schedule or to do replanning, if necessary.

An effective method of planning clearly indicates the interrelationship between activities or planned elements of work making up the project, and shows as to how delays at any particular stage will affect the remaining project, and hence the total completion time, thus enabling reallocation of resources in time to overcome the last minute delays, problems and *panics*. It is an accepted fact that traditional methods of planning by *Bar Chart* or Gantt Chart are good methods only for illustrating progress. They do not provide any direct indication of the relationship that exists between the various activities, nor do they show the extent to which the delays affect the remaining project. These limitations of the traditional methods created a felt-need for more effective tools for planning and controlling.

The *programme evaluation and review technique* (PERT) was evolved by the Programme Evaluation Branch of die US Navy in 1958 for the Polaris Missile System. It enabled the completion of Polaris Missile Project two years ahead of schedule. At about the same time, the critical path method (CPM) was developed by the Du Pont. In the UK, this method was developed a little later, and is known as the Central Electricity Generaling Board's Network Technique, which is used to plan and control the overhaul of capital equipment.

The critical path method has received universal recognition for the planning and controlling of everyday projects, both large and small. Though the construction engineering industry leads in the use of CPM, it has also been extensively employed on research, design and development projects. It is safe to say that this method can be applied with advantage to almost any type of project which requires careful planning. The only limiting factor is the capability and desire of the individual applying this technique.

Network technique is not new. Most managers have been using it in some form or the other, and have called it *commonsense*. However, today they face an enormous increase in the *complexity* of their work. Further, they face the problems of *uncertainty* about future developments. Network specifically deals with these two situations. It provides the help a manager needs, when he is defining the complex relationship that exists in *sequencing* and *time* between so many jobs or planned elements of work in a project. Thereafter, during the execution of the project, the inevitable deviations occur from the planned schedule. The network technique then helps him understand and determine the importance of these deviations and to take most cost-effective corrective actions.

BASIC STEPS IN A PROJECT

The managers are concerned with the planning and control of the project. A project can be defined in different ways—any task that has a beginning and an end and during which each job only occurs once. It has a well-defined objective, a stipulated time schedule and monetary constraint. The management has to carry out two functions—to plan the work schedule, and to ensure that it is executed according to the plan.

The plan has to produce a time-table of work, with each job allocated a specific time-frame for start as well as completion, ensuring that the resources necessary to execute each task or job will be available when required. The main steps are as follows:

(i) *Logic:* In the first step, all the jobs that make up the project are arranged in their correct and sequential order. Here no consideration is given as to how long the job will take or what resources will be required. The methodology used to place the jobs in their right order is mainly graphical and involves the construction of network or arrow diagram. Thus, the management will have its first graphical representation of their project plan.

(ii) *Timing:* The next step is the estimation of time that each job will take and then placing of this information against each job on the network.

(iii) *Analysis:* In every project there is a sequence of jobs, specifying the longest project. This determines the time required for project completion. The first requirement is to find out through analysis the sequence, and to identify it. It is called the *critical path* and all the jobs on this critical path are known as *critical* jobs. The analysis will also indicate the starting and finishing (completion) dates for the critical jobs. All the other jobs are classified as *sub-critical* or *non-critical* as they will have varying degrees of time to spare. The analysis will also determine the stipulated time within which these non-critical jobs can be executed.

(iv) *Scheduling:* Here the resource requirements are considered in detail. Scheduling means deciding the starting and finishing dates of all the remaining non-critical jobs. Thereafter, a time-table is prepared for the project to complete it in time.

(v) *Controlling:* This involves the physical progress of the project against the schedule, and taking corrective actions when necessary. Deviations always occur from the original plan. If the deviations are so large that they cannot be corrected, replanning becomes inevitable because of the changed circumstances.

Once the network has been completed, it is possible to assign details of duration, cost, and resources to each activity. It is also possible to examine schedule and control the rate of progress, costs, and the allocation of resources. The main advantage of network analysis is the logic and discipline it brings to bear on the planning and control functions. It is a valuable technique, particularly for planning and overhaul of equipment, and its control within the time-frame and cost as planned. The details of network planning—critical

108 *Maintenance and Spare Parts Management*

data method (CPM) and programme evaluation review are given in Chapter 35 on overhauling spares and PERT.

Here the photos of 2 lathe machines (Figure 9.1 and Figure 9.2), Benchtop milling of machines (Figure 9.3) and horizontal boring machine (Figure 9.4) are given. These machines are used in most of the factories for turned around. You can understand the complicated nature of overhauling after stopping these basic machines.

Figure 9.1 Lathe machine.

Figure 9.2 Lathe machine.

Maintenance Turnaround **109**

Figure 9.3 Benchtop milling machine.

Figure 9.4 Horizontal boring machine.

Section II
Related Auxiliary Functions

10. Inspection and Lubrication
11. Calibration and Quality
12. Maintenance Training and HR
13. Safety and Maintenance
14. Computers and Maintenance
15. Productivity and Industrial Engineering
16. Activity Sampling for Work Measurement
17. Energy Saving by Maintenance
18. Facilities Investment Decisions (FID) and Life Cycle Costing (LCC)
19. Evaluation of Maintenance Function

Chapter 10

Inspection and Lubrication

WHY INSPECTION?

Inspection is the examination of certain critical machines or parts thereof, of equipment or machines to determine their state or condition. This information is useful for planning and programming of actions which help to maintain accuracy of the machine, and the prevention of unwarranted breakdowns.

A sudden failure rarely takes place as there are always signs of gradual wear and tear noticeable long before an actual failure occurs. Periodic inspection helps detect extent of deterioration and plan for its repairs or rectification, or if need be, even make replacements before an actual breakdown occurs, It is, therefore, a means of preventing breakdowns, and reduces loss of production and the cost of expensive repairs, and many other allied expenditures.

Daily inspections are performed to ensure that, critical equipments are in a good condition so as to continue to perform satisfactorily, and that no such situation develops, which might cause a failure or breakdown suddenly.

The most important question while deciding about the inspection is that which machine should be inspected and when, for one cannot inspect all machines everyday. Hence, criticality analysis would be required to arrive at a decision. Such an analysis would include the following questions about the machine:

(a) Will its failure hold up the production?
(b) Will its failure be a threat to the health and safety of the worker?
(c) Will its stoppage cause wastage of material being processed?
(d) Will it cause damage or danger to other equipment?

Take, for example, the case of a fan or a pump which has been considered critical and will be inspected daily. The inspection routine would involve carrying out checks as per the following checklist:

(a) Any abnormal vibrations, or any abnormal noise;
(b) The temperatures of all the bearings to ascertain that they are at acceptable levels and that they are not at overheating levels;
(c) Leakages from the glands and gauge to see whether they are excessive;
(d) Oil levels in cups;
(e) Grease nipples to ensure that they are not dry.

This is a simple example of an inspection checklist, which indicates exactly what to look for and do. Similarly, a daily checklist for a capstan lathe gear box would include:

(a) Physical inspection of gears for teeth condition and smoothness of meshing;
(b) Check for noise and vibration;
(c) Check for oil leakage or seepage;
(d) Check for any visible cracks or dents on the box.

Different checklists are required for the same machine, depending on whether the machine needs a weekly, monthly or an annual check. Depending on the decrease in frequency, each of these checklists will become more and more complicated and detailed. In the annual check for an air compressor, one has to check the piston rings for cracks and wear, the cylinder for oval shape and scoring, the connecting rods for the condition of their bearings. However, for the same air compressor in the weekly checks, one has to check the condition of the air filter by taking it out for dusting and cleaning. The fan belt too should be checked frequently for its condition and tension and if not perfect, then it is required to adjust the tension.

Thus, checklists are to be prepared for each machine depending on the frequency of inspection, e.g. daily, weekly, monthly. In the field of aviation, the concept of inspection is very critical, and has to be implemented very seriously.

FREQUENCY OF INSPECTION

The frequency of inspection is determined by an engineering analysis, which considers the following parameters:

(a) Age of the machine, its condition and value.
(b) Severity and intensity of service.
(c) Hours of utilisation—Are they prolonged—or intermittent (for 8, 16 or 24 hours a day)?

(d) Susceptibility to wear and tear—Is the machine subjected to dirt, friction, fatigue, stress, corrosion, smoke, ash?
(e) Susceptibility to damage—Is the machine subjected to severe vibration, overloading, abuse, heal, freezing cold?
(f) Susceptibility to losing adjustment during use—Will the maladjustment or non-alignment affect the accuracy or functioning? Will the lack of proper balancing affect performance?
(g) Safety requirements and considerations.
(h) Criticality of item—If very critical, then the item may need daily inspection.

As time goes by, the deciding parameters may undergo change and then the frequency will need to be changed as per one's experience.

PLANNING FOR INSPECTION

The Master Schedule for inspection can be prepared once the inspection requirements of the machines and their parts have been determined, and the frequency need has been established. Attempts should be made to prepare a schedule for the entire year, which can be then broken up into monthly, and then into weekly and daily schedules.

Master inspection schedules are invariably kept in the Maintenance Control Centre (MCC) and serve as guides in starting the work. A sample of the Inspection Master Schedule is shown in Table 10.1. The "×" mark indicates, work not done as scheduled, and a ✓ (tick mark) indicates work completed as per schedule.

Table 10.1 Master Schedule for Inspection

Department (General Engineering)	Master Schedule for Inspections		Month/Year
Equipment	Work to be Done	Frequency	Dates
Air Compressor	1. Dismantle air filter element and clean	Daily	✓ ✓ × ✓ ✓ ×
	2. Check fan drive belt for its condition and tension adjustment	Daily	✓ ✓ × ✓ ✓ ✓
	3. Drain oil and check for metal particles and contamination	Weekly	✓ ✓
	4. Remove oil strainer, check for rupture, clean and put back	Weekly	× ✓
	5. Check value disc for wear, spring for functioning—replace if faulty	Monthly	✓

INSPECTION AND TESTING FACILITIES

It is needless to emphasise that the inspection process has to consider the inspection equipment and testing facilities. It is well known that it is lopsided, as there are plenty of equipment available in research laboratories and testing houses, in universities, and National Laboratories, which are grossly underutilised. However, the industrial organisations are not ready for investing in inspection equipment, and this affects the maintenance process/performance adversely. For instance, it is known that only very few airports have ground equipment needed for proper inspection and identification of faults even for the newly acquired Air Bus A 320, resulting in avoidable delays for the readiness of aircraft and thus for passengers.

Similarly, in the thermal power sector, some plants do not have adequate facilities to check the water system. Unpurified water contains impurities like calcium, iron, silica, sulphates, chlorides and fluorides. These impurities can cause serious damage to the turbine rotor blades and boiler tubes by scaling, and the valve seats by fouling, if used in its impure form. The standard method of obtaining good quality water in terms of alkalinity, conductivity, hardness, pH, etc. is by using a DM water treatment plant. But to ensure quality DM water, checks have to be carried out at periodic intervals. This requires adequate inspection or testing facilities to check silica and chloride content in DM water, pH of DM water just before being filled into boiler, residual CO_2 content in degasified water, etc. For all these, one needs adequate laboratory testing facilities.

Many maintenance-oriented process industries do not have proper testing facilities for even carrying out acceptance inspection to differentiate genuine bearings from spurious ones, recycled in original packing, resulting in repeated failures.

Many organisations do not have proper engine test bed inspection facilities for checking overhauled engines for their operating parameters, efficiency, output and performance. In fact, a proper calibrated engine test bed is required to do justice to this task.

An inspection work order form is used for the authorisation of work and its documentation by the supervisor or the maintenance controller. A sample of inspection work order is given in Table 10.2.

Inspection is a serious maintenance function and responsibility, as it can indicate, through observations, the necessary conditions for planned corrective action so as to avoid unscheduled breakdowns. It may also be necessary that minor repairs/replacements/adjustments may have to be done by the inspector. Therefore, the inspector must be given discretionary powers to decide on his own, if he feels the seriousness of the situation so demands. He should be in a position to:

(a) Ask for a machine to be stopped
(b) Set it right and to let production continue
(c) Ask for immediate attention and help from higher authorities.

Table 10.2 Inspection Work Order

Scheduler	Inspector	Work Order Number		Date
P.G.	A.K.B.	2135/H.I./IPCL		
Equipment	Work to be done	Inspector's Report		
		Observed conditions	Corrections required	Rectified by................ Date ....
Air compressor	Check fan belt for condition and tension adjustment	1. Fan belt frayed	1. Replace fan belt by original fenner (Model)	A.S. Ramana
		2. Tension loose	2. Adjust tension	30th September 1989
		3. Fan belt of non-standard manufacture	3. Check adjustment bolt, put lock-washer	
Remarks of Production Head		Remarks of Maintenance Engineer		Signature of Inspector
Satisfactory P.V. Ganesan		OK R.C. Garg		A.K. Banerji

Therefore, the inspector must have a sound judgement and responsibility, and should be a mature, experienced and skilled person. Hence inspectors have to be chosen very carefully, and given due respect for their actions and decisions. Finally, the inspection schedules must be integrated into the total maintenance planning and then carried out, monitored and controlled with proper documentation. Successful implementation of inspection can show excellent results. Laser measuring tools, surveying equipments, environmental measuring instruments electronic telescopes, must be checked for magnification, focusing ranges accuracy levels, weight, dust/water protection and memory capacities.

LUBRICATION

Lubrication plays a very important and effective role in planned or preventive maintenance and rightly, therefore, it is basic to maintenance planning. Ensuring proper lubrication can and does reduce a large number of breakdowns. More so, in those industries where machine or equipment is exposed to an abrasive or corrosive environment. The role of lubrication where machinery is exposed to the ravages of dust and grime, or to moisture and salt laden sea-breezes, or the punishment inflicted upon machinery and plant in the chemical industry as a whole cannot be overemphasised. Grease and oil are the usual lubricants, as they prevent metal contact with each other and machine movement is easier.

Proper lubrication helps to (a) prevent rust formation; (b) reduce friction, thus reducing wear, scoring and seizures, and economises on power consumption; (c) reduce heat; (d) wash away worn material and particles; (e) increase equipment life (f) increase accuracy etc.

The points/places that need lubrication, the quantity to be used, the type of lubricant, and the required frequency should normally be indicated by the original equipment manufacturers in their operating and maintenance manuals. However, at times, it is seen that such information is either insufficient or not complete and comprehensive. The OEM should also indicate the substitute lubricants, oils and greases. The design of the equipment must provide easy access and the means of lubrication; if not, lubrication piping or channels, which eventually terminate in oil cups, or grease nipples must be provided with maximum and minimum marks indicated to ease the lubrication. Sometimes it is preferable to seek help of the oil companies who will provide help, if asked for, in the hope that the organisation will buy their lubricants sooner or later. They have a special advantage, for they can apply knowledge gained from die problems and experiences of their many users, to the specific problems of the concerned organisation. Another valuable and helpful source is the first-hand experience gained by other users working under similar conditions or by the personnel of their own organisations. Lubricants are usually derived from crude oil.

The quality and quantity of lubricant used have an important bearing on any lubrication programme. Lubricant properties have to be carefully selected to meet specific needs of the machine and its operating conditions. For example, the use of very little amount of lubricant is worse than its excess use. However, excess too can cause problems such as overheating and churning. Hence, the amount of lubricant needed is an important criterion. It can range from total immersion in an oil bath to a few drops only. Suppliers manual on the machinery usually prescribes when, what, how of lubrication.

LUBRICATION PROGRAMME AND PLANNING

Proper lubrication requires a sound technical design for lubrication and a set of proper management and control systems to ensure that every item is properly lubricated. The following steps are recommended in developing a lubrication programme:

(a) Identify and list every item/equipment that needs lubrication. It can be in the form of a register, punch-card or tape.
(b) Provide identification number/code for every equipment so that it can be recognised and cross-referred to in the register or punch-card, including its location in the factory.
(c) Designate every part, point, or location on each of the above equipment, which needs to be lubricated.
(d) Earmark the lubricant to be used for each of the locations/parts by either type or code or as per lubricant manufacturer's code, and indicate substitute lubricant.
(e) Lay down the correct method of lubrication to be followed.
(f) Establish the frequency or periodicity of lubrication.
(g) Identify those equipment that can be lubricated while in operation and those for which a shut-down is mandatory. Use colour codes for such identification, and these should be prominently displayed on the machine itself.
(h) Decide who are the persons responsible (by names preferably) for lubrication.
(i) Lay down lubrication routes by oilers.
(j) Standardise lubrication methods and lubricants to be used to as small a number as possible.
(k) Create proper storage and handling facilities for lubricants, and lay down procedures for lubrication to help in safety and to avoid spillage and waste.
(l) Take recommendations of lubricant manufacturers seriously and evaluate the newer lubricants proposed, to take the best advantage of the latest developments.
(m) Analyse the failures due to too much, too little or no lubrication, and take immediate corrective actions to avoid their recurrence.
(n) When in doubt, use suppliers' manual or get help from machinery marketing people.

The following planning effort can help in formulating a well-thought out lubrication plan:

(i) *Survey:* In establishing a lubrication plan, the first step is to carry out a survey of the plant/equipment/machine to determine the type of lubricant required and the frequency of lubrication. Here the help of oil companies can be made use of; they will be able to provide information about the correct type of lubricants to be used, and the frequency of application. They can also help in the standardisation of use and thus bring about a reduction in the number of lubricants, after the completion of the survey. Needless to say, the fullest of cooperation must be extended to them by the maintenance and the operating personnel.

(ii) *Lubricant requirement:* It will be known after the survey has been made as to which lubricants are required. Then a comparison has to be made between the lubricants that are in use and those which have been recommended by the lubricant manufacturers. The next step will be to rationalise the requirements by reducing the variety of lubricants after due consideration of suitability and compatibility of the alternative brands/types available in the market.

(iii) *Lubrication system:* Develop lubrication charts for each equipment, clearly showing all the lubrication points, the types of lubricants required, the method of lubrication, the quantities, and the frequency. These charts can be affixed on the equipment itself for easy reference. Use a colour scheme or colour code to indicate each spot that needs lubrication. Have a distinguishing code to clearly indicate the spots to be lubricated by the operator of the machine and by the maintenance personnel.

(iv) *Schedule:* Having determined the lubricant to be used and the correct frequency, it is necessary to lay down a schedule of operations. This can be done by having check sheets made for daily, weekly, and monthly work-loads. This check sheet can use standard time or expected time to indicate the time required for lubrication. It must consolidate all information, like parts to be lubricated, type of lubricant to be used, the amount and method of lubrication, and so on.

(v) *Training:* Proper training is the key to success in this field as elsewhere. It should be ensured that each operator and maintenance man is properly trained to do his specified task. He must know and recognise the system of colour or coding used to distinguish and differentiate types of oils, and grease the points to be oiled/greased, the method of lubrication and the exact quantity to be put in and the frequency. A lubrication manual prepared by the Maintenance Manager, indicating the full details will go a long way in helping personnel to do their jobs properly. The oiler must be literate, intelligent, trained and capable of proper work to ensure success.

(vi) *Storage/Handling:* This is of great importance. Keep sealed oil and grease drums under cover in a dry place and observe laid down safety and fire precautions. Transfer of oil from drums to cans, etc., should be done under supervision and clean conditions, using strainers if needed. All lubrication appliances and handling equipment must be kept very clean to prevent contamination or ingression of moisture or of foreign matter. The operator should daily check oil levels. Analysis of machine parts, their material composition are by the mechanical engineer will help the situation. Free running of the machine before loading is another way to check oil. Some industries keep oil man for the purpose.

TRIBOLOGY

A well designed, efficient and effective lubrication plan, when implemented, can give excellent results, by appreciably reducing the down time. However, such benefits may not always be in evidence immediately, but they will surely result in enhancing the life of the correctly lubricated equipment. An important task of the Maintenance Manager is to establish a feedback system which will assure completion of assigned lubrication tasks and a proper follow-up on any deficiencies noticed.

Friction of two materials or parts moving relative to each other causes heat and wear. According to a researcher in the UK, friction-related problems cost their industry over a billion dollars each year. Hence a new term was coined in the UK, known as *tribology,* derived from the Greek word 'tribos' which means 'rubbing'. Tribology attempts to find newer approaches to solve and reduce the problems of friction, heat and wear with correct lubrication. Great improvements have taken place in the recent years with newer discoveries and technologies being extended to improve wear resistance of metals, plastics, and other surfaces in relative motion. Another technique employed towards this end is *surface coating* which is done with the required alloy deposits that can be used to counteract wear. This technique, which provides a highly alloyed cover resistance surface, is being widely adopted today. Apart from standard lubricants, air bearings are being used in gyroscopes and other very sensitive devices, where friction has to be kept down to the minimum. But the best solution to the problem still lies in planning, scheduling and control of a well-thought out lubrication plan and its proper implementation.

Chapter **11**

Calibration and Quality

STANDARDS OF CALIBRATION

Calibration has long been an area often neglected, for it has not received the attention and planned action it deserves. In fact, calibration should form a major area of responsibility for the Maintenance Manager because it is a preventive practice which ensures that measurement and control instruments and other test equipment/facilities are kept within specified tolerance limits and are always kept in a fully serviceable state. It is a *must* to set up adequate calibration facilities and to actually carry out the calibration of equipment at prescribed intervals, because if this is not done, then an uncalibrated test equipment may give outputs or results which are inaccurate, thereby causing a very high level of rejects and rework at a later stage.

The need for calibration cannot therefore be over-emphasised. A well-run and fine-tuned calibration facility will provide a reliable means for the prompt detection of defects, and thus enable the maintenance personnel to take timely actions to rectify the same. Calibrations of measuring instruments, equipments, gauges and production aids, as per national and international standards is indeed a stupendous task.

In order to calibrate, it is necessary to decide on and lay down a *standard*. Once the parameters have been fixed, the maintenance personnel should endeavour to calibrate in accordance with these standards. There are two types of standards for calibration, namely, the *primary standards* and the *secondary standards*. (i) The primary standards are laid down by the relevant body of standardisation in all technologically advanced countries. In India this is the Bureau of Indian Standards (BIS) which till very recently was known as the Indian Standards Institution (ISI). The primary standards are actually the parameters for the standards. (ii) The secondary standards are available with the National Test and Calibration Laboratories. These are designed to the closest possible tolerance to the primary standards and put into actual use.

The mechanical engineer must be aware of ISO 9001–2008 and ISO 9001–2008 and ISO TS 16949 calibration principles, instrument handling and protection, selection of calibration laboratory, knowledge of NABL and ISO.EC17025-2005 standard, requirements for the company to testing and calibration laboratories, metrology principles and preparation of calibration records.

CALIBRATION SYSTEM

For this purpose every organisation must have a manual or document describing its calibration system, which must contain the following information/details:

(a) A list of all equipment measuring instruments, gauges and production aids that need calibration
(b) The calibration procedure to be adopted for each of the listed equipment
(c) Calibration procedure for each of the standards used
(d) Calibration intervals for each of the above equipment
(e) The necessary environmental conditions for carrying out calibration, e.g. controlled temperature and humidity condition, or dust-free environment
(f) Method of implementing the calibration system and its coordination with user departments
(g) Method for providing timely corrective actions
(h) System of ensuring that calibration tags and records are affixed for proper implementation and control
(i) Designated facilities for calibration within and outside the organisation.

The inspection interval or periodicity has to be laid down carefully, based on either the hours of utilisation or the calendar time, e.g. every 1000 hours of use, or weekly, monthly, quarterly, or half yearly use. Obviously, it is based on the extent of usage, and criticality. The system once adopted stabilises over a period of time. Initially, the periodicity of checks has to be at closer intervals, but if the equipment maintains accuracy during successive checks, then the interval between two successive calibration checks may be increased. As a rule, on a scheduled calibration check, not more than 5 per cent of a particular type of equipment should be found beyond the specified tolerance limits at the end of the interval. Then the interval between checks should be reduced until a stage is reached when less than 5 per cent of the total is found to be defective. It must be noted that only by strict adherence to the laid down calibration schedules can one guarantee success. It is essential, that a habit be formed among the users so that they insist on using only properly calibrated and labelled checking and testing equipment.

CALIBRATION DOCUMENTATION

Documentation plays an important role in the process of calibration. A list of essential documents to be maintained in the Maintenance Control Centre for each and every item of equipment or instrument is given now:

(a) Equipment Record Card, indicating the history of its use and present location
(b) Calibration interval or frequency and the next due date
(c) Calibration procedure, checks and tolerances
(d) Actual calibrated values of latest calibration checks
(e) Certificate of acceptance or rejection
(f) History Record Card indicating all the maintenance, repairs, modifications and replacements carried out from the day the item was put to use.

All the test and measuring equipment must carry a label or a lag indicating date last calibrated, calibrating authority (signatures), status of serviceability, and the next calibration due date. A format to illustrate this is given below:

Identification No.
Last calibrated date   ....
Calibrated by engineer   PGK
Status in department   S
Next due date   ....

Label

If for some reason the calibration label cannot be attached to the equipment then an identifying code ought to be affixed to reflect the serviceability status and next due date for calibration.

Calibration is expensive but essential for every piece of equipment that provides a variable or an output which is significant to success. Hence, the entire calibration system has to be planned carefully, taking into account the cost effectiveness. Thus, a mix of internal facilities and dependence on external sources has to be designed. There should be a possibility within the system so designed to cater to the needs arising from any possible problem/mistake/misuse from the user's point of view, e.g. the user makes wrong electrical connection to the meter. It is highly probable that the meter reading would be inaccurate and it needs to be repaired, checked and recalibrated before its next use. Similarly, if a test equipment is dropped or mishandled, or has suffered severe vibration, it will need a recheck and calibration.

In this age of advanced technology, it may be worthwhile considering the setting up of mobile calibration vans, well equipped with their entire calibration equipment, manuals and teams of competent personnel to calibrate equipment at the location of use, rather than to bring them to a centrally located facility. For example, in the field of aviation, often a transport aircraft is fitted with all the required calibration facilities and completes its calibration task as per schedule and is more or less like a flying laboratory.

MAINTENANCE QUALITY

Quality of maintenance like the quality of a product must be designed and built into the system, process, or method of maintenance for *quality cannot be inspected into it.*

There are three principles of quality audit in maintenance system:

(a) Prevention is not only better but easier than cure.
(b) Defects should be detected at the start and not at the time of inspection.
(c) Quality should be built into the process or system; this is better since inspecting it is difficult.

The total quality of maintenance, depends upon well-designed plans, systems, and procedures, and the use of proper tools and test equipment, the adoption of correct technical practices, and the creation of a conducive environment for good maintenance, catering to the needs of trained and motivated manpower, with the top management giving support and realising the importance of maintenance. However, it is well worth remembering that it is always the man behind the machine who really matters. One may have the best technology, the most sophisticated machines, and the best possible system for maintenance, yet one has to remember that it is eventually the man behind the machine who will have to operate and maintain it in order to obtain quality outputs. Therefore, it is his attitude to work, and the quality consciousness, which is ultimately important.

Achievement of consistent quality outputs over a period of time should be the main objective of the maintenance function. To carry out this function effectively, the worker must know and realise, what is expected of him. Regular training at different intervals will enable him always to perform his duties efficiently. But, apart from that, monitoring must be such that the laid down procedures are being followed properly and adequate time and facilities are being provided to him to do his job well. To ensure success, the only right way to carry out the maintenance function is to ensure dial the work is standardised, checked, monitored, and controlled properly.

INNOVATION FOUNTAINHEAD

The supervisor working under mechanical engineer has a major role to play in the maintenance function. In order to achieve success, he has to be physically present at the site, where his men are at work, and motivate and guide them, pointing out their mistakes to improve their performance. He has to be alert and thorough with his inspections and should not allow work of inferior quality, only then he can ensure quality outputs.

For an organisation to be truly successful, quest for excellence should motivate its workers. A properly motivated and trained person enjoys his work. He can reach beyond himself and become the *fountainhead of creativity and innovation* in his work sphere. This is the potential which has to be tapped by the managers in order to achieve the desired level of quality and excellence in work. There can be no better precept than the example set by oneself to motivate the workers, by showing one's own commitment to work, and one's dedication to quality and excellence. This will contribute immeasurably to the progress of the organisation. We will discuss more about this in the next chapter.

BUILDING UP OF QUALITY

Quality can be built into the design of a system or procedure. This can be achieved by taking the following steps:

(a) During any scheduled maintenance activity, always dismantle all items strictly in accordance with the work specification or the dictates of the manufacturer's maintenance manual.

(b) Place dismantled parts/components in correct sequential order on a clean tray/receptacle.

(c) Replace worn out parts only after obtaining certification by superior/ inspector, stating that it has reached the stage of wear beyond acceptable limits.

(d) Ensure correct replacement of rejects by original spare parts.

(e) Segregate items which have to be discarded/disposed.

(f) Ensure such items/spares are mutilated before being sent to the salvage yard so that they may not find their way back into the stores after minor repairs/modifications/reclamation by unscrupulous vendors.

(g) Send rotables to workshop.

(h) Carry out repairs/rectification as required.

(i) Reassemble, and test as laid down.

(j) Carry out testing schedules using proper test equipment. Ask supervisor/ inspector to check and certify serviceability.

(k) Make entries in the documents concerned.

(l) Return item to service.

(m) Form quality circles or voluntary groups of maintenance personnel, analysing defects and suggesting improvements.

(n) Try to achieve zero-defect maintenance.

It is obvious therefore that with care, planning, and nurturing, maintenance quality can be achieved and the desired results obtained.

Chapter 12

Maintenance Training and HR

NEED FOR MAINTENANCE TRAINING

In the present day automated and complex continuously operating plants, it is essential to have as many stream-days as possible, as down time costs are prohibitive. Thus, the availability of such plants for continuous productive utilisation is dependent upon the effective and efficient performance of the maintenance function. Modern plants are precise and capable of sustained output but, as a consequence they have become very complex pieces of machinery, with high degree of technical sophistication (this could not even be dreamt of before a couple of decades back). These machines are in need of a very special kind of caring. These advanced technology equipment used in the automated plant itself make extremely harsh demands on the maintenance personnel. In order to be effective and efficient under these conditions, they have to be trained accordingly. Such a training, however difficult or expensive it may be, is essential in today's context.

The maintenance personnel who were trained earlier have become totally outdated to meet the present-day needs. Today's maintenance personnel need a totally new set of skills and considerable knowledge in a variety of complex fields such as Electronics, Instrumentation, Servo-mechanisms, Computers, Data Acquisition Systems, NDT, Diagnostic Instruments, Robotics, etc. Hence they require better education to be able to absorb the constant upgradation of technology, and to understand the equipment, its functioning and controls, and its preventive maintenance tasks, including overhaul, trouble shooting analysis, and calibration. The training should include operators, fitters, foreman, mechanical engineers and compact precise coverage relevant to the process, position, period, macro- micro-background directed towards availability of the machine.

PLANNING BACKGROUND

Training is a continuing process and it can be successfully planned and implemented, only if the organisation has a proper *training policy* and has the support of its top management. The Chief Maintenance Manager is responsible to plan and train his personnel. He does this usually by making a detailed document, clearly projecting the maintenance requirements of the organisation well in advance (short-term and long-term projections) which are based on long-range corporate plans of the organisation and producing likely forecasts of the types of equipment and their design for the creation of the necessary facilities.

Maintenance is the least glamorous job and a low profile activity in all organisations, as compared to finance, marketing, operations, human resources and materials management! No wonder, you do not have any institutions, universities offering degree/post graduation in the world! It is only touched in a passing manner in industrial engineering, production engineering and mechanical engineering, even though without maintenance the plant cannot function. As a matter of fact machines are purchased and operating systems developed without considering the maintenance with the accountability resting on maintenance. Rarely you witness the maintenance at general managers' level and maintenance professions NEVER reach the board of any company. Maintenance always receives the blame for machine stoppages and never of kudos! In many organisations the maintenance has to be at the beck and call of the production supervisor or even the machine operator for fault repairs. Maintenance manager's stature is never commensurate with his responsibility and accountability. Often the problem stems out of frustration, demotivation, and non-recognition by the shortsighted top management.

The chief mechanical engineer should take the initiative with human relation manager problems relating to advertisement, interview, selection, induction, training, motivation, financial incentives and non-financial incentives of the maintenance staff of his organisation, after shedding off his ego, sensitiveness, emotion, pride, personal differences, likes and dislikes, etc. Training depends upon level of education, skill mix, level of IQ, motivation levels, etc. Less skilled staff requires greater training and supervision.

TRAINING LEVELS

From the maintenance requirement projections, a list of various trades or skills required to meet the maintenance needs can be prepared. Then a demand projection will have to be made, listing the number of each type of tradesmen (according to skill) needed yearly to meet all the maintenance requirements. These requirements can be plotted as a curve, and against this curve one can plot the current physical strength, tradewise and yearwise, making allowances for retirement, deaths, resignations, etc. The gap between the two curves will clearly indicate the number of younger men who need to be trained and inducted in, to fill the gaps created by retiring older people and the additional load created by newer work due to the induction of newer equipment. Development of subordinates and skill mix must form part of training. The training should adequately cover role of supervisors, HR practices, communication skills, team building, leadership skills, self-development, stress-relieving, time management, motivation, team work etc.

An analysis similar to the one mentioned above has to be done with regard to supervisory and executive cadre as well. It would, therefore, be clear that for self-development, proper planning of maintenance training, a very careful analysis of various planning documents of the company has to be done. The whole exercise has to be dovetailed with the human resource development plans of the company. Here the need for advance planning is a *must*, particularly when the organisation is embarking on an expansion and modernisation phase. The forecast and the planning have to be as accurate as possible, if the gap between the *need and the availability* has to be narrowed down and bridged. HR must emphasise on developing positive attitude for self work, interactions with colleagues, motivation, assertiveness, communication skill, group working, psychological instruments for each participant, personality development.

The three levels at which training has to be imparted:

(a) Worker level,
(b) Supervisory level,
(c) Executive level.

The focus should be on 'distressing' the mind as every one says, "I am too busy".

(i) *Worker-level training:* Here we have to train workers for two different kinds of need fulfilment. In the first instance, along with general training, specific trade knowledge and skills have to be imparted, and in the second case, specialised knowledge of a plant machine or process will be the focus of training.

   (a) Along with a general, all-purpose training, specific trade training can be imparted to newly recruited personnel in trades like those of electrician, instrument mechanic, milling machine operator, etc. This is the basic step in developing the skills of the newly recruited personnel.

   (b) Specialised training on a particular machine or plant, on the other hand, has to be imparted to an already trained trade technician. And the training in this instance has to be on the fundamentals of a newer technology or process, or to impart the know-how on a particularly critical machinery, e.g. a machine which is numerically controlled by a computer, or to impart a special skill such as the one required to balance a turbine blade. In order to acquire specialised knowledge, the worker has to undergo intensive and specialised training, which perhaps can be best achieved by sending him to be trained at the Original Equipment Manufacturer's (OEM) works. And on return he can be employed to further train others.

Similarly, workers have to be separately trained for preventive maintenance and overhauls, as skill levels for these two jobs are different in nature.

(ii) *Supervisory training:* The Supervisor, to be effective, needs the cooperation and respect from his men who are invariably becoming more and more skilled in their jobs. Therefore, it is essential that he has a good knowledge of the job, and in fact be the best man on the floor in terms of problem identification and problem solving. Besides, he should have a humane approach in understanding people and their problems which helps to motivate and communicate with the workers in coordinating various activities. These qualities, along with the requisite technical know-how, have to be nurtured and cultivated during his developmental training. Obviously, the training needs will have to be clearly understood, and the training inputs formulated, keeping in mind the objectives of the *supervisory role.*

(iii) *Executive training:* Executive training must focus on decision-making commitment and communication. It has to be divided into separate categories, depending upon the hierarchical segment at which it is being directed. This is usually conducted outside the company by professional agencies.

(a) The training which is aimed at the *top level management,* if it is well designed, is bound to prove to be extremely valuable and beneficial for the well-being of the organisation as a whole. Hence, a short Appreciation Course designed for giving board level executives a macro-level view of the role of maintenance, its costs, and its operational problems is a must, for it deepens their understanding of and involvement in this crucial function.

(b) *The middle level maintenance executives* need more of a Management Development Programme, which will have a balanced mix of inputs like Finance, Industrial Relations,

Spare Parts Management, Material Requirement Planning, Leadership, Conflict Resolution, Group Working, Motivation, Down Time Cost concept, Cost Reduction concepts, technical knowledge on NDT, Calibration, etc. Besides, they also require inter-disciplinary and inter-branch training, e.g. mechanical maintenance executives being trained on Instrument Maintenance and vice-versa. This, besides broadening the scope of the executives knowledge and expertise, improves mobility and removes the many barriers and prejudices which are built into departments kept in water-tight compartments.

(c) *The junior maintenance executives* need a mix of technical (in higher proportion) and management training. Basic problem solving techniques like brainstorming, pare to model, cause and effect analysis, data collection and analysis, bar charts, histograms, pie charts, scatter diagrames, quality control charts, quality circles, innovative power team spirit, participative management, motivation to inspire the team must be included. They can have short courses on the technical aspects of Maintenance Engineering like Plant Safety, Principles of Planned Maintenance, Documentation and its Analysis, and Non-Destructive Testing, The management training can comprise subjects like Management Controls, Decision-making, Man-Management, Industrial Relations, and Cost Control.

TRAINING METHODOLOGY

Depending upon the class composition and the type of training to be imparted, a variety of training aids/methods can be made use of. These include the following:

(a) *Simulators* which can simulate faulty conditions in the system for defect diagnosis and their rectification. This is a very powerful, though expensive, training aid available with the trainer/teacher today: PowerPoint presentation, practice examples, knowledge transfer are also used.

(b) *Sectioned models* of machine or equipment which can explain the entire internal functioning like the sanctioned models of a new engine, or a turbine.

(c) *Wall diagrams and charts* for lubrication, or the fuel flow system.

(d) *Maintenance manuals,* publications, illustrated schedules of spare parts and specific training manuals.

(e) Training by suppliers/AMC/ASS

(f) *Video films.*
(g) *Top management interaction.*
(h) *Computer simulation games.*
(i) *Multiple choice questions.*
(j) *Job rotation, problem solving.*
(k) *Trouble shooting charts.*
(l) *Class-room teaching* by experienced and knowledgeable personnel
(m) *Case studies* discussions, and their presentation.
(n) *Exercises, inspection interchange of experience*
(o) *On the job training* under expert supervision
(p) Lectures, safety simulation, negotiations, mock sessions, business games, social, political, economical micro macro problems technological upgradations. Problem solving situations and interdepartment conflict resolution.

It has to be remembered that the training needs have to be designed and catered for, depending upon the size of the organisation, technology level, and capability in terms of training costs. All the training facilities need not be, and cannot be located within the organisation itself. Certain training facilities available outside can be made use of, like those available at the Indian Institutes of Management. The National Productivity Councils and such other training organisations who can offer training courses to meet a particular organisation's management development needs. Similarly, an organisation should make use of training facilities available with the OEM for technical training.

In medium-sized organisations, one cannot usually afford to have separate executives for each and every specialised function, like Mechanical, Electrical, Electronics, Maintenance, etc. The answer to this problem lies in imparting inter-disciplinary training to their personnel. Moreover, with ever increasing complexity, there is a tremendous need, as each day passes, to have multiskilled workers, except of course for the super specialised areas which require highly specialised skills. Thus, the training should be so designed as to meet all these ends.

Another point worth mentioning is the *rotation* concept for executives, which should be adhered, to, wherever possible. A maintenance executive can be rotated, subject to suitability Material functions, or to Production functions and vice-versa. Of course, this should be implemented after imparting due inter-disciplinary training. This will generate the need to know the 'other' man's job, and to appreciate and understand the other's problems. This growing awareness and understanding are bound to foster a spirit of cooperation and coordination among the different disciplines in an organisation. The training must ensure that the chief mechanical engineer becomes a real leader like

Lord Vinayaka whose small eyes symbolizes the idea that despite reaching great heights in wealth and wisdom one should perceive other to be bigger than oneself.

MAINTENANCE INCENTIVES

It is a common complaint that the maintenance workers often work below expected levels of performance. However, it is possible that the management have not provided the proper environment and motivation to match their expectations of a reasonable performance level. It is noticed that in well-run organisations, where there is a good rapport and understanding between the maintenance and the production workers, and the work environment is good, people are motivated and the down time of the machines is low and maintenance effectiveness is high. Some organisations do provide an incentive plan for maintenance workers, which is aimed at motivating the worker to achieve that *extra* bit or to achieve more and more. On the other hand, in return, the organisation rightly expects higher productivity, better quality of workmanship, and a reduction in maintenance costs.

An *incentive* is a plan to compensate workers in proportion to their output on a reasonably direct and continuing basis. Incentives can be financial or non-financial. The non-financial incentives are designed to provide psychological motivation and have an emotional appeal. They usually help in boosting pride in his work and aspiring towards excellence, quality of workmanship, and achieving outstanding performance. The management responds by giving recognition to the worker by giving a write-up on him in the organisation's news letter, by displaying his photograph at various important locations, by making him "man of the month" or the head of the organisation issuing a commendation letter/card in the presence of the whole department.

Financial incentives are of two kinds: direct and indirect. The indirect incentives are based on profit-sharing of the reward for cost reduction. These incentives usually do not encourage a worker to the desired extent for, in his perception, his own contributions have not been individually recognised and he feels that even though he has worked so hard, in return he has received only what the general run of workers, who had contributed so little, were given. The direct incentive provides monetary compensation or additional pay, proportional to the individual's output or efficiency to be measured according to certain laid down criteria. The underlying idea is extra money for extra effort or work. Such plans do create a tremendous interest for doing sustained hard work for further achievement. In the area of maintenance such plans are designed for individuals or for groups.

For any plan to be successful, it must not only be properly designed and have an adequate basis, but has to be well administered thereafter. Each organisation has its own problems and priorities which have to be considered. Therefore, there can be no standard plan to fit in all organisations. The plan ought to be of such versatility that it provides for the best use of the manpower and of the production facilities. It must also be balanced enough not to create industrial relations problems, but must be designed to improve such relations. This is possible only if the plan is equally fair to all, and no discrimination is practised by those administering it. It should also meet the needs of the worker and is clearly understandable to him.

It should be remembered that the most important aspect of a plan is to ensure that the worker feels and knows that a proper and unbiased evaluation of his work or performance will be done. If the incentive is attractive enough to motivate him and is so designed as to provide equivalent opportunities, it will spur him on to still better performance. The organisation, on the other hand, expects, in return, higher production outputs and better capacity utilisation without sacrificing quality and safety so that they can put a product of acceptable quality at a fair price in the market. If well implemented, this will result in a good industrial relations climate.

INCENTIVES IN MAINTENANCE

To achieve organisational objectives through financial compensation, an incentive plan for maintenance is introduced. For this, attention should be focussed on these management objectives falling within the worker's purview, to which he can contribute directly. However, unlike the production functions, the maintenance functions are rather intangible, and it is not easy to identify or precisely locate the contributions of the individual worker towards these functions.

Our main objective and requirement is to obtain improved work performance through incentive schemes. A substantial lowering in the cost of production can be brought about by extra vigil, initiative, skill and care by the maintenance team, for this will result in lower machine down time, higher quality output and lower and lower levels of rejects and rework. Therefore, it is imperative that, apart from the amount of work completed/achieved, a critical evaluation be made of the effect of this improved maintenance work, and the impact it is having on related areas of work, like the general condition of the machines, down time reduction, frequency of breakdowns, improved accuracy, etc. An area worth mentioning is the area of capacity utilisation which needs to be improved. As more complex and expensive equipment comes into industry, the need for avoiding and reducing expensive down time can hardly be overemphasised. The maintenance personnel can and will do their best to achieve this goal, under an inspiring incentive scheme.

There are two types of incentive plans: direct and indirect, which are presented below. The greatest happiness of the greatest number is the foundation of morale and motivation.

(i) *Direct incentive plans:* These are usually based on the measurement of maintenance work, by any of the measurement techniques in use, like work estimating, activity sampling (or work sampling), analysis of past performance, and standard data, etc. These are now briefly dealt with.

  (a) *Work estimating:* The time required for each job is estimated by a knowledgeable person, which is then used as a standard. It is difficult to say if this can be used to correctly measure performance for incentive payments for, after all, it is only a judgement based on the experience of the estimator. It is a fact that administrative cost of using estimated standards is low, but it may entail increasing the cost of a job/work.

  (b) *Activity sampling:* It is a technique based on observing whether a worker is working or not working at random intervals of time and immediately recording his work performance. The sample performance is used as a basis for measuring his performance for the full working day. This method has two basic constraints. First, the worker learns when he is being watched and pretends to work. Second, the evaluator does not always know if the work being done is necessary or not. Hence activity sampling, though inexpensive, is of doubtful value, if used by itself for incentive payment. It can be used to support more direct measurement methods. This is discussed in Chapter 16.

  (c) *Statistical data from past performance:* At times incentive plans are based on the use of certain factors or coefficients which are obtained from statistical analysis of recorded past performance, e.g. allowances such as those given as relaxation allowances or rest allowances for repetitive work or heavy physical work, machine interference time, etc. Now these coefficients were calculated based on conditions prevailing in the past and, therefore, do not include excess time consumed for operations by factors which are important but are rather vague and difficult to quantify, like improper methods/procedures used, ineffective organisation structure, poor skill level due to inefficient training, and so on. But, today when improvements have been made in these very same areas, the organisation may find itself bound to pay incentive schemes which provide for the inefficiencies of the past, though the reasons for those past inefficiencies have been taken care of today. Hence, there would be a real need for moderating the coefficients/figures according to the present situation and generalised judgement cannot be made by just accepting the statistical data of past performances as a standard. At best the data must only serve as a guideline.

(d) *Standard data:* This data is available in organisations where actual measurements have been done for various types of maintenance jobs. In this instance, various elements of jobs and their times are added on, to calculate the standard time for a specific operation or a complete task. The standard data provides an accurate and comprehensive measurement method for maintenance work. Incentives designed on the basis of standard data have a sound plan, where monetary benefits accrue according to measured performance of the worker.

(ii) *Indirect plans:* Indirect incentives are based on broad indications of maintenance costs. They depend upon cost increases of total maintenance cost, including labour, material, and overhead costs. Such indices reflect *ratio changes* in maintenance costs over a period of time stretching to a few years.

Certain ratios which have been used as maintenance cost indices are:

(a) Maintenance cost as a percentage of manufacturing cost
(b) Maintenance labour cost as a percentage of direct labour cost
(c) Scrap and rework cost as percentage of manufacturing cost
(d) Overtime as a percentage of total maintenance labour cost
(e) Actual operating hours as a percentage of available hours or scheduled operating hours.

INCENTIVE ADMINISTRATION

Reduction in maintenance cost can be an excellent basis for an incentive plan, provided it is possible to make available the maintenance costs incurred over the years with figures indicating the changes that have been taking place. The extra pay is proportional to the cost reduction or its effect quantified in terms of monetary value.

Profit sharing is a form of incentive which needs to be considered, but in the maintenance function this does not influence the workers sufficiently or attract enough attention from them as they are unable to appreciate how their own inputs influence these profits. It is true that reduction in maintenance cost can result in higher profits, but as long as it merely remains a vague quantity, its ability to influence the maintenance worker will continue to remain negligible.

It is to be understood that an excellent administration of the incentive scheme is essential to get the best out of the plan. To implement the same, it is desirable to have a policy document on the incentive plan, which must include:

(a) The objective and rationale of the plan
(b) Measurement methods to be used
(c) Procedure for application of measurement
(d) Unmeasured work assessment
(e) Quality standards requirement
(f) Delay allowances
(g) Time-documentation and procedures
(h) Incentive pay calculations
(i) Methods of appeal
(j) Development of subordinate, commitment and transparancy.

Apart from methods and methodology, incentive administration will always take into account that very sensitive factor, viz. the human emotions, while deciding on a course of action. The formulation and application of good incentives policy will surely prove to be beneficial to the organisation. The training must also cover quantitative techniques like Poison distribution, MTTF, MTBF, and other reliability parameters, discussed in Chapter 30.

Chapter 13

Safety and Maintenance

SIGNIFICANCE OF SAFETY

Safety is an attitude, a way of life, that has to be practised. Safety is an important factor in industry and it is the responsibility of maintenance management to ensure it. A non-existent or even an inferior safety system will invariably lead to accidents which are totally undesirable and unwarranted. Very often avoidable mistakes do occur involving the personnel or machines, resulting in loss, injury or damage. In the case of an accident involving human beings, apart from the grief, the loss, and the disability or pain, the psychological suffering that the injured person undergoes cannot be quantified in monetary terms. The cost to the employer can be computed under many heads, but the medical, legal, and the compensation costs alone can become formidable. The other costs are machine down time, and tremendous worker demotivation leading to loss of production, machine damages and their repair/replacement costs, and the recruitment and training costs for new workers, if the need so arises. It is obvious, therefore, that proper attention to *safety* is extremely important, both for the employer and die employees.

For the employee, it does the following: (a) Prevents having a feeling of physical and mental anguish and suffering to himself and his family. (b) Prevents loss of earning and protects him from losing his earning capacity in future. (c) Protects him from being labelled *accident prone*. (d) Removes the fear which may prevent him from enjoying his work by instilling a sense of fear about a hazardous piece of machinery.

For the employer, it does the following:

(a) Prevents waste and loss of production.
(b) Reduces insurance costs and compensation liabilities.

(c) Prevents unwarranted equipment/machine replacements, which is an expensive exercise on multiple counts.
(d) Prevents criticism from all quarters and thus improves reputation.
(e) Improves the industrial relations climate in the organisation.

Some of the basic human characteristics and altitudes which affect safely are: (a) Indiscipline; (b) Recklessness; (c) Stubbornness; (d) Slow learning abilities; (e) Physical and mental conditions (perhaps influenced by social pressures or by family and personal problems); (f) Fear.

All accidents are preventable and safety is an attitude of life. Life is precious. These characteristics coupled with certain acts/actions lead to the following: (a) improper handling of equipment; (b) ignoring laid down rules; (c) not using protective clothing or equipment, or not wearing safer working clothes for operating a particular machine; (d) operating machines with *unserviceable* safety devices, or even rendering safety devices which are non-operative; and (e) operating a machine without authority.

POTENTIAL HAZARDS

A few of the potential hazards that can cause accidents and their possible remedies are indicated now. This should be read with reliability discussed in Chapter 30.

(a) *Slipping, tripping and falling:* Well demarcated and painted walk ways with *treads* on incline, non-slip floors with guard-rails for handhold and, if need be, safety belts.
(b) *Bumping injuries caused by falling objects and impact:* Proper overhead clearance screen to catch falling objects, elimination of blind comers, proper use of material handling equipment, warning signs for moving equipment, and even painting them with fluorescent paints.
(c) *Catching:* Elimination of projections that can catch clothing, proper passages for entry and exit, and better layout design.
(d) *Burning:* Protective clothing, emergency showers, alarm mechanism, first-aid kits, and training for their use, strict system of checks for maintenance of such equipment that are carrying molten metal, liquid glass, steam, etc. Adequate fire-fighting equipment facilities and training for correct handling in case of an emergency.
(e) *Blinding:* Using goggles and protective shields.
(f) *Explosion:* Proper and careful handling of industrial explosives, chemicals, gases, O_2, H_2, helium, nitrogen, their proper storage, taking

precautions, and safe transportation. Having inside and outside safety distances, fire-breaks for storage, and using automatic sprinklers.

(g) *Fire:* Fire fighting symbol display, having as well as using correct types of fire fighting appliances to fight/extinguish fire, e.g. water, sand, foam, soda-ash, etc. In particular, fire caused due to electrical short-circuiting needs smoke-sensing devices and their correct installation and periodic checks for serviceability. Petrol, oil and lubricants and their storage need special attention as they are all potential fire hazards. Taking safety precautions (their display), and actions in case of fire, training of personnel, disposal system of cotton-waste and oily rags. Safety instructions are usually ignored by people belonging to all professions, for instance, the car insurers give safety pamphlet which are usually ignored by drivers.

SAFETY PRINCIPLES

The top management, particularly the Chairman (i.e. the executive head), should take the initiative in developing a safety culture, and they should be seen (by the employees) to be giving importance to this aspect. The Chairman can be involved by having an internal safety department, with the safety officer directly reporting to him. If the Chairman becomes also the head of the safety department, then he will be seen to be taking interest in the safety of his employees. This will also ensure that all the recommendations are implemented with far greater care and speed, reducing the recurrence of accidents. Thus, safety culture can be inculcated in the entire organisation, only when it emanates from the higher levels of the hierarchy, and percolates down to one and all. When such a situation prevails, higher levels of safety are bound to be achieved.

Using cell phones often removes your membrane in brain and children below 15 are prohibited to use cell phones in US, Europe. Stringing cell phone affect your brain.

An accident is an *unexpected* happening or is an *unforeseen* course of events which occurs by *chance,* whereas an incident is an event that occurs in the normal course. Before an accident takes place, quite a few forewarnings are available. If they are noticed, heeded, and taken care of, most of the accidents can be avoided. The principle of incident reporting and its thorough investigation is to be strictly adhered to know what has gone wrong. To quote David Miller, the former Aviation Safety Director of the US National Transportation Safety Board: "The complexity of modern technology is such that I defy anybody, however good, to predict all operating errors. The role of incident reporting is to find these errors before they become accidents".

Automation which has replaced most of the manual functions has brought about a high degree of reliability. This has led to certain amount of carelessness and this in turn has generated *automation complacency* which results in accidents.

Simulators have been developed and are increasingly being used for basic training, retraining and for conversion training purposes in the field of power engineering and, in particular, in the field of *aviation*. Simulators also help, even if the running of an actual power plant is different and not as it is on the simulator. It has the advantage of being able to simulate all the different kinds of crises one may suddenly face while handling a highly sophisticated piece of machinery which runs by itself. The actual facing up to a situation, even if it is a mock situation takes away the edge of complacency which, if not checked, can be disastrous. So, it makes one alive to the dangers and to realise that it is necessary to remember to follow all the standard operating procedures as laid down and to perfect them.

Disciplinary action should not be taken against those people who confess to having committed mistakes because, if that is done, then the operators will never report their mistakes, knowing that they will be blamed for these, and action taken against them. Waiving of action in such cases brings about more honest communication and thereby encourages a good incident reporting system.

Training has to be given a pride of place in the organisation. Hence, the teacher/trainer is the best possible man for the job and his selection should be one of the more important and major tasks of the management. He should be respected on his own merit in the organisation and there should not be a general feeling among the workers that he has been entrusted with the job as he does not fit well into other roles.

In selecting the person, the consideration should be given to seniority, age, expertise and experience. A senior, experienced supervisor can in no way be, replaced by a few diagnostic instruments. The experience, critical judgement and the capabilities he has acquired over the years can help detect many faults even through a cursory glance, which these diagnostic equipment may not be able to detect. Many car insurance agencies provide safety book to avoid accidents, but the letters are too small that they themselves cannot read. Safety belts are never worn by drivers of car and helmets are never worn by many Indians.

As said earlier, safety is a top management responsibility. Hence, a safety department has to be created, which ought to be headed by a fairly senior executive with sufficient experience. Besides, there should be a safety council or committee which ought to include all the heads of the operating departments, as members, presided over by the Chairman, and the safety officer as its Secretary. The committee must meet at least once a month and deliberate on all past actions and incidents and their investigations, as also the implementation of its recommendations. It must also chalk out a safety plan for implementation in the organisation.

A safety culture or consciousness has to be developed through lectures, slides, pamphlets, programmes, demonstrations, poster displays, suggestion schemes, and the observance of a safety week from time to time. Safety training is an important factor which can help an organisation to move in this direction.

It is also essential that a safety manual be prepared and a safety schedule designed to be strictly followed. The contribution of the Maintenance Manager in these activities is considerable, as he has to be a member for the compilation of the safety manual and schedule and its implementation. For example, he must ensure that the 'Permit To Work (PTW)' is always supported by a safety document materials, at least one, related to work on electrical, chemical, mechanical equipment and hot work safety permits, etc. are scrupulously followed:

HOT WORK SAFETY PERMIT

Hot work safety permit on any oil line, oil tank, lines transporting hydrogen gas after completely isolating and washing by water and/or stream proper venting of lines or tank must be ensured. They should be locked so that they are hot in operating condition and will not be available for operation. After isolating oil pipe, pump, tanks for H_2 pipe line or inflammable materials, at least one fireman with extinguishers and other fire fighting equipment will be stationed near work spot. The Officer-Incharge of fire station will be present with others standing by.

The Hot Work Safety Permit (Table 13.1) will have a safety document on fire safety which will be strictly adhered to during operation.

A permit is valid only for 8 hours shift period. If work extends beyond that period, fresh permit should be issued.

Hot Work Safety Permit is issued on any plant/equipment on which hot work may lead to a *fire hazard,* etc. in a thermal power plant. The potential areas of fire hazard are:

(a) Hydrogen generation plant
(b) Area near turbine, oil system
(c) Main oil tank
(d) Turbine lubricating oil pump
(e) Seal oil system
(f) H_2 gas feeding system of turbo-generator

Table 13.1 Hot Work Safety Permit

Hot Work Safety Permit	
Permit No.	Date ....
Location and details of work and machinery to be used ......…...............................	
Details of precautions to be taken..	
Permit issued to	
	Name ....
	Signature ....
This permit valid from …….. to …….. hours	
Certified that site has been inspected and Isolation and Safety arrangements have been made.	
Signed by	Signed by ....
Officer-Incharge Fire Station	Shift Incharge/Engineer
Certified that all men and materials have been removed and equipment is safe to be put into service.	
	Name ....
	Signature ....
Permit Closed by Officer-Incharge Fire Station	
Officer-Incharge Shift Engineer	

  (g) Turbo-generator
  (h) Boiler feed pump lubricating oil system
  (i) Fuel oil pump house
  (j) Fuel oil tanks in main fuel oil pump house and auxiliary boiler area
  (k) Fuel oil lines from fuel oil pump house to main boiler and other fuel firing floors
  (l) Fuel oil lines of auxiliary boiler fuel firing system.

Each of these is indicative and has to be modified to suit the needs of the concerned industry, its criticalities, dangers involved in the process, and technology and safety needs. The samples of PTW, etc. which have been shown here, are from the thermal power sector.

FAULT TREE ANALYSIS

Fault tree analysis is an analytical tool, which helps in closely examining critical and hazard-prone areas to find out the exact and specific reasons for accidents taking place and in suggesting corrective actions to avoid their recurrence in future. This is an inexpensive means of improving safety standards in an organisation, but it needs trained maintenance engineers.

A 'fault tree' is a diagrammatical representation (Figure 13.1) of inter-relationships in a system, which comprehensively describes a specific operation.

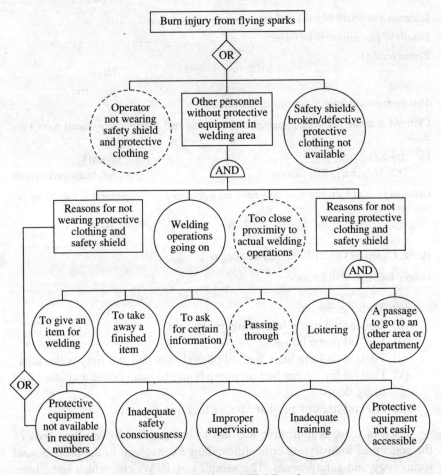

Figure 13.1 Fault tree: Welding shop.

It was originally designed for carrying out reliability studies. It has been adapted, with modifications, to depict interdependence and relationships of different factors that lead to an accident. Such an analysis provides a means for predicting the probability of an accident taking place by unravelling those aspects of a system, which may be contributing factors to this end, but are not visible otherwise.

Such a safely study is indicated whenever we are faced with unacceptable rates or levels of accidents taking place in any particular area or in a subsystem.

From a list of known accidents, one is picked up for a detailed study. This in technical terminology is known as the *head event*. It is likely that one subsystem may have quite a few such head events. Each head event will have a fault tree of its own. Each fault tree is then connected together to have a total picture.

Factors which are responsible for the occurrence of the head event are located and identified down to the smallest of events, known as *basic* events. These are recognised by carrying out a top-to-bottom analysis starting from the head event. This tree is developed by progressively analysing and making critical examinations of even the most casual events through an 'AND'/'OR' concept.

SAFETY SYMBOLS

Certain symbols are used to represent the criteria for the analysis. These symbols are as follows:

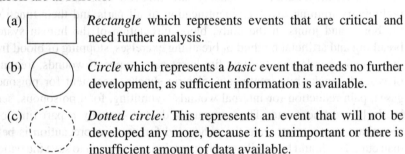

(a) *Rectangle* which represents events that are critical and need further analysis.

(b) *Circle* which represents a *basic* event that needs no further development, as sufficient information is available.

(c) *Dotted circle:* This represents an event that will not be developed any more, because it is unimportant or there is insufficient amount of data available.

(d) *Semi-circle:* 'AND' gate which indicates two or more inputs to an output. All inputs must occur for the output to be produced.

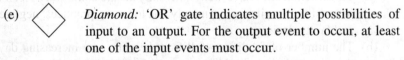

(e) *Diamond:* 'OR' gate indicates multiple possibilities of input to an output. For the output event to occur, at least one of the input events must occur.

To explain the concepts, we shall take an illustrative example from welding. The *head event* is burn injury from flying sparks, which is indicated at the top within the triangle. Straight down is the 'OR' gate within a diamond, indicating three ways the head event could occur.

The most critical of the three events is within a rectangle, as has been indicated by analysis of accident investigation records; the maximum number of burn injuries are suffered by those who are not operating any equipment and do not wear protective equipment and clothing/shield. This aspect is then rejected to in-depth analysis and critical examination. Six conditions that must occur before a burn injury takes place are shown below the AND gate on the right-hand side. Five conditions/reasons are defined in the OR gate. It may be noted that AND gates indicate what must happen to cause an accident, whereas the OR gate brings out what could be the causes.

Firstly, it should be remembered that whatever be the techniques or method, safety can be fully ensured only when it becomes the concern of every single person in an organisation. But the example, as stated at the very outset, has to be set from the top, from where it percolates down, making everyone safety conscious and creating a safe environment for work where deeds speak louder than words.

FIRST AID KIT

It is absolutely essential that first aid kit must be kept in all the shop floors and the foremen must be trained in first aid provision to ensure the accident involved victim should be reasonably safe, till '108' ambulance arrives. The preliminary training involves identification of all parts and their functions, the bones and joints in the body, blood circulation of the human system, breathing and artificial method of breathing exercises, stopping of blood from the wounds, bandaging, treating fire burnt parts, types of wounds, treatment of wounds, snake-bite treatments, electric shocks, treatment for poisonous gases, pain reduction for internal wounds—vomiting, food poisonous, sense headache treatment, etc. The company's doctor—usually a part time or a consultant doctor must train the concerned foremen, as precaution is better than cure. He should be familiar with auspicious respiration to save the patient.

SAFETY CULTURE A–Z ISSUES

(a) Construction sites have areas inherently unsafe conditions. Accidents in construction sites occur mainly due to wrong practices, negligence and faulty decision.
(b) The number of fatalities—fall from heights—are increasing day by day, particularly in building sites and factory sites.
(c) None of site construction carries first aid kit/boxes.
(d) Every project sites, as it advances towards completion, changes its nature continuously, involves different agencies and increasingly prove to accidents with workers under tremendous pressure to complete the fast track projects.
(e) Accidents demotivate and traumatise not only the workers but the total staff. Machinery and tools can malfunction, leading to injuries.
(f) Exposure to toxic chemicals, five dangerous debris falling objectives or hazardous material may be another cause.

(g) You may remember Bhopal gas tragedy by Union Carbide/now Dow Chemicals; where lakhs were seriously affected even now, with the then Chairman Anderson allowed to leave the country by the then Congress Government. Even now lakhs continue to suffer for decades without any compensation.

(h) Officials of the disaster management cell of the Navi Mumbai Municipal Corporation (NMMC) were alarmed by reports of toxic ammonium fumes emanating from the gunning bags in Rebale on July 9, 2011. Locals complained about a strong order, irritating in the eyes. The fallen gunny bags were covered with mud, before disposing them off in Talora.

(i) Accidents occur, when tools are not properly placed in their slots or supports are not safely secured.

(j) Unprotected duct openings and partially constructed building components result in fatalities.

(k) Everyone have to wary of cylinder explosions, and electrical shocks to improper laying of cables.

(l) Electric motor windings must be reliably protected against high temperatures as the windings of three phase motors are coated by an insulating film that can get damaged or destroyed due to high temperatures.

(m) Motor protection at high ambient temperatures, motors used at high switching frequency, for long starting up and breaking procedures, motors used together with frequency converters, irregular intermittent ducts, in all such cases the thermistor motor protection devices can be used.

IN THE WAKE OF A QUAKE

(n) Due to the movement collision of titanic plates under Indian Ocean and Pacific Ocean, Indonesia and Japan are the most frequent earthquake affected areas. The behaviour of Japanese was incredibly excellent. The recent experience of Fukushima Daiichi resulting in earthquake, tsunami, floods, and radiation leading Japanese devastating disaster in 2011—leading to downgrading of its currency is well known. This imbalance affects India, periodically due to 'Overdevelopment'—in Bombay islands sky scrappers, leading to earthquakes. New York and Coastal America experience periodic floods/hurricanes. The earthquake-experienced Japanese did not panic, but moved to a safer area within seconds by seeking refuge beneath tables, cots or strong furniture.

(o) If the furniture is not available, stand near a strong wall with the hands pressed against the floor for balance.
(p) Do not stand against doors, windows, book racks, mirrors etc.
(q) Never use the lift.
(r) The body must be covered with pillows and bedsheets.
(s) People living in old-vulnerable houses move to open place taking care not to stand near trees or electrical poles.
(t) In case the house or building has suffered damages people must collect food, water, medicines, valuable documents, and move to safe location.
(u) Post quake tremors often build down the buildings.
(v) Use only torches as lamps and candles might cause a fire.
(w) People travelling must park their vehicles at open space avoiding trees and bridges.
(x) Venture out with footwear when the quake is over.
(y) Care must be taken to evacuate women, old people and children first.
(z) Keep away from chemicals, gases, electric poles and trees turn off power and gas connections.

It is crucial to keep ready a first-aid-kit, bottled water, eatables that a lust a few days, battery operated radio and other items.

In spite of compulsory wearing of helmets, you see even kids driving without helmets.

Drivers are forced to wear seat belts for safety side, but you rarely see anyone driving wear seat belts.

UNSAFE METROS

Not only due to terrorist attack or natural disaster or greater pollution of all kinds, but because cities are floating on water as sea water has gradually crept into the river to fill that space and is into ground water, making an increase in salinity in water used for human consumption. There is no breathing space in many cities. The cell phone towers and cell phones are unsafe and affect your brain, so that all children up to 14 years have been advised not to use cell phones in advanced countries. The tsunami warning systems setup after tsunami in 2004 in south east of India have also been eroded. Mumbai's Mithi river, Kolkata's Hooghly river, Chennai's Coovam river, Delhi's Yamuna river, Bangalore's Ulsoor lake, Hyderabad's Musi river, etc. are occupied by hutment—*Jhuggi-Jhopri* dwellers turning them into hutments and toilets. The entire Sunderban Delta basin is vulnerable to raising sea levels and high tides with southern Kolkata, 180 km away may 'sink' if ground water level is not arrested immediately.

Most dams are built without a sustainable flow of fresh water into the river. Illegal encroachments of mangroves encouraged politicians are a common feature. Low lying areas are prone to floods, tsunamis, and cholera/ mosquito breeding centres, leading to natural disasters, affecting the *aam aadmi* or vulnerable poor.

To minimise the above damages and minimise inter state water dispute Dr. K.L. Rao, Congress minister for irrigation in 1960s, suggested linking of all rivers, which was continuously ignored by successive Congress governments since 1947. The cause of linking rivers will not only provide employment for millions, but also provide a cheap navigation system besides giving water to thirty crores of people and will cost the government a small fraction of the corruption money swallowed by politicians and bureaucrats.

SAFETY NORMS SWITCHED OFF

In 2010–2011 alone, live wires claimed 72 lives, including 22 EB staff in 'educated', Coimbatore region of "progressive Tamil Nadu". The incidence of electrical accidents can be minimised as most of them are sheer negligence. The major reasons for safety violated accidents include snapping of conductors, accidental contact with live electric wire or equipment, violation of all safety measures, lack of supervision, defective appliances or apparatus, inadequate maintenance, unauthorised work, substandard construction, stealing of electricity by putting a hook from the 400 kW cables by political parties and urban poor, etc. The incidence of electrical accidents can be minimised as most of them are due to sheer negligence of the proven principles of electrical safety, poor, quality of cables and poor awareness among the public. Most of the transformers in India are installed without proper fencing and accounts for 407 of electrical accidents. Poor maintenance of transformers, transformer oil, faulty cables, inferior carriers making life threatening mistakes in a hurry, workers carrying iron rods accidentally touching live electrical wire, workers talking on cell phones while carrying out maintenance jobs, overconfidence of maintenance engineers, shortage off staff with sanctioned vacancies remaining unfilled, buildings closest to high tension wires and transformers earth leakages, etc.

ELEVATED RISKS

Tragic deaths of kids trapped between doors of lift is very common in India. All the lifts in residential buildings are poorly maintained. Builders have to get permission from inspecting staff before elevator is fixed, but no one bothers. While considering the lift installation the inspector has to check all aspects before certifying its use. The feasibility of mandatory service after sales or AMC should be examined. The electrical inspector (without corruption) should have the authority to enter in any multistorey complex and inspect the lifts. Only trained personnel must operate the lift, open the lift under emergencies and use emergency keys for opening/closing lifts, children and senior citizens are prone to more risks in lift. Do not open the gate if the lift is not at the level. Do not force to open the lift gate under any circumstances. Do not switch off the cabin lights. Do not force your hand through the gates. Do not open the door when lift is in motion. Do not panic under any circumstances. Do not allow kids to misuse the lift. Ensure water does not accumulate in the basement under lift. Do not stop button unnecessarily. Do not overload the lift. Have enough lights in the lobbies for your safety.

FOOD SAFETY

The number of food borne diseases have increased much in recent times and stands at a whopping 2 billion cases per year resulting in 10 million deaths per year, with many food producers, hoteliers and suppliers becoming insolvent. In advanced safe country like US, 1 out of 6 is affected by food poisoning. Indian hotels, the same oil is used again and again as it is not thrown causing digestive problems to many. In view of serious nature of food diseases and finds eating junk foods, Chinese noodles, Italian pizza, Manchurian sauce, Arabic shouerma, American hot dogs, etc.—leading to stomach disorders, certification of safety by a third party that demonstrates compliance to international food standards in a transparent way unnecessary for India, particularly for street vendors, where foodstuff are occupied by mosquitoes, flies and cockroach.

Chapter 14

Computers and Maintenance

COMPUTER SYSTEM

Due to development, friendship, parenting and indeed the very fabric of society is changing fast. Thanks to technology and social networks, words like Microsoft, Google, Facebook, Twitter, laptops, personal computers, skype chatting, video kiosks, satellite, television, online trading, internet watching, investment banking on line ticket booking. Computer simulations, graphic efforts in cinema, have invaded every sphere of activity in all corners of India. It is hence inevitable for mechanical engineer to completely immerse in computerisation in order to catch up with his professional colleagues. It is based on the principle GIGO/TITO—meaning garbage in garbage out, trash in, trash out implying you will reap what you sow.

An excellent information system is essential for effective maintenance performance functions, and maintenance executives can be far more effective if they have the information required for decision making. In many organisations, quick, scientific and effective decision making is still not forthcoming because of the lack of requisite information in terms of quantity, quality and speed, despite the best efforts made by them, e.g. having a total planned maintenance system, proper schedules, trained manpower and the backing of a proper organisational structure and top management support. One solution to this complex and seemingly intractable problem is to have computers which, with their prodigious memory, can handle a vast variety of information. These machines will be a great boon, particularly in organisations, dealing with data, information which would be programmed to get the desired results.

Paperless maintenance using computer is witnessed in many firms. The information should be so designed as to serve the needs of the Maintenance

Manager. Computers can be of great help, but success to a large extent depends upon whether or not the manual systems are comprehensive and fully developed as well to supplement them.

The *inputs* of the computer system are exactly the same as the inputs of the manually entered records of the specific task or operation. The programming of the computer has to be developed by trained programmers in consultation with the user. It is not necessary that the Maintenance Managers have to be expert programmers, but they ought to be sufficiently familiar with the computer system to know and appreciate the basic fact that the accuracy and usefulness of output are totally dependent on the accuracy of *input* data fed to the computer. They should also know in advance what exactly they want in terms of information.

There can be no hard and fast rules about the size of an organisation to justify the installation of a computer. Each case needs an indepth examination about the size of plant, its complexity, and the importance attached by the organisation for correct and timely retrieval of information and the way it is used. The organisation should also realise the saving potential of the computer through better coordination and reduction in volume of manual work against its initial cost and cost of operations.

The computers are useful tools for carrying out repetitive tasks and difficult calculations quickly, to cater to speedy and accurate information needs. The data fed into the computer can be so programmed as to summarise, reorganise and combine the maintenance data as required for decision making.

Some areas of computer application in the maintenance function are now indicated. Maintenance software programmes are readily available, but they need to be developed and modified to suit specific requirements of the company.

COMPUTER APPLICATION AREAS

The first important area of application is documentation, i.e. to keep a permanent record of each piece of equipment according to its needs. These records could be stored under heads such as equipment record cards, history record cards, record of each repair carried out, man-hours and spares consumed, etc. If stored in the computer, they can save considerable amount of clerical effort and manpower. From these records the requisite analysis can be done effectively and quickly to improve maintenance efficiency.

Down time reports can indicate the quantum and causes, and give a comparative data of, say, the last three months. These at a glance can apprise the Maintenance Manager of his performance level in terms of down time, and whether corrective actions have been taken and, if so, whether they show a reduction in down time over a period of time. A cause-wise down time analysis can be available to the concerned person for the last three or six months at a glance and can provide him with many interesting facts to guide him in his future decision making. It can pin-point if the main cause, as identified earlier, still remains the same, and if so, its rectification at a later stage has reduced down time or not. On the other hand, if the prime cause for down time has shifted to some other area, it would definitely need fresh investigation. Similarly, many such comparisons and analyses can be carried out very quickly, which can be helpful in decision making and taking remedial actions to improve maintenance effectiveness.

(i) *Schedules* can be developed for various kinds of maintenance activities. The inputs need to have the specific maintenance operation for planned, preventive or predictive maintenance work, its details, frequency and time estimates to complete the task. The computer can then add up the time needed to carry out these tasks each week. If it is more than what is available in terms of maintenance man-hours, then extra jobs should be deferred to other weeks except for critical equipment which can brook no delay. Thus, a balanced and equitable schedule can be created which will be acceptable, and implementable in terms of *need, costs* and *time*. This exercise would be very time consuming if done manually. Similarly, every tradesman's work can be scheduled and planned.

(ii) *Inspection and lubrication:* To ensure smooth and proper lubrication, the lubrication route card and lubrication schedule can be computerised. Similar details can be prepared for preventive maintenance and inspection schedules. Details desired can be added on, like the individual worker (who has to perform the task) by name, his job card, etc. These details can be prepared in advance for all the weeks in a month and each week divided into daily needs, and kept ready in the Maintenance Control Centre (MCC) to be issued to each worker through his supervisor in the morning of die actual day of work.

(iii) *Turnaround management:* This includes reconditioning, retrofits and overhauling of plants. Here, of course, a CPM/PERT package should be used. This is an extremely useful area for turnaround planning, implementation, monitoring and control and is being done by some organisations in India.

(iv) *Allied areas:* These include areas such as the keeping of records of the backlog, maintenance work, and pending workloads for critical equipment maintenance. These can help the management to take proper decisions with regard to maintenance manpower needs, and to decide whether to go in for overtime or to ask for contract maintenance, or to continue to do the maintenance work with the existing manpower/infrastructure.

MAINTENANCE COSTS

(i) *Cost comparison.* Estimated repair costs versus actual costs can be shown weekly, fortnightly or monthly. This projection will indicate if actual costs are more than, less than, or equal to the estimated costs. If not, by how much has it deviated? Keeping in mind the actual costs against the funds provided for in the budget is a control function. The computer can provide exception reporting where the deviation is more than say 20 per cent and is above the budgeted provision.

(ii) *Planned/Preventive maintenance cost:* Periodic reports can be provided, giving details of the total maintenance costs, equipmentwise, or for a group of equipment or for types of equipment. It can also include down time of machine, cost of lost production, or the number of requests received by maintenance from production to attend to that particular equipment.

A monthly report must be brought out, indicating those machines or equipment which in descending order of priority have incurred losses. This should cover the maximum repair cost for the month, the maximum down time for the month, and the machines that have been down most frequently during the month. If concerted corrective actions are directed to these machines, then they will be restored to normal condition and thereby improve both production and maintenance results.

Forecasting about spares and other material requirement planning needs can be improved tremendously by computerisation. Consumption records of spares and other data are shown in each job card, which can then be added up machinewise or equipmentwise or as desired. With the help of trending and other forecasting techniques, a near enough to exact requirement of spares, etc. can be forecasted and indented for.

The preparation of five or 10 off-lists by analysis of past recorded data is of tremendous help in planning for future overhauls. This task, if done manually, is cumbersome, but the same job can be accomplished by the computer quickly and accurately.

(i) *Manpower:* Each job card has a column indicating estimated man-hours or the standard man-hours for that job, and after job completion, the actual man-hours spent on dial job are indicated.

$$\text{Manpower effectiveness ratio} = \frac{\text{Actual man-hours}}{\text{Standard man-hours}} \times 100$$

This ratio can be indicated in weekly, monthly or six-monthly time segments for each individual, by name, trade, or department. Additionally, other information, such as overtime and idle time, can also be shown. If the monthly manpower effectiveness ratios are analysed for, say three or six months periods, then they can indicate the trend. If the ratios are satisfactory, then no actions need be taken, but if they begin to show a decrease, the causes for this have to be closely studied and methods found to rectify the mistakes.

(ii) *Work progress* can be analysed in two separate ways. In the *first* method we use the knowledge of the estimated time to complete a particular job. At the point of time when we wish to assess the progress, we know the amount of time that has already been spent on that job. The ratios between the time already spent on the job to the estimated time of completion becomes the indicator of work progress. In the *second* method, the assessment is based on the work *actually* completed. But in this case, one must have the capability to assess the amount of work done correctly. A computer can be of immense assistance while estimating work progress, for it involves the obtaining of a very large number of job cards and extracting information from them, which has to be then compiled and put together in terms of ratios.

Computers are very useful in establishing the level of plant maintenance. Usually, the two methods that are in use are the *cost appraisal method* and the *elemental analysis method.*

COST BENEFIT

The cost appraisal method consists of computing the ratio of maintenance cost during an established reference period to maintenance cost in a similar current period. The computer would quickly add up the various costs like down time cost, scrap and rejects cost, and rework cost, for the two periods, and then give the ratio by dividing the two computed costs. The ratio of one indicates that the same amount of money has been spent in the current period as in the reference period. If the ratio is less than one, it indicates that the maintenance level is deteriorating and shows the need for investigation. When the ratio is more than one, an improving trend is indicated.

In the elemental analysis method, the ratio of the reference period to the current period is used, but it is not in terms of cost, but on the basis of the value established by the inspection of equipment. This method assigns point values for each equipment malfunction or undesirable condition, and compares these points with the reference base period. During periodic inspections, equipment and machines are rated for their overall condition by assigning penalty points for each observed defect/malfunction. Point values are then assigned to each segment of each piece of equipment on the Facility Register. This is in strict accordance with the relative importance of each segment for its operation in the totality. This is done by a committee consisting of industrial engineers, Maintenance Managers and Production Managers. For example, the equipment segment can be the mechanical condition of a milling machine, and the point values assigned for bearings would be, say, (5), gears (7), shaft alignment (3), and appearance and cleanliness point value (4). Similarly, during inspection, penalty points can be awarded, e.g. for defective bearings (2) each, worn out gear each teeth (1), dirty machine (1), paint peeling-off (1), lubricating oil smeared on machine (1), and so on.

These are mere illustrations, and have to be developed by each organisation on its own. This method of rating can be made as elaborate as desired or as concise as one would like to have. It has its limitations, namely, the complexity and variety of factors involved in it, and the subjectivity involved in making a judgement. However, this method can be employed by the computer very accurately and quickly.

Apart from the foregoing examples, there are many other areas of computer application, like fault diagnosis and trouble shooting, replacement decisions, standardisation, condition monitoring systems and for kinds of spares inventories.

The computerisation of the maintanance system can achieve many objectives, like improved decision making and better utilisation of resources from a single source of information. But, unfortunately, it has been noticed that quite often computerisation does not yield the expected results. Therefore, to get the best out of the computer, the change has to be planned and introduced gradually and then personnel have to be prepared for the changeover and trained. The maintenance engineer must be fully involved in the development of software and fully trained to face his new responsibilities.

MANAGEMENT INFORMATION SYSTEM (MIS)

The most important use of the computer is the maintenance information system (MIS), which can and should provide all the information the Maintenance Manager needs and warns to provide to the management. They include the following:

(a) Pending jobs report and work-order completion report
(b) Planned maintenance schedule status report
(c) Material availability report
(d) Failure analysis report

- (e) Delay analysis report
- (f) Cost analysis report
- (g) Equipmentwise repair cost analysis
- (h) Equipment down time (percentage) report
- (i) Cost of maintenance as percentage of total sales
- (j) Number of stock-outs recorded in a defined period
- (k) Ratio of planned work to emergency work
- (l) Man-hours actually spent on maintenance as compared to estimated planned man-hours.

It should, however, be clearly understood that the computer has to be programmed correctly and data fed into it to obtain the desired output. The Maintenance Manager should, therefore, work in close coordination with the computer specialist to get the best out of the computer to meet his needs. Monthly, quarterly and annual reports on the above can be produced, compared and action taken.

MAINTENANCE BUDGET

Any investment consists of fixed investments and working capital, about which the maintenance executive is concerned with. He has to be involved in the formulation of the maintenance budget, relating both to the fixed assets and working capital. In a developing economy like ours, this is a neglected area, as the Finance Manager is often the final authority for budget finalisation. Quite often, it is seen that the Maintenance Manager is ignored in the budget finalisation process, while the source documents and the technical competence for the formulation of the maintenance budget is with the Maintenance Manager, who unfortunately plays a secondary role. The maintenance budget usually comprises up to 10 per cent of the value of the equipments, including the value of spares consumption.

For the preparation of the operating budget, the period of one year is normally broken down into quarterly and then monthly budget figures. To prepare this budget, the Maintenance Manager must consider the following factors:

- (a) Salary and all statutory payments for the entire maintenance personnel
- (b) A portion of overheads to be allotted to maintenance (the basis for the same can be obtained from finance)
- (c) Materials like consumables, lubricants, oils, greases, cotton waste, and all categories of spares likely to be used during the financial year. (The quantity required for each category of material will have to be estimated based on past records and experience and the task ahead.) The unit current price for these items can be obtained from

(i) original equipment manufacturer (OEM), (ii) material planning cell, (iii) previous work orders, (iv) history record cards, and the (v) finance department. Admittedly, the spares component will be a major percentage (about 60 per cent of the overall maintenance operating budget).

(d) Depreciation of tools, tackles, test equipment, calibration and inspection facilities have to be included in the maintenance operating budget. The amount of depreciation to be included will be as per the laid down depreciation policy and as approved by the Income Tax authorities. Capitalising insurance spare parts also comes in this processs.

At the end of the year, it is advantageous to review the maintenance budget against the actual expenditure, in order to identify the deviations and their causes and for taking appropriate corrective remedial measures. Based on the experience gained and by intense interactions with the finance department, the Chief Maintenance Manager can introduce sophistication to the budgeting process, like performance budgeting, zero-based budgeting, etc.

A performance budget is an operational document which translates the aspirations of an organisation into meaningful and feasible action programmes and activities, for realising the objectives by integrating financial as well as physical targets of performance on major items of business or service.

Zero-based budgeting is a process which requires a manager to justify his entire budget request in detail, starting from scratch. In other words, he determines the minimum basic requirements to perform the functions of his department. Any costs above this requirement are identified as increments, that must be justified before they are funded. The world wide web is one of the most baffling paradoxes of our times. On one hand, it has created nine billion dollar corporations out of the college start ups and on the other created business which sounded great on the face, but have not been able to figure out a way to monetise their operations in some cases!

Chapter **15**

Productivity and Industrial Engineering

PRODUCTIVITY CONCEPTS

The national productivity council (NPC) and local productivity council (LPC) are doing excellent training in maintenance. In this chapter we shall discuss some concepts which have been researched and formulated into techniques for enhancing productivity. We may note that their application is not confined only to maintenance.

Productivity, being one of the major aims and objectives of the maintenance function, includes the following topics, usually coming under industrial engineering.

(a) Work study
(b) Method study
(c) Ergonomics.

Productivity is the ratio between output to input in any activity. This activity may pertain to office work, the maintenance of a vehicle, the assembly of fans, confirming a seat in the airline, or a room in a hotel. It is the function of achieving the maximum possible with minimum resources. The resources are: (a) manpower, (b) material, (c) equipment, (d) spares and building, (e) capital, and (f) time.

The responsibility for achieving higher productivity rests on those who are entrusted with the task of management. They obtain the facts and then plan, direct, coordinate, control and motivate in order to produce goods and services.

Different organisations have different objectives and their necessary resources. These have to be balanced and coordinated to achieve the best results. While thinking of productivity, we have to think of it in relation to time. The yardstick by which this is measured is the output of goods/services in a given number of man-hours or machine-hours.

The total time required for a job is determined by (a) basic work content, (b) additional work necessitated due to defect in design and specifications and inefficient methods of production/manufacturing, and (c) additional efforts put in and time spent due to ineffective management and work. If time spent on such additional work is reduced or eliminated, productivity can be improved, which is the responsibility of the management.

Productivity is talked about too frequently, and treated as if it is a panacea for all ills. Seen in the correct perspective, productivity is not a measure of production quantity, but simply the ratio of output and input. Also, it is not a measure of profitability, but only indicates the efficiency of operations, and suggests the profitability; however, inefficient operations too can be profitable if a product enjoys a favoured market status. It is a sure way to reducing inflation; it may be a moderating factor and one of the many economic factors that determine the general price trend.

WORK STUDY

The basic activity of industrial engineering is work study, aimed at most effective and efficient work systems. *Work study* is a term that denotes the techniques of method study and work applying their specialised education and experience measurement which are employed to ensure the best possible use of all the available resources—human and material—in carrying out a specific activity. It is specifically concerned with productivity since it is used to increase the amount produced from a given quantity of resources without further capital investments. Hence, work study is of interest to a manager, because it is concerned with all the resources a management processes, and their systematic improvement so that they are effectively and efficiently utilised.

To take decisions one must have facts. It is only the analysis of facts collected impartially, objectively and systematically that can help one to take the right decisions. Work study provides the technique for the collection of all the facts (which governs an activity) and the critical examination of these facts, and makes recommendations in a balanced way.

Method study is concerned with how work is done and how it can be done more efficiently. The major objective is elimination of wastes—waste of skills, time, capacity, material, capital, investments, etc. and this will be dealt later.

The subject of work measurement is time. It involves fixing a reasonable period of time for a given work. It is the systematic determination of the proper amount of time for the effective accomplishment of a defined task by a qualified person by a specified method. It aims at having a yardstick for determining the effectiveness of the human work involved in any activity. It allows for a (a) portion of rest time and (b) definite and minimum rate of work. Work is ubiquitous and is performed by all human beings—as individual or teams at every place and tine. While you study any work, the system is improved.

To appreciate the aims and potentiality of work study, the following points need to be emphasised:

(a) The aim of work study is to improve productivity, and not to make the worker work harder or make him redundant.
(b) Responsibility for its introduction and application rests with the management, and not with the work study practitioner.
(c) As conditions change and experience and knowledge are not static, work study needs to be applied continuously to update and bring in further improvements.

Areas of application of work study are diverse, and can be summarised as follows:

(a) Better utilisation of equipment, machinery, materials and spares.
(b) Improved methods of doing work, by laying down the best method (by method study).
(c) Improved factory and work place layout.
(d) Higher functional and operational efficiency.
(e) Increased administrative efficiency, better and speedier production, thus cutting down delays.
(f) Allocation of required time for a task, thus providing a scientific means of improved manpower planning (manning) and control, and a basis for sound incentive schemes.
(g) Improved productivity.

METHOD STUDY

Method study is a scientific analysis of all inputs and searches for more effective procedure by asking questions of what, when, where, why and how.

Method study is concerned with the best way in which a specified activity can be carried out or a process or procedure followed, its philosophy being that there is always a better way of doing things.

Method study consists of a systematic analysis of the present method and bringing about improvement through critical examination.

The basic steps of method study are given below:

(a) *Select* the work to be studied.
(b) *Record* all the facts pertaining to that work by using any recording or charting techniques.
(c) *Examine* the facts critically in ordered sequence and apply the critical examination questioning techniques to all the key facts.
(d) *Develop* the best method(s) under the present prevailing conditions.
(e) *Install* the method so chosen as standard practice.
(f) *Maintain* this standard practice by regular checks.

After the work to be studied has been selected, all the facts relating to the present method of doing that work or activity are recorded. This forms the basic ingredient of the critical examination and, therefore, the success of a study depends on the accurate, impartial and systematic recording of all facts.

The need for symbiosis between world of knowledge and world of work, is felt by professionals technology is upgraded by R&D personnel.

The easiest way to record the facts would be to write them down, but describing a process in writing is tedious and the written word is difficult to be visualised. Hence, symbols are used which combine a system of analysis allied with shorthand.

RECORDING TECHNIQUES

To record facts from observation, several recording techniques are available. These are as follows:

(a) String diagram : This shows the paths of movement.
(b) Models : These are two or three dimensional.
(c) Floor diagrams : This shows how to scale location of activities, and routes followed, with distances.
(d) Outline process chart (OPC) : This indicates the sequence of key operations.
(e) Flow process chart (FPC) : This follows the activities till the end product FPC (Man) charts the activities of operator and FPC (Material)—movement and processing of material.
(f) Two-handed process chart : This is used for highly repetitive work in short cycles.
(g) Multiple activity chart : Man and machine against common time scale.

(h)	Cyclograph	: Here, small lights are attached to the hand of operator, thus tracing the path of movement.
(i)	Chronocyclograph	: This is similar to the above, but here pear-shaped dots are obtained by making lights flicker and by taking time exposures of the same.
(j)	SIMO charts (Simultaneous motion charts)	: Study of the motions of two or more inter-related tasks at the same time in order to reduce the total time.
(k)	Memo motion studies	: These are spread over long periods: single frame exposed at regular intervals.

Contribution of critical examination is the most important aspect of method study. It consists of a set of questions about the key facts and answers obtained with respect to:

(a)	Purpose	: What is done? Why is it done? What else might be done? What should be done?
(b)	Place	: Where is it done? Why is it done there? Where else might it be done? Where should it be done?
(c)	Sequence	: When is it done? Why is it done then? When might it be done? When should it be done?
(d)	Person	: How many people are employed, what is the skill mix, age mix, etc.?
(e)	Means	: How is it done? Why is that done in a particular way? How else might it be done? How should it be done?

All these questions and the respective answers obtained result in elimination, modification, simplification, combing, or shortening of activity.

The critical evaluation provides the ideal methods. This has to be developed into an implementable and practical method, taking into account the prevailing human, physical, economic and technical limitations.

Also, all this effort will be wasted if this new method is not installed correctly.

Casual changes take place and a drift sets in the new method. Periodic reviews focus on the changes, if they are useful, then accept them, otherwise reject them.

Method study contributes to improved efficiency through the following:

(a) Better layout of office, factory, and work place
(b) Improved utilisation of resources

(c) Improved handling of material
(d) Effective flow of work
(e) Standardisation of procedures/methods
(f) Improved working conditions/environment
(g) Better safety: Due to job enlargement and enrichment, it is both a challenge and opportunity for mechanical engineer.

ERGONOMICS

Consciously, unconsciously or even sub-consciously, man has perhaps always endeavoured to match machines with himself. By evaluation, this concept has blossomed and become a full-fledged science which is known as *Ergonomics*, The word 'Ergonomics' is derived from two Greek words, 'ergos' meaning *work* and 'nomics' meaning *laws*. Though a new discipline, it has a long list of names by which it is known. Some of these are:

(a) Ergonomics
(b) Applied and Human Engineering Research
(c) Applied Experimental Psychology
(d) Human Engineering
(e) Engineering Psychology
(f) Human Factors in Engineering
(g) Bio-mechanics.

However, quite often the concepts, perhaps in a less evolved state, are already there. Thus, the basic concept of ergonomics was already present, but the concept of an integrated approach adopted by specialists in these fields towards the problems was new. These problems arise when man undertakes the performance of any activity with a machine, or without its help. The need was felt for full-time research work to be carried out, based on the accumulation of basic information, and this led to the evolution and growth of the new scientific/discipline of Ergonomics.

When equipment and machinery were small, wieldy and simple in design, and limited efficiency, it was possible to train the worker to suit the demand of the equipment. However, with the rapid strides made by man in the field of science and technology, more versatile, efficient and complex machines have been introduced. It has become imperative under these circumstances, it has become imperative that the areas of contact between man and machine be properly matched, if the man-machine system is to operate smoothly. The physiologists and psychologists have joined hands with the design engineer, and through coordinated efforts, they have analysed in detail the design aspect of machines/equipment, so that man can operate them more easily, more accurately, and with greater speed.

The earliest application of Ergonomics can be traced to the days of F.W. Taylor (1856–1915). His experiments sought to arrive at designing of the *optimum design equipment* which would be required for specific types of work, e.g. the design of different types of shovels for shovelling different types of materials. He also laid great stress on the selection and training of the worker and developing him, rather than allowing him to train himself haphazardly. The next major step towards the same goal was taken by F.B. Gilbreth and his wife Lillian, who together enunciated and elaborated the principles of "motion economy". The Gilbreth's laid special emphasis on lightening loads, introducing rest pauses and spacing out work, so as to reduce fatigue and eliminate stresses. Researchers of Ergonomics delve deep into the human make-up and orients it towards the study of the effects of working conditions and environment, and the collection of *anthropometric* data (measurements of the human body in relation to work), and the study of the permissible limits of loading and methods of acquiring higher skills.

The International Labour Organization (ILO) defines Ergonomics as

> the application of biological sciences in conjunction with engineering sciences to the worker and his working environments so as to obtain maximum satisfaction for the worker which at the same time enhances productivity.

However, it should be noted that a professional ergonomist is primarily concerned with the worker's job satisfaction, and the increase of productivity accruing out of it is just incidental to him. A simple definition of Ergonomics could be "an endeavour to study the effect of a work situation upon a worker".

Motivating the man at work and meeting the work suit him, is the prime objective of Ergonomics. It attempts to bring about the most effective accomplishment of work by adapting the work system, which includes human tasks, working equipment, working space and working conditions, to suit the requirements and capabilities of the human operator.

Ergonomics is a science that can have real application in any activity involving human effort, whether it is at home, office, factory, during travel by land, sea, air or even in outer space. It has been successfully applied in several spheres of activity. These include:

(a) Plant and work-place layouts
(b) Selection, training and placement of personnel
(c) Design of equipment, power tools, displays, jigs, fixtures, and even furniture
(d) Motivation of the worker

(e) Working conditions and environments
(f) Computation of relaxation allowances.

The areas where ergonomic investigations can be profitably carried out are shown in Figure 15.1.

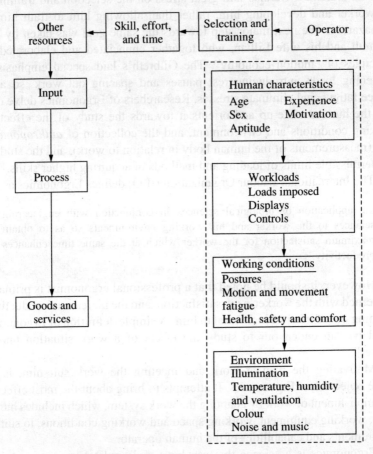

Figure 15.1 Areas of ergonomic investigation.

Man and His Work

Before having a detailed discussion on various factors relating a person to his work, it is necessary to understand the part played by him.

In any activity the man receives and processes information and then acts on it. Figure 15.2 shows man as a component in a closed-loop system and the factors which may affect his efficiency.

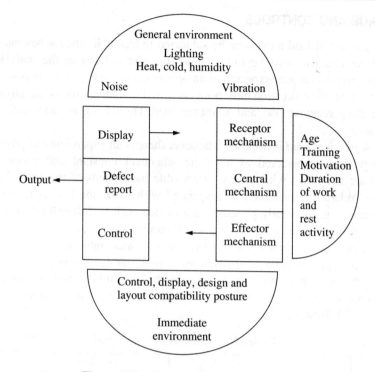

Figure 15.2 Man in a closed loop system.

In the above, man is shown as a component in a closed-loop system and the factors which may affect his efficiency.

The receptor functions are performed by the eyes, the ears, the nose, and the touch. Through these receptors, the sensations of heat and cold are conveyed to the *central* nervous system, where information is processed to arrive at a decision. These new inputs are then integrated with previously stored information and finally the decision is taken through an *effector* mechanism, which involves muscular activity based on the skeletal framework of the human body.

Thus, we can see that man forms part of a closed-loop servo system and forms that part of the system which makes decisions. Hence, he plays an important role in the efficiency of the system.

A man-machine system will have to be designed as a whole in order to achieve maximum efficiency. A situation must prevail where man will be complementary to the machine, and the machine in turn will be complementary to the abilities of the man.

LOADS AND CONTROLS

The amount of load a man can be subjected to has its limits, for beyond that load, his characteristics begin to change. In order to facilitate the analysis of inputs into the outputs expected of an operator, the loads that are imposed by a particular work situation have been classified into three groups: (a) physical load, (b) perceptual load, and (c) menial load. These loads are both static and dynamic:

A *physical load* is imposed whenever there is an expenditure of physical energy. It is considered to be static whenever constant and continuous pressure is applied without any appreciable movements of the limbs being made. When die movement is associated with work, the load is treated as dynamic, e.g. (a) steady pressure on a device such as call-bell (static), and (b) loading and unloading of a toaster (dynamic).

A *perceptual load* is caused by sensory inputs into an operation. The different types of perceptual loads are visual, aural, olfactory, tactile, and taste. When the perceptual load is at a steady level, it is termed as *static*, and *dynamic* when there are differences in the intensity of the sensory inputs. Table 15.1 illustrates this:

Table 15.1 Static and Dynamic Load

	Static	Dynamic
Visual	Observing a watch dial making a steady movement	Watching the movement of the needle when the tyre pressure is being gauged
Aural	The humming sound of a refrigerator	The ringing of a telephone
Tactile	Holding a pillow close to oneself	Holding a child in one's arms
Taste	Letting a cube of sugar melt in the mouth	Biting into and chewing a hot green chilli

Both physical and perceptual loads cause corresponding mental loads. However, mental loads could even occur in their absence. Mental loads can also be classified as static and dynamic.

Static loads help remember facts or figures so that actions can be initiated, making use of the same when need arises, e.g. memorising multiplication tables.

Dynamic loads aid in evolving new ideas, designs, plans, tactics, strategies, e.g. writing a play, composing music, and planning an overhaul.

The permissible limits of loading under each category are determined by psychological tests and by making use of psychometry. The Defence Institute of Physiological and Allied Studies (DIPAS) is conducting some research work in this field in India.

The aim behind analysing each kind of load is to

(a) try and eliminate, if possible, or at least reduce physical load, thereby reducing fatigue, and increasing comfort and health;
(b) simplify and reduce the number, type, range, and the layout of the display panels and warning systems to reduce perceptual load;
(c) reduce and simplify the controls and their layout for easy response and speedy control; and
(d) to enable the operator to fit the equipment at the design stage itself.

WORK STUDY AND ERGONOMICS

As the aim of work study is to provide job satisfaction and to increase productivity, the work study practitioner becomes dependent upon ergonomic research data which sheds light on many facets of human reactions to a given work situation. Some of these are:

(a) Measuring the limits of physical endurance, the normal speeds of move-ments, and the optimisation of the methods of handling controls.
(b) Ascertaining the receptivity to sensory inputs and measuring the time required for the perception of deviations.
(c) Gauging reaction time for motor output, and the time required for perception and deviations.
(d) Design and layout of equipment, work place and furniture to be guided by anthropometric data.
(e) The effects of environmental conditions.

Knowledge which is useful to work measurement is kept up-to-date through Ergonomics by contributing to the study of energy expenditure, factors influencing fatigue, and so on.

ERGONOMICS AND MAINTENANCE

Ergonomics is a science that can be put to the service of management from the initial stages, that is, at the planning and design stages of a work system. The organisation and control should appreciate for the human factors involved in a work system. And this need is very well catered to by ergonomic research. The availability of data on all perceptual and mental loads and its limits helps in planning the work. This allows for a more effective application of the managerial talent to the solution of complex problems, over a wider range of activities. This single improvement, wherein loading and the other factors relating to it are understood clearly, leads to substantial increases in the overall productivity of the organisation without the need for making any changes in the methods on the shop floor.

Optimum productivity, as we all know, should be the aim and objective of all industrial activity. We have also observed, in the course of our discussions, that productivity on the part of the maintenance function has far-reaching positive consequences on the productivity of any organisation as a whole. Therefore, the concepts which we have discussed, though applicable in any sphere of activity where work is carried out, have applications in the field of maintenance, for efficiency and productivity. In this segment of industrial activity, it has much to offer towards the overall health and well-being of an organisation, which is the most important corporate goal.

Chapter 16

Activity Sampling for Work Measurement

Activity sampling also called snap reading is a management technique which employs a statistical approach for collection of complete information about the activities of men, machines, and process equipment. It is a practical compromise between the extremes of purely subjective opinion, and the certainty of continuous observation and detailed study. This is useful to assess maintenance workload.

The technique is based on the laws of probability and the random sampling theory which is a statistical technique similar to that used in total quality management (TQM).

SNAP READING

Sampling is the process of drawing inferences concerning the characteristics of somewhat smaller number of items drawn at random from a large mass. (*Population* or *universe* is the term used for denoting a large mass, group or total.)

(i) *Activity sampling* is a work measurement technique for the quantitative analysis in terms of time of the activities and delays of men, machines or any other observable state or condition of observation.

(ii) The principle can be illustrated with a simple example. Imagine a worker working on a machine for 7½ hours out of 8 hours (excluding half hour lunch break). We have to find as to how much productive work the worker puts in. The two activities in that case would be working and idling. Results of a continuous study can show that he works for 6 hours and idles for 1½ hours; the distribution of idling is as shown in Figure 16.1. His working time will be $(6 \div 7\frac{1}{2}) \times 100 = 80\%$ of total working hours, excluding lunch interval. Watching the worker advertantly will irritate him.

Instead of a continuous study, if activity sampling is carried out, then 50 instantaneous observations would have been made, as indicated by the arrows in Figure 16.1. If it is observed that the worker has been *working* for 42 times out of 50 observations, then we can say that, over 7½ hours he has been working (for 7½ × 42/50 hours), and the rest of the time he has been idling. Actual working time as indicated by activity sampling would be 42/50 × 100 = 84%.

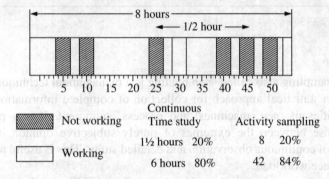

Figure 16.1 Distribution of idle time.

It is evident that the accuracy of this result as predicted will increase as the number of observations increases. It can be statistically worked out as to what should be the number of observations required to predict within a certain limit of accuracy.

NORMAL DISTRIBUTION

Based on the theory of probability, the probability of occurrence of a certain change phenomena can be represented by a Normal distribution of Gaussian curve. This curve has a characteristic bell shape, i.e. the data is clustered in the middle and dispersed at the extremities (Figure 16.2). The curve is symmetrical about its mean value, and the area under it corresponds to probability.

In normal distribution, the area is measured in intervals of (mean ± standard deviation). In practice, the interval containing mean and standard deviations will include the following proportions of values:

Mean ± 1σ includes 68.27% of total area

Mean ± 2σ includes 95.45% of total area

Mean ± 3σ includes 99.73% of total area

This is called *six sigma effect*, covering most items.

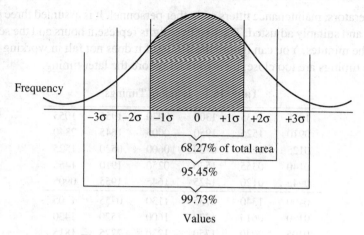

Figure 16.2 Normal distribution curve.

These percentage values are also referred as *confidence levels* in maintenance management parlance. Depending on the confidence level and limits of accuracy of the required prediction, the number of observations to be made varies. The formulae, giving the number of observations required for 68.27%, 95.47%, and 99.73% confidence levels are, respectively, the simplified formula are given below:

$$N = \frac{P(100-P)}{L2} \text{ for } 68\%$$

$$N = \frac{4P(100-P)}{L2} \text{ for } 95\%$$

$$N = \frac{9P(100-P)}{L2} \text{ for } 99\%$$

In the formulas,

N = number of observations to be made
P = percentage of occurrence of a particular activity (to be studied in detail by activity sampling) as decided by past experience or pilot study
L = absolute limits of accuracy expressed as percentage of N

Table 16.1 gives random times for conducting snap study.
Table 16.2 gives the areas under symmetric continuous distribution.

RANDOM TIMINGS

Table 16.1 gives the activity sampling person, in 5 minutes gap random timings generated by computer to start taking rounds for observing the work of operators, maintenance fitters and other personnel. It is assumed three shifts work and suitably adjusted. The first two digits represent hours and the second two the minutes. You can ignore the timing if it does not fall in working shift. If the timings are too close, then you can ignore the later timing.

Table 16.1   Random Timings

1015	2035	1305	0930	1955	1055
0910	1525	1950	0005	1645	2340
0125	0550	0720	0600	0520	1525
0140	0355	1530	0250	1910	1455
0945	0120	1425	1645	1955	0905
0400	1340	0735	1130	0455	1705
0130	0615	1320	1600	1520	1420
0405	2040	1750	1230	2225	1815
1115	0025	1655	1020	0540	0620
1400	0510	1930	2145	0830	0900
1000	2035	0155	2240	0335	0225
1955	2145	2240	0140	2310	1220
1140	0300	0140	0825	2020	0755
0035	2355	1515	1110	1810	1140
0045	1200	1150	1250	1754	1710
0835	1515	1725	2025	1650	1215
2320	3400	2000	1745	1235	0440
1830	2340	2215	2270	0420	0910
1950	1620	0705	0245	1015	1125
1750	2155	0935	1300	1750	1345

All numbers 0, 1, 2, 3, 4, 5, 6, 7, 8, 9 are likely have same frequency in the long run.

AREAS UNDER NORMAL DISTRIBUTION

Proportions of observations lying above a distance of "t" from the mean—positions are measured in terms of the total area is one and at mean –0–, the area will be equal on either side. Areas are less than the total area is one under the curve and hence all one.

Six Sigma

Normally the frequency distribution of a large number of observations/characteristics always form into a normal curve as it normally happens. Every distribution approximates to normal curve in the long run. Published tables are available to know the areas covered if the average and standard deviation are known.

Six sigma has become a central part of business success to measure the inevitable variation in all parts of business, committed to excellence. It provides the road to success. Several books have been written on the applications of six sigma.

For normal studies, the value of confidence level assumed is 95.45%, and absolute limits of accuracy are between 2.0 and 3.5%.

Table 16.2

t	Area	t	Area	t	Area
−4.0	0.99997	−1.3	.9032	1.4	.0808
−3.9	.99995	−1.2	.8849	1.5	.0668
−3.8	.99993	−1.1	.8643	1.6	.0548
−3.7	.99989	−1.0	.8413	1.7	.6446
−3.6	.99984	−0.9	.8159	1.8	.0359
−3.5	.99979	−0.8	.7881	1.9	.0287
−3.4	.99966	−0.7	.7850	2.0	.0228
−3.3	.99952	−0.6	.7257	2.1	.0179
−3.2	.99931	−0.5	.6915	2.2	.0139
−3.1	.9990	−0.4	.6554	2.3	.0107
−3.0	.9987	−0.3	.6179	2.4	.0082
−2.9	.9981	−0.2	.5793	2.5	.0062
−2.8	.9974	−0.1	.5793	2.6	.0047
−2.7	.9965	−0.0	.5000	2.7	.0035
−2.6	.9953	+0.1	.4602	2.8	.0026
−2.5	.9938	0.2	.4207	2.9	.0019
−2.4	.9918	0.3	.3821	3.0	.0013
−2.3	.9893	0.4	.3446	3.1	.0010
−2.2	.9861	0.5	.3085	3.2	.0069
−2.1	.9821	0.6	.2743	3.3	.0048
−2.0	.9772	0.7	.2420	3.4	.0034
−1.9	.9713	0.8	.2119	3.5	.0023
−1.8	.9614	0.9	.1841	3.6	.0016
−1.7	.9454	1.0	.1587	3.7	.0011
−1.6	.9452	1.1	.1357	3.8	.0007
−1.5	.9332	1.2	.1151	3.9	.0005
−1.4	.9192	1.3	.0968	4.0	.0003

The random sampling technique is used for activity sampling. For success of random sampling, the following three conditions must be satisfied:

(a) Each item selected must be completely independent.

(b) Each item must have equal likelihood or opportunity for selection, i.e., there should not be any bias.
(c) The characteristics of each item being observed should remain constant throughout the process of sampling.

In order to avoid bias, which may be intentional or unintentional, random observations are to be made by selecting timings at random. These can be obtained either from random number tables or by drawing lots from a hat. Each minute over the complete period is represented by one lot. The picking up of lots gives the random timings, at which the observer should make his observations.

The conduct of the study will consist of three main steps:

(a) Prepare a plan for study
(b) Perform the study
(c) Evaluate the results.

APPLICATIONS OF ACTIVITY SAMPLING

Activity sampling can be used in the following areas:

(a) Inexpensive overall survey of an office, workshop or any work situation
(b) Analysing non-repetitive or irregularly occurring activities
(c) Preliminary survey which helps define the exact problem for further study
(d) Determining the nature and extent, both cyclic and peak load variations, of various activities
(e) Analysis of men and machine utilisation
(f) Planning for manpower requirements
(g) Working out shop effectiveness and efficiency
(h) Checking observance of specific management policies and safety regulations
(i) Helping in collection of data for operations research studies.

ADVANTAGES OF SNAP READING

As compared to time study, activity sampling has the following advantages:

(a) It is convenient, economical, time saving and suitable for observing many operators and machines. (Reduction in cost ranges from 5% to 60%, depending upon type of study.)
(b) The observations can be spread over a period of days/weeks to allow for fluctuations.

(c) The observers need not be given elaborate training. Given sufficient briefing, supervisory staff can act as observers.
(d) Study may be interrupted at any time without affecting the results.
(e) Study can be modified to the pre-assigned requirements of reliability and accuracy.
(f) It is less tiring to observers and preferred by operators, as a stop watch or any other special equipment is not required for study.
(g) It consumes less time in compiling results.
(h) It does not irritate the worker, who dislikes some one sitting on him with a stop watch and hence used universally.

DISADVANTAGES OF SNAP READING

Activity sampling has the following disadvantages:

(a) It is not economical for studying single operator or machine.
(b) The results will be less accurate as compared to time study.
(c) Time study with a stop watch permits finer breakdown of elements.
(d) An activity sampling made on a group will present only average results of the group.
(e) In certain studies, no record of method is kept, and a new study has to be carried out even if the method is changed slightly.

There is a tendency on the part of some observers to underestimate the importance of fundamental principles of activity sampling, such as taking observations at pre-assigned random timings, making required number of instantaneous observations, and carefully defining elements. This would lead to erroneous results.

For the past 60 years activity sampling has been used in industry for measuring activities of men and machines. Slowly, the emphasis is shifting from studying direct labour to indirect labour, such as measuring working time and non-working time of office staff and factory workers employed in indirect activities. As industry becomes highly mechanised, it is imperative that machine down time be eliminated or reduced to bring down operating cost of machines and process equipment. As a fact-finding tool, this technique can provide valuable information about men and machines in lesser time, and at a lower cost to the management.

WORK MEASUREMENT

Work measurement is the application of techniques designed to establish the time for a qualified worker to carry out a specified job at a defined level of performance. It provides the basis (a) for quantitative assessment of the human work in a specified task, and (b) for establishing proper time for the effective performance of the task.

According to the accepted Bureau of Indian Standard Institution definition, work measurement is the application of techniques designed to establish the *proper time* for accomplishing a *specified task* by a *prescribed method* by determining the time required for carrying it out at a *defined level* of performance by a *qualified worker* under a given set of environments and working conditions.

Proper time should take care of the following:

(a) The worker is helped to keep up a normal steady pace throughout the working day.
(b) He can sustain this pace all through his working life.
(c) Depending upon the type of work, he can be allowed the relaxation allowance to get over his physical and psychological fatigue.

Specified task and prescribed method means that each job must be fully defined in relation to:

(a) exact method of performance;
(b) details and type of operations;
(c) place and its environments, where work is done;
(d) specification of the quality of output;
(e) state and specification of equipment and tools.

A qualified worker must have the following attributes and capabilities:

(a) *Physical fitness*—having physical attributes and characteristics as required for the job.
(b) *Mental fitness*—adequate education and intelligence for performing the task.
(c) *Experience*—having been long enough on the job to develop:
    (i) sufficient knowledge of the job;
    (ii) adequate skill to perform the job;
    (iii) satisfactory standard of output in quality and quantity
    (iv) proper regard to safety for himself and the equipment he is operating.

Defined level of performance is the average rate at which a qualified worker would work, provided: (a) he knows the specified method of performance; (b) he adheres to the method while working; (c) he is properly motivated to apply himself to the job.

A qualified person who works at this pace is said to work at a rating of 100.

Work measurement contributes to the following:

(a) Correct pre-planning of work
(b) Effective planning of work load
(c) Having a basis for deployment of manpower
(d) Having a basis for effective control of work
(e) Compiling of data for estimating costing and budgeting
(f) Building up a basis for sound financial incentives.

TIME STUDY

Time study is a work measurement technique for recording the times and rates of working for the elements of specified job carried out under specified conditions. The data so obtained is analysed to calculate the time necessary for carrying out the job at the defined level of performance. The steps followed are:

(a) Work is examined and broken up into a series of small elements of half minute duration.
(b) Each element is timed with a stop-watch and an assessment made of the rate of working. This is done by comparing observed rate of working with the observer's concept of standard rate.
(c) Derive, for each clement, the time required when work is done at the standard rate of working.
(d) Calculate the basic time as follows:

$$\frac{\text{Observed time} \times \text{Observed rating}}{100} = \text{Basic time}$$

   (i) *Work content:* Work content is the total of basic time + Relaxation allowance + Any other allowance for work, e.g. part of contingency allowance which represents work.

  (ii) *Relaxation allowance:* Relaxation allowance is an addition to basic time, which is intended to provide the worker with the possibility to recover from the physical and psychological effects of the job performance under specified conditions, and to allow the worker to attend to personal needs. It depends on the nature of work. It is computed as a percentage of basic time; standard tables are available for reference.

 (iii) *Contingency allowance:* Contingency allowances are small allowances, which are given to meet legitimate and expected items of work or delay, precise measurement of which is uneconomical because of their infrequent occurrences. These allowances are never more than 5%.

(iv) *Interference allowance:* Interference allowance is provided to a worker operating several machines. These machines are liable to cyclic or random stoppages which give rise to cyclic or random machine interferences.

(v) *Standard time:* Standard time is the total time in which a job should be completed at standard performance, which is equal to the sum of work content, the contingency allowance for delay, the unoccupied time, and interference allowance.

The allowances can be categorised as follows:

(a) Rest and personal allowances
(b) Process allowance
(c) Special allowance
(d) Contingency allowance
(e) Policy allowance

(i) *Relaxation allowance:* (a) *Constant allowances.* Personal allowance for washing hands, drinking of water, visit to toilet, etc. Women workers need more time than men; and allowance for overcoming fatigue.
(b) *Variable allowance.* Variable allowance is provided, depending upon the posture adopted at work and the working conditions. The factors involved are abnormal position, standing, use of force or muscular energy, bad light, temperature, humidity, cold, close attention, high noise level, concentration, monotony, etc.

(ii) *Process allowance:* Process allowance is provided to compensate the worker for 'enforced idleness' due to characteristic of operation, or process needs. This is applicable where piece rate payment is in force.

(iii) *Special allowance:* (a) *Periodic activity.* This allowance is provided for work carried out periodically in the course of manufacturing of a given batch. It includes allowance for activity carried out at *definite* intervals, e.g. regarding tool, machine cleaning, resetting machine, periodic checks, and allowances for activity carried out *once* in the course of a batch or order, like setting the tool at the beginning of a batch, and preparing plant to manufacture paint of a fresh colour and mixing its ingredients. (b) *Interference allowance.* This depends on the skill and effort of operator, the number of machines he is looking after, and on the time taken to rectify machine stoppages.

(iv) *Contingency allowance:* This is meant to cover irregular occurrences, measurement of which is uneconomical.

(v) *Policy allowance:* It is given at the discretion of management, as a result of agreement with union, enabling the workers to earn a higher percentage of bonus.

Chapter 17

Energy Saving by Maintenance

NEED FOR ENERGY SAVING

Energy systems of one kind or the other like electricity, steam, water, diesel, oil are in use in every organisation. These systems use some or the other form of energy, whether it be in factories, laboratories, schools, colleges, hospitals, shops, cinemas or the railways. The consumption of energy is a must. Maintenance of energy systems comprises regular schedules of inspection, followed by the necessary cleaning, lubrication, adjusting, rectifying and calibration of the inspected machinery. It also includes specific as well as general tasks, like the cleaning of fins on compressors, or the repainting of walls to improve effectiveness of lighting, and the plugging of water/steam or any other variety of leakage, like the cold air leaking out from the doors of an air-conditioned laboratory and insulation monitoring.

Poor maintenance leads to the inefficient operation of a system, which ultimately can lead to higher consumption of energy for its operation, and this means the incurring of additional cost. A well-thought out and good *energy system maintenance programme* can save an organisation considerable sum of money, by reducing energy wastage due to losses in water, steam, electricity, petroleum, diesel, oil, and air and gases, etc. Further, it can reduce expenses incurred due to breakdown of energy systems and the consequent loss of production.

Unfortunately, maintenance remains even today as one of the most neglected areas in industry. This sad state of affairs has been perpetuated, because regular maintenance provides no immediate economic benefit, nor poses any immediate threat to motivate management. So far, management has paid little attention to this vital aspect. But today, with rapidly escalating energy costs, the threat has become real, and serious efforts are needed to improve maintenance productivity. In view of the energy crunch, the Government has announced that any energy saving device can be depreciated at 100% in a year. Energy audit, energy monitoring and energy saving measures are advertised in internet as well.

ISSUES OF MAINTENANCE AND ENERGY

The following actions can help in formulating a plan for creating a viable maintenance programme for energy saving:

(i) *Listing of facilities:* This includes the following steps:

   (a) Determine the present condition of the facilities. This presupposes that a Facility Register be maintained, listing all those equipment, plant, machines, buildings which have to be maintained. From these a detailed list of transformers, motors, airconditioners, lights, and other components, which make up each of the systems, together with a report on the *condition* of each of these, can be prepared.

   (b) Furthermore, the listing must indicate the *condition* of each item at the time it is entered in the Facility Register.

(ii) *Preparing a schedule of routine maintenance inspection and checks:* It indicates frequency, types of tradesmen, needed materials and spares, test equipment, special tools and gauges, and the estimated time required to complete the task.

   This involves the following steps:

   (a) Indicate the frequency of checks required for different items.
   (b) List the tradesmen/craftsmen required for each job.
   (c) Prepare a list of materials and spares required and make arrangements for holding adequate stocks in store.
   (d) Make arrangements for providing specific tools and gauges for each task, and for the maintenance of these items.

(iii) *Designing a maintenance plan:* It includes planned and/or preventive schedules incorporating the above steps, and taking into account the manufacturers recommendations, the extent of utilisation, and environment of operation for achieving the desired level of maintenance.

(iv) *Instituting a monitoring system:* A monitoring system has to be designed and instituted to ensure the implementation of the maintenance plan, as per schedule.

While stating the present condition of the facility, the following things are necessary:

(a) A complete list of all equipment in the building, showing the name, location, and condition of each item.
(b) A folder containing manufacturer's data, regarding operating specifications/conditions, limits of temperature/pressure etc., equipment-wise.
(c) Diagrams showing location of all important pieces of equipment on hydraulic/lubrication charts for major systems.
(d) A comprehensive list of maintenance activities required for each piece of equipment, with details.

This information constitutes an 'equipment data bank' unique for the equipment in an organisation. This should be updated and centrally maintained in the Maintenance Control Centre. The compilation of the data bank can be simplified, if separate energy-related systems are defined within the plant, for easy reference and examination.

A suggested classification would be as follows:

(a) The building requirement
(b) Boiler and steam distribution system
(c) Air-compressor systems
(d) High voltage air-compressor system
(e) Electrical system
(f) Lighting system (lights, reflective walls, ceilings).

Each of these systems should be inspected and the condition of each part noted.

Adoption of energy conservation methods and energy audit schemes will lead to better maintenance. Economical and safe operation of electrical equipment with specific reference to power factor is a step towards good maintenance for the maintenance of H.T. breakers earthing, switch gears, starters motors, transformers—it's a must.

BUILDING PROBLEMS

The building requirement consists of facilities that let air into or leak out of any building, workshop or laboratory. From the maintenance point of view, the most important aspect is to provide a highly insulated environment to reduce heating or cooling losses. The vulnerable points should be described by a blueprint of the building, showing locations of all the outside walls, windows, ceilings, doors, etc. The primary malfunction of the system is the leakage of air that can be detected by sight (a crack) or sound or feel (a draft). The benefits accruable are: reduction in the amount of air that must be heated or cooled, and increased comfort due to decreased amount of hot or cold air. For some of the system components, the problems encountered and the initial maintenance actions required are as indicated in Table 17.1.

Table 17.1 Maintenance Problems and Steps

System Component	Problem	Initial Maintenance Action
Doors	Loose fitting	Hinge to be repaired or frame to be replaced
	Does not close properly	Check door-closing mechanism, rectify door fit, balance the intake and exhausting of air
Windows	Air leakage	Replace broken panes, put fresh patty strips
Walls	Drafts from wall cracks/openings	Seal opening on outside of wall, patch/seal cracks with sealant or cement
Ceilings	Draft around exhaust piping	Repair
Roofs	Leakage	Patch, cover, repair or renovate by tarfelt

BOILER AND STEAM DISTRIBUTION

The boiler is the biggest fuel consumer in any organisation. Hence, any maintenance improvement will be reflected in decreased energy consumption and in decreased energy costs. If the steam distribution system has leaks or improper insulation, the boiler will generate more steam than is really needed, thus consuming more energy. Leaks lead to wastage of energy and excessive noise. Steam leaks can be detected by using acoustic probes or just visually. Leaks must be checked, particularly at valves, stuck traps, stuck by-pass valves, and condensate systems. Routing of pipelines ought to be designed in such a way that if leakages develop, they should be visible and accessible for immediate repairs.

Figure 17.1, which is in the form of a graph, indicates the annual heat loss from steam leaks.

In an operating steam plant it is commonly noticed that the insulation is damaged or there is no insulation at all along significant lengths of piping. This takes place when continuous but haphazard repairs are carried out over the years in the pipes, or they are re-routed without adequate care being taken to re-route the insulation along with it. Such a condition could arise from a lack of awareness of the need to take care or due to sheer carelessness.

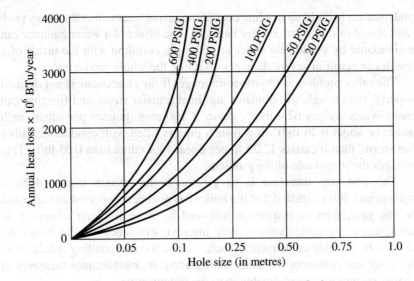

Figure 17.1 Annual heat loss from steam leaks.

Figure 17.2 indicates energy losses that accrue due to the bare steam pipelines.

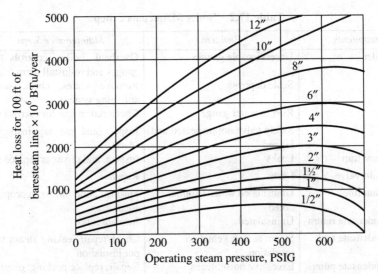

Figure 17.2 Energy losses due to bare steam pipelines.

BOILER MAINTENANCE

Another area to be looked into is the steam traps, which have to be always maintained in a good operating condition because, if a blocked trap causes condensation to build up in a line containing active steam, the steam may push water ahead of it to form a water hammer. The effect of a water hammer can be simulated by visualising water at 100 mph, colliding with the inside of a pipe. It can result in severe damages affecting the whole production.

The other problem is improper drainage. If the condensate is not drained properly, then weight accumulates and heat transfer stops, and freezing can result. Water weighs 62.4 lb/ft^3, steam at 100 psig (pounds per square inch gauge) or about 0.26 lb/ft^3. If a 6 inch pipe is filled with condensate rather than steam, then it carries 12.25 lb per linear foot rather titan 0.05 lb/ft. This indicates the magnitude of the problem.

The first step therefore is to get the boiler system inspected by a professional. It is estimated that if a boiler has not been inspected and adjusted for one year, then an inspection followed by reconditioning according to maintenance recommendations, will improve efficiency of the boiler by 12–15% by suitable adjustments. There will be a corresponding reduction in fuel costs and consumption. Thus, taking adequate maintenance measures in a boiler system will be of great help to the organisation.

Table 17.2 shows the initial maintenance actions pertaining to the boiler system components. It is well known that government boiler inspections periodically examine the boilers of factories.

Table 17.2 Boiler Maintenance Steps

Components	Problem	Maintenance Steps
Boiler	Unserviceable gauges	Overhaul boiler controls and gauges and re-install them
	Scale deposits	Remove scales, check water-softening system
	Rust in water gauge	Check return line for corrosion
	Safety valve not inspected and lagged	Inspect and put tag inspection (date)
Steam trap	Leaks	Inspect, and repair or replace
Steam valve	Leaks	Repair
Steam line	Uninsulated water hammer noted	Put insulation on; fix appropriate steam-trap
Condensate return	Uninsulated	Insulate if hot
Condensate tank	Steam at tank vents; no insulation	Check/repair leaking steam trap; put insulation
Condensate pumps	Excessive noise; leaks	Repair; replace packing; overhaul or replace pump

AIR COMPRESSOR SYSTEM

Compressed air is used as a medium for cleaning by blowing away various kinds of dirt or to dry material, or as an energy source for tools or machines.

As leaks lead to yearly operating loss due to wastage of energy, they are a big financial burden on the organisation. Air leaks take place at fittings, valves, and air hoses, which are easily detectable by swabbing soapy water, as bubbles form when the air leaks out.

Table 17.3 shows annual heat losses from compressed air leaks.

Table 17.3 Heat Losses (yearly) from Compressed Air Leaks

Hole Diameter (in inches)	Free Air Wasted*: ft³/year, by air leak at	Fuel Wasted**: (million BTu/yr)
	100 psig	
3/8	79,900,000	2190
1/4	35,500,000	972
1/8	8,880,000	243
1/16	2,220,000	60.6
1/32	553,000	15.1
	at 70 psig	
3/8	59,100,000	1320
1/4	26,200,000	587
1/8	6,560,000	147
1/16	1,640,000	36.6
1/32	410,000	9.2

* Based on nozzle coefficient of 0.65.
** Based on 10,000 BTu fuel/kWh.

Table 17.4 serves as a guide to initial maintenance actions for air compressor systems, which will reduce the energy losses.

Table 17.4 Maintenance Steps for Compressor Systems

Component	Problem	Maintenance Action
Compressor	Low suction pressures	Check and rectify leaks on low pressure side
	Gauges u/s	Repair, overhaul, replace
	Excess vibration	Check mountings
	Loose or frayed wiring	Repair with insulation tape or replace
	Leak on low-pressure side	Examine compressor, change gaskets, connection etc.

HEATING VENTILATION AND AIR CONDITIONING

Every organisation has certain sections which should be dust free and/or an air-conditioned insulated environment where precision work is carried out, as in the case of instruments, or electronic repair, the overhaul or manufacturing sections of delicate components, standards rooms, calibration rooms, explosive workshops, etc.

The heating, ventilation and air-conditioning systems are expected to supply enough air at the right temperature to keep working people comfortable and to exhaust the contaminated air. A complete description of this system existing in an organisation must include a blueprint indicating the location of all dampers, fans, ducts, and the control system, showing the location of all gauges, thermostats, valves, etc. and manufacturers manuals, operating manuals, engineering diagrams, spare parts catalogue, etc.

Significant savings can be effected by proper maintenance of this system. The savings accrue by reducing the energy used by the system, by decreasing the amount of unanticipated repair costs, and reducing down time, caused by non-availability when the system does not work and conditions become uncomfortable to work. Further, a good maintenance programme can spot the deterioration taking place in the equipment well in advance, and can thus schedule repair at a time, which will cause the least amount of disruption of work. Large savings can be made simply by adjusting the controls, e.g. by using a lower temperature on weekends, when no one is likely to be inside the facility. Hence, great care must be taken to bring the control system to a fully efficient operating condition.

CONSERVATION GUIDELINES

Table 17.5 lists certain problems and their initial solutions.

Table 17.5 Maintenance Actions Needed for Heating, Ventilation and Air-Conditioning System

Component	Problem	Initial Maintenance Actions
Filter	Dirt	Replace and/or clean
Damper	Blocked or linkage disconnected	Check damper controls
Duct work	Leaks, open joints	Clean and overhaul; repair with duct tape
	Loose insulation in duct work	Replace/attach firmly
	Water leakage or rust spots	Repair
	Crushed	Replace
Grill work	Dirt	Clean
	Blocked by other equipment	Remove equipment; allow air-flow
Fan	Excessive noise	Check bearings, belt tension
	Insufficient ventilation	Check fan and surrounding duct and gril-work
	Belt too tight or loose	Adjust motor mount
	Pulleys misaligned	Correct alignment
Pump	Hot water, pump cold	Inspect valve, check direction of flow

(Contd.)

Table 17.5 Maintenance Actions Needed for Heating, Ventilation and Air-Conditioning System (*Contd.*)

Component	Problem	Initial Maintenance Actions
Blower	Not throwing enough air	Check direction of rotation and clean
	Noise	Check bearings
	Wrong direction of rotation	Check wiring
	Shaft does not rotate freely	Check lubrication; repair pump
Cooling tower	Scaling on spray nozzles	Remove by chemical or mechanical means
	Leaks	Repair
	Cold water too warm	Check fans, pump and piping for blockage
Compressor	Temperature reading inaccurate	Calibrate
Thermostat	Leaks—water or oil from mounting	Check lines for breakage and ensure free-flow

The following simple guidelines can be used for conservation in heating, ventilation and air-conditioning system operations:

(a) In workshops, laboratories or specific sections which require heating or cooling, the controls should be examined and set depending upon the outside temperature.
(b) Ventilation, air and exhaust requirements must be ascertained. A reduction in air flow will save energy, because pumping power varies as the cube of air flow rate.
(c) Constantly review air-conditioning and heating needs. Seal off those sections that do not need these. During non-working hours, the equipment (heating, ventilation and air-conditioning system) can be shut off or reduced, depending upon the comparative cost effectiveness.

ELECTRICAL SYSTEM

Electricity bills can be reduced considerably by carefully attending to some simple problem areas. If a motor is operating at a lower voltage than it is designed for, then it is using more amperage, thus causing losses in transmission. If the wire diameter is too small for the load i^2r losses can be large. It can also become a source of an additional fire hazard. Voltage imbalances in a three-phase motor, and leaks to ground are other major sources of waste. These must be checked by a professionally qualified and experienced electrical engineer, as all these factors cost money and are safety hazards too. At the same time, a check must be carried out to ensure that the wiring, the transformers and the switch, etc. are of the appropriate size for the load they are expected to carry.

Equipment that contribute to the lowering of power factors are welding plants, induction motors, power transformers, electric are furnaces, etc. A low power factor can be corrected by installing capacitors.

The following steps can be taken for saving electricity.

(a) Use the highest voltage which is practicable, as we know that for a given application, doubling the voltage cuts down the required current to half, and thereby reduces the i^2r losses by a factor of four.
(b) Eliminate unnecessary transformers as they waste energy. It is better to order equipment with motors of current voltage, even if this costs more than to instal a special transformer.
(c) Check overall power factor of the plant for low power factor. This can be improved by
   (i) reducing inefficient loading (Motors running at full load have much higher power factor);
   (ii) making use of power factor correction capacitors;
   (iii) using synchronous motors instead of induction motors.

At the same time, basic maintenance actions must be initiated, such as having a well-ventilated and clean transformer room, with no oil leaks, no burnt spots on contacts, and no flickering or arcing of switches.

LIGHTING SYSTEM

For the personnel in the organisation to take energy saving seriously, the management will have to set an example. If lights, fans, and room-coolers, air conditioners, computers, are switched off in the Manager's room, when not needed, it will have a positive psychological impact on others.

Many factors affect the lighting system; e.g. the condition of lights, cleanliness of bulbs, cleanliness of walls, ceilings and floors, and in particular the windows. Energy saving bulbs must be used instead of old ones.

The level of illumination, or the quantity of light which reaches the work surface, is measured in units of Lux (lumens per square metre, lm/m^3) by a lightmeter, which is a simple portable device. The recommended lighting intensities for different jobs are compiled by the Illuminating Engineering Society (IES) or the ISI, both professional bodies. Hence, the first task to be undertaken is to measure the levels of illumination at work places, i.e. on shops, storehouses, offices, laboratories, etc. Many work places are perhaps overlit and some underlit.

In India, there is abundant bright sunlight practically throughout the year. Undoubtedly, few factories can be lit with day light alone, but a judicious use of day light and electric light can be made. General offices and drawing offices, etc. can benefit immensely by this judicious combination. But, seating arrangements usually place the working desks in the middle, and the filing cabinets/cupboards are placed near the windows, thereby hiding the sunlight. This needs to be looked into. Workshops can also use the sunlight to advantage. However, this would necessitate window-pane repairs and cleaning to be done on a regular/frequent basis. Solar energy can also be used.

Thus, it is possible that those people working near the windows will not need artificial light for at least part of the day. But, the problem arises when the 'switching on' circuit is centralised. Hence, for convenience and economy, it is necessary to have two separate switching devices, one for the group which usually requires lighting, and the other which requires it only when day light is insufficient.

Very often one sees that yard lights, security lights, and corridor lights are kept on even in broad day light, as either the switches are out of order or someone has forgotten to switch them off. By fitting a light-cell switch, which controls the lighting according to the intensity of day light, this problem can be solved. Contrary to belief, switching lights 'on' or 'off' does not appreciably reduce the life of the light. This happens in many cities where street lights, sometimes, are not switched off in day time.

CYCLIC REPLACEMENT OF LAMPS

The illuminating capacity of light decreases with use, and the dust build-up on the lights further adds up to this loss. The fluorescent fittings in factories lose approximately 60% of their luminosity after 12 months of use. The loss accruing from the dirt which collects on the lamps (if not cleaned regularly) adds up to another 25%. At the end of two years, the figures move up to 80% and 75% of loss in each case. These statistics have been verified in the United Kingdom. Therefore, the loss in luminosity due to dirt and grime must be much greater in a country as dry and dust-laden, as ours. It can well be argued that, if lights are allowed to become dirty, and the illumination level allowed to dwindle and deteriorate, then no extra cost is being incurred. But this would be a wrong assessment, because either extra lights would have to be provided or one will have to accept higher rates of rejects, rework, and scrapping. The replacement of lights is a relevant factor as the major cost in the lighting of an installation is not the hardware itself. This is because the energy cost which is around 70 to 80% of the total cost far surpasses the hardware cost. After 75% of the rated life-span of the lamps is over, 5 to 10% of them normally fail; thereafter the occurrence or probability of failure becomes more frequent with each passing day. Therefore, it is often economical to replace all the lights when they have reached somewhere between 60 to 80% of their rated life span. While doing this, the fixtures and fittings should also be given a thorough cleaning, making use of the same manpower. Group replacement of bulbs is more economical.

If a room is painted totally white, then the least amount of lighting will be needed, as white reflects the maximum amount of light. Rooms and ceilings painted in dark colours will need more lighting. Pastel shades are more pleasing and are cooler. Shades such as light blue, grey and green have the added advantage of not absorbing a great degree of light.

ECONOMISING ON LIGHT

The recommended illumination levels are available in Illuminating Engineering Society Handbook. Some of these levels are given in Table 17.6.

Table 17.6 Recommended Illumination Levels

Type of Building	Recommended Range in Foot-candles (USA)	ISI Recommended Values in Lux
Banks	50–150	300
Offices	30–100	150–300
Stores	20–50	150
*Library	30–100	300–700
Schools/Colleges	30–150	300–700
Manufacturing area		
Ordinary tasks	50	450
Different tasks	100	
Very difficult task	300–500	see IS 3646 Part II 1966
Exterior		
Building security	1–5	
Floor lighting	5–30	

*For details see 1.S. 2672 1966 Codes for Library Lighting.
To convert foot-candles to Lux multiply by 10.7. IS 3646 (Part II)-1966 Code of Practice for Interior Illumination.

Efficacy of a light source is the ability to convert power (watts) into light (lumen) expressed as lumens/watts. The lamps can be incandescent, fluorescent, energy saver, high-intensity discharge, mercury, metal halide, and high and low pressure sodium. Some ideas can be had from Table 17.7 for selecting an appropriate lamp for a specific application. But before making a final selection, consult the lamp manufacturers literature, asking for full technical specifications and details.

In the final analysis, there are four basic checks to be made and actions to be taken:
(a) Removal of excessive or unnecessary lamps.
(b) Ascertaining whether relamping is possible. Installing lower wattage lamps.
(c) Fixtures, lamps, reflectors periodically, as it improves light output.
(d) Devising better controls of lighting.

Table 17.7 Efficacy of Light Sources

	Incandescent	Fluorescent	Mercury Vapour	Metal Halide	High Pressure Sodium	Low Pressure Sodium
Wattage	15–1500	15–219	40–1000	175–1000	70–1000	35–180
Life (hr)	750–12,000	7500–24,000	16,000	1500–15,000	24,000	18,000
Efficacy	15–25	55–100	50–60	80–100	75–140	Up to 180
Lumen maintenance	Fair to excellent	Fair to excellent	Very good	Good	Excellent	Excellent
Comparative fixture cost	Low-simple fixtures	Moderate	Higher than earlier two types	Higher than mercury	High	High
Comparative operating costs	High-short life and low efficiency	Lower than incandescent	Lower than incandescent	Lower than mercury	Lowest of HID type	Low
Auxiliary equipment needed	Not needed	Needed medium cost	Needed high cost	Needed high cost	Needed high cost	Needed high cost

Life and efficacy rating as per US practice.
For Indian practice, check with indigenous manufacturers latest literature and Illuminating Engineering Society of India, Calcutta.

LIGHT SOURCE CHARACTERISTICS

Fluorescent lights enjoy the largest percentage share of commercial utilisation. In higher wattages their efficacy is better than that of mercury vapour and some metal halide lamps. Their low cost makes them the most competitive type of light. However, a planned maintenance schedule for cleaning lamps/fixtures and timely replacing can only ensure maintenance efficiency.

It is neither possible nor necessary to address ourselves to the variety and diversity of industrial operations (and their myriad problems) that exist in our country. Each industry, facility or plant is unique, and hence needs to be examined individually on the basis of its particular merits and demerits. An attempt has been made here to present certain general concepts and principles with their applications. It should be possible to make up to 10% of savings easily. These simple concepts so easily applicable do yield very good results with very little added expenditure.

Chapter 18

Facilities Investment Decisions (FID) and Life Cycle Costing (LCC)

ASSET MANAGEMENT

The Government of India has opened up its economy since 2004 onwards in most of the sectors. Inspite of huge inflation, heavy investments are made by public sectors, private sectors, foreign direct investment, and foreign institutional investors. The investment is touching one trillion per year recently particularly because the other economies except China, are not doing well. Investments of such huge magnitude are being made in public as well as private sectors for the creation of capital assets like plant, machinery and equipment each year. Only their adequate production utilisation can ensure the planned development of our economy and the raising of the standard of living of our countrymen.

DECISION VARIETIES AND FID

However, these investment decisions are extremely important, because those organisations that have invested in technologically inferior machinery or those who have not replaced their ageing plants and equipment in time are most likely to lose their competitive edge in the present-day dynamic market environment. Investments in infrastructure industries has increased and high tech machines are put into operation in the sixties, since reinvestment decisions and replacement decisions.

On the other hand, overinvestment in providing facilities is another damaging aspect of decision making, because it leads to underutilisation of the created capacity/facilities. Capacity underutilisation is a serious problem in our country today. Some of the important reasons for this are:

(a) Incorrect choice of technology
(b) Lack of adequate demand for the product
(c) Shortage of power
(d) Shortage of input raw materials
(e) Poor maintenance.

These are due to managerial shortcomings and inadequacies, which can be corrected only by broadening the perspective of our decision makers and creating in them the capabilities for proper assessment of the total environment, market surveys, technological needs and improved analytical skills.

It is therefore necessary to understand the analytical framework within which such decisions ought to be made, the focus being on economic analyses like the concept of time value of money, the economic life of an asset, and the impact of taxes and depreciation, and the salvage value.

The managers responsible for investment decisions have to face a multitude of issues and problems due to the availability of various alternatives. The decision variables may be as follows:

(a) What is the appropriate technology?
(b) How much capacity is to be created?
(c) What kind of machines are required—general purpose or specific?
(d) Should all the machines be bought or taken on lease? What could be the correct mix?
(e) When and how should the machines be replaced?
(f) What about handling facilities?

These questions are difficult to answer, and we must first try and understand the various factors that affect these decision variables before attempting to seek solutions.

FACTORS INFLUENCING FACILITIES INVESTMENT DECISIONS (FID)

Figure 18.1 illustrates the various factors involved, and no analysis can be complete without taking into consideration all the factors. Let us take the example of investment in producer/consumer goods which are power and basic raw material intensive. These must only be made after fully knowing and analysing the government's policy and plans for growth in power and other concerned core sectors. The technical analysis cannot only be confined to process and product efficiency, but must also deal with the availability of inputs, raw materials, the spare parts availability, and the obtainability of appropriate technical skill levels to maintain, etc. Further, no analysis is complete without considering the 'human element' as it can have serious implications on the industrial relations climate and even bring in political repercussions. The facilities include industrial processing machines, handling machines like E07, Grab cranes, Grab bucket, single ropes, Electro mechanical, hydraulic systems, laddle, vertical coil tongs, hydraulic filter, hooks, hydraulic platforms, coil lift estackles, fork lift etc.

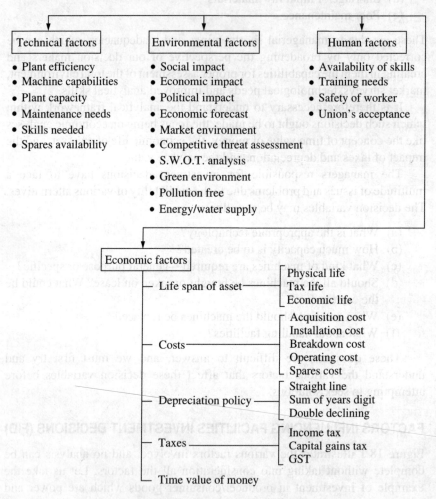

Figure 18.1 Factors influencing FID.

Because of the large quantum of investments involved, only on being assured of long-term benefits arising from such investments, should any organisation commit its scarce resources for investment. Mistakes committed here can be fatal to any organisation. In case, the organisation is not assured of a potential market, and wishes to launch a new product, it will be prudent to take the machinery or equipment (favourable) on lease rather than to buy it. On the other hand, if it desires to expand the existing capacity, then increasing the number of shifts, working overtime or subcontracting are viable alternatives, to making immediate capital investments.

SELECTION/REPLACEMENT PROBLEMS

The proposals for buying equipment to cater to a new or expanding market have differing characteristics and organisational implications compared to requests for replacing existing physically functioning but economically inferior, aged machines. These are shown below, and are categorised under two heads, namely, Selection Problems and Replacement Problems (Table 18.1).

Table 18.1 Selection/Replacement Issues

Selection Problems	Replacement Problems
These arise due to the need for new or expanded capacities, to fulfil the desire to improve product capabilities.	It is essential to replace worn out or inefficient equipment; no significant addition to capacity takes place.
Proposals come from the marketing research or corporate planning divisions.	These emanate from the production or maintenance departments.
The problems have major implications for the direction in which the firm wants to grow in future years; necessary market uncertainties must be faced.	These pertain to activities in which firm has expertise and an established market; large data on machine cost characteristics are available; less uncertainty is involved.
Decision alternatives include different machines with varying initial and recurring cost characteristics.	Decision alternatives are the number of years (economic life) a particular machine should be retained before replacement.
Top level decisions that may require the help of experts from within and outside the firm.	Can be dealt with by lower level managers and decisions can be routinised.

In spite of the existence of such structural differences, the framework for economic analysis is surprisingly common for these two categories of problem, since concepts of time value of money, enonomic life, and impact of taxes and depreciation are equally applicable to both. These are now explained.

ECONOMIC LIFE CONCEPT

The physical life of an asset covers its entire period of utilisation until it is scrapped. It will not be hypothetical to assume here, however, that the operating costs which comprise power and other input costs, and preventive maintenance costs will escalate over the years. The probability of breakdowns also will increase as machines will be worn out due to usage. However, the notion of physical life is very static and is confined to the single dimension of time with no reference to costs. The tax life of an asset is defined as the number of years over which its capital costs are amortised, and varies from the physical life only to the extent that the former is based on a fixed time period. Depending on the type of facility tax laws permit, a life of 10, 20 or even 50 years is ascribed to an asset for the purpose of depreciation. In contrast to both these, the notion of economic life of an asset is a dynamic one, which recognises the time and cost dimensions related to the asset. It is defined as the "service life of an asset which will minimise the average cost per period of service".

OPTIMUM REPLACEMENT MODELS

Many replacement models are available from Cooperations Research and we will discuss some of the simpler:

Let P = initial investment in the asset (acquisition cost).

C_j = sum of operating and breakdown maintenance costs in period j.

TC_n = total cost of owning and using the asset for n years.

$AC_n = TC_{n/n}$ = average cost per period.

The problem is to choose "n" such that AC_n average cost per period is minimised. Let the optimal value of n be denoted by N. Then

$$TC_n = P + \sum_{j=1}^{n} C_j$$

$$AC_n = \frac{TC_n}{n} = \frac{P}{n} + \frac{1}{n}\left(\sum_{j=1}^{n} C_j\right)$$

For the cost minimising optimum in N, the following conditions must hold:

$$AC_{n+1} - AC_n \geq 0$$
$$AC_{n-1} - AC_n \geq 0$$

We can solve these inequalities and obtain the rule as

$$C_n \leq AC_n \leq C_{n+1}$$

Hence, the rule is: Replace it at the end of any period for which the operating and breakdown maintenance costs in the next period exceed the average cost up to the time of replacement. Do not replace as long as the recurring costs in a period do not exceed the average cost to the end of the period. Let us use the following example to explain this rule: Let

$$P = 5000$$

Year	1	2	3	4	5	6	7	8
C_j	100	210	330	460	600	1000	1700	2500

Then the analysis will yield the result given in Table 18.2.

$$AC_n = \frac{P}{n} + \frac{1}{n}\sum_{}^{n} Cj = \frac{5000}{4} + \frac{1}{4}(1100)$$

Hence, $N = 6$ and $AC_n = 1284$. Note that $C_6 = 1000 < AC_n < C_7 = 1700$, which satisfy the rule developed earlier. The physical life of this asset may be long, but its economic life is only six years and it must be replaced then. Such a decision contradicts the notion of thrift, held by many decision makers who would like to extract every hour of service available from the machine. Yet they must be warned that their decision defies the concept of economic efficiency.

Table 18.2 Optimum Replacement Decisions

n	P/n	$1/n \sum C_j$	AC_n
1	5000	100	5100
2	2500	155	2655
3	1667	213	1880
4	1250	275	1525
5	1000	340	1340
6	834	450	1284
7	714	629	1343
8	625	863	1485

In the foregoing analysis, using the mathematical model, we assumed that a replacement is available at the same price as the original one in all future periods. It may not be valid in many real-life situations. There is no cause for alarm since varying initial costs can be easily accommodated in the calculation of average costs in the model with minor modifications. Similarly, we can include the salvage value of the machine when it is disposed of and evolve modified decision rules. These are extensions of the model only, and do not invalidate the basic principle involved in the concept of economic life.

The foregoing analysis, however, suffers from a serious drawback in that it has not taken into account "the time value of money". Being a scarce commodity, money has its market price and a rupee borrowed today must be returned with interest in any future period. We shall pursue this in detail in the following section but may note here that the concept of average cost discussed in the derivation of economic life of an asset must be replaced now with a similar concept of equivalent annual cost which would take into account the time value of money also. But for this change, the fundamental proposition of the notion of economic life remains intact.

TIME VALUE OF MONEY

Let the prevailing interest rate of i per year and the funds available at present period be P_0. Its value at the end of one year period is

$$P_1 = P_0 + iP_0 = P_0(1 + i)$$

Similarly, its value at the end of t years will be

$$P_t = P_0(1 + i)^t$$

In other words, any shrewd investor must be willing to accept an amount P_0 now or P_t after t years as equivalent amounts. This means that the present value of P_t is P_0 which can be expressed mathematically as

$$P_0 = \frac{P_t}{(1+i)^t} = P_t \left[\frac{1}{(1+i)^t} \right] = P_t \text{ (SPPWF)}$$

The term $1/(1 + i)^t$ is called the *"single payment present work factor (SPPWF)"*. Note that the present value of any future amount P_t can be obtained by multiplying it with this factor. From the relation for SPPWF it is seen to be a function of interest rate i and the time period t. We need not evaluate SPPWF every time since it is precalculated and readily available in tabular forms for various i and t values.

Instead of considering single payments at the end of t periods, we can visualise an alternative scenario where a specific amount R is available at the end of each year up to year T. The present worth of such a series of payments is

$$P = \sum_{t=1}^{T} \frac{R}{(1+i)^t} = R \text{ (SPWF)}$$

where SPWF is the *series present worth factor* which is a function of i and t as in the case of SPPWF. We can refer to SPWF published tables and determine its value also.

Note: Mathematically,

$$\text{SPWF} = \frac{(1+i)^{t}-1}{i(1+i)^t}$$

and it can be shown that for an infinite time period it simplifies to ($1/i$).

CAPITAL RECOVERY FACTOR

There are instances when we are interested in the reciprocal nature of relationships such as the one described above. A typical example is a lender who wants uniform end of period payments, necessary to repay his debt with interest charges. We use a capital recovery factor (CRF) here which is the reciprocal of the series present worth factor (SPWF):

$$\text{CRF} = \frac{1}{\text{SPWF}} = \frac{i(1+i)^t}{(1+i)^t - 1}$$

The relationship is uniform end of period payments $R = P$ (CRF). In other words, R represents the equivalent annual amount of P for a given i and t.

We are in a position now to extend the economic life concept from the average cost to equivalent annual cost estimation. Consider the case where the initial investment in an asset is P, its operating costs amount to C_t in year t, and the salvage value at the end of T years equals S_T. This is represented in a time framework as in Figure 18.2.

Present worth of such a stream of expenditures and revenue equals:

$$P + \frac{C_1}{1+i} + \frac{C_2}{(1+i)^2} + \cdots + \frac{C_T}{(1+i)^T} - \frac{S_T}{(1+i)^T}$$

This complex expression can be simplified if

$$C_1 = C_2 = \cdots = C_T = C$$

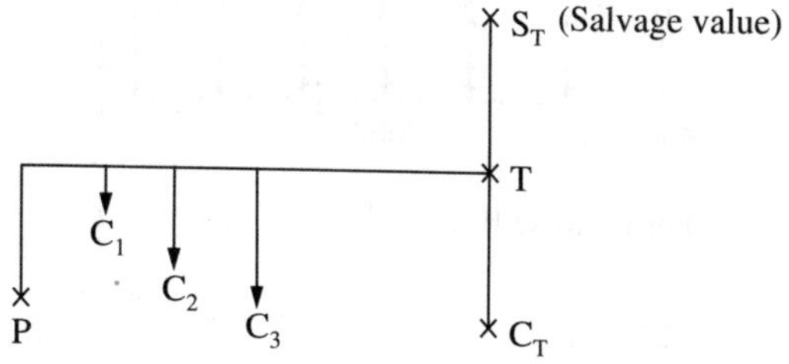

Figure 18.2 Time frame of costs.

Then,

$$P + C\left[\frac{1}{1+i} + \frac{1}{(1+i)^2} + \cdots + \frac{1}{(1+i)^T}\right] - \frac{S_T}{(1+i)^T}$$

$$= \left[P + C\,(\text{SPWF}) - \frac{S_T}{(1+i)^t}\right]$$

$$\text{EAC} = (\text{present worth}) \times \text{CRF} = \left[P - \frac{S_T}{(1+i)^t}\right]\text{CRF} + C\,(\text{SPWF}) \times \text{CRF}$$

This equation can be simplified to get the expected average costs as

$$EAC = (P - S_T) \text{CRF} + i \times S_T + C$$

Apart from its use in determining the economic life of an asset, this equation has been widely used in comparing a new machine to an existing one in replacement decisions. This will be clear from the following example: Suppose machine A was purchased four years ago for ₹22,000 and it was established at that time to have an economic life of 10 years and a salvage value of ₹2000 at the end of its life. Its operating costs amount to ₹7000 per year. Presently, a salesman is offering machine B which has an estimated economic life of six years and is priced at ₹24,000. Salvage value at the end of six years will be ₹3000 and the annual operating costs equal to ₹4000. Further, he offers to purchase machine A for ₹6000 if machine B is bought. Assume $j = 15\%$.

Solution. The original cost of ₹22,000 spent four years ago is irrelevant for our analysis at present since it is a sunk cost. This will become apparent if we consider the problem from the point of view of an outsider. He has at present the option of buying machine B for ₹24,000 or A for ₹6000, which is the price offered for it by the salesman.

Then the comparison would be:

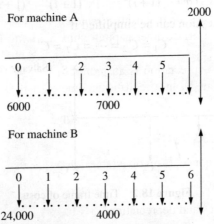

Figure 18.3 Time value of money.

Using the formula

$$EAC = (P - S_T) \text{CRF} + iS_T + C$$

with $i = 15\%$ and $t = 6$ years, for machine A,

$$EAC = (6000 - 2000)0.2642 + 0.15(2000) + 7000 = ₹8357$$

and for machine B,

$$EAC = (24{,}000 - 3000)0.2642 + (0.15)3000 + 4000 = ₹9998$$

Hence, in spite of its lower operating costs, it is not economical to buy machine B at this time.

EFFECT OF TAXES AND DEPRECIATION

So far, we have confined our analysis to cost considerations alone. However, the picture is not complete without including the effects of taxes including GST—goods and service tax and depreciation policies in our discussion. For example, the difference in operating costs for machines A and B described above is ₹3000. If machine B is bought, not all these cost savings are available to the company since it has to pay taxes on all income earned. (The income may come from increased sales revenue or through cost savings; in both cases it is taxed.) Similarly, when an asset is disposed off, it may be sold at a price different from its book value and thus incur either capital gains or losses. If we gain, then there is a tax to be paid on this (at a different rate than the income tax) which reduces the total net income obtained through transfer of such capital stock. For example, let the book value of an asset be ₹6000 and the capital gains tax rate be 50%. If we sell it for ₹10,000, the net realization due to this transaction is only ₹10,000—(10,000 + 6000)/(0.5) which is ₹8000. On the other hand, if we incur losses, say by selling the same asset for less than its book value, the loss on disposal is treated as a long term capital loss. This loss can be used to offset other capital gains resulting from other activities during that year.

The effect of depreciation is an indirect one since the pattern in which assets are depreciated can affect the income earned before and after taxes every year.

CASH FLOW ILLUSTRATION

Let us explain the concepts involved through an example.

		General Model
Income before taxes and depreciation	₹10,000	x
Depreciation	2000	y
Income before taxes	8000	$x - y$
	4400	$(x - y)0.55$
Income after taxes	3600	$(x - y) \times (1 - 0.55)$
		$= (x - y) \times (0.45)$
Add depreciation to determine cash flow	2000	y
Final cash flow	5600	$(0.45x) + 0.55y$

The general formula for final cash flow is $(1 - t)x + ty$, where t is the income tax rate.

We can deduce the following results from the above formula:

(a) Additions to the after tax cash flows occur as the tax rate times depreciation value of assets.
(b) The more the depreciation and in a given year for a fixed income, more is the cash flow. (Hence, it would be advantageous to switch over to accelerated depreciation methods wherever possible since we stand to gain more in earlier years.)

In all our discussions so far, we have introduced the term *cash flow* without explaining it. This only means that the way accountants keep their books and calculate profits is not exactly proper for the net present worth calculations in comparing different alternatives. What really counts is the pattern in which actual cash flows occur. Cash flow is concerned with expected future cash receipts and disbursements. For example, capital assets are depreciated over a number of years and profits calculated in the books of accountants. But, in the cash flow approach the entire amount paid is debited to the time period in which it is incurred and depreciation is taken into account in calculating future cash flows only. Earlier, we learnt that final cash flow comprises both income after taxes and the depreciation amount. In the ultimate analysis, this is the proper framework for evaluating investment decisions.

DEPRECIATION

Depreciation accounting sets a pattern for recovering capital invested in an asset and it relates the cost of owning a machine to its output. In effect, a portion of the income derived from the operation of a machine is set aside in book keeping accounts towards the cost of replacements.

The simplest method is a straight line depreciation by which a constant amount is charged each year of asset life to recover the revised capital.

Thus, annual depreciation charge (ADC) = $\dfrac{P - S}{n}$

where
P = purchase price of an asset
S = salvage value at the end of the economic life of the assets
n = economic life in years

Example
P = ₹60,000
S = ₹5000
n = 10 years

$$\text{ADC} = \frac{60{,}000 - 5000}{10} = 5500 \text{ per year}$$

The machine, therefore, has a book value that decreases uniformly from ₹60,000 to ₹5000.

Accelerated depreciation methods are popular as they recover more of the invested capital during the early years of the life of an asset. By the double declining method, ADC = $2/n$ (book value), where book value = purchase price – accumulated depreciation. The maximum rate factor of depreciation = 2/10 = 0.2. This means that 1/5 of the book value at the end of the preceding year is the depreciation charge for the following year. Therefore,

$$\text{First year charge} = 60{,}000 \times 0.2 = 12{,}000$$

$$\text{Book value} = 60{,}000 - 12{,}000 = 48{,}000$$

$$\text{Second year depreciation charges} = 0.2 \times 48{,}000 = 9600$$

$$\text{Book value at the end of second year} = 48{,}000 - 9600 = 38{,}400.$$

Another fast write-off method is *sum of digit depreciation*, where

$$\text{ADC} = \frac{\text{(Digit associated with assets age)}}{\text{Sum of digit for the economic life of an asset}} (P - S)$$

Example
Sum of digits $(1 + 2 + 3 + 4 + 5 + 6 + 7 + 8 + 9 + 10)$

$$= n\frac{(n+1)}{2} = \frac{10(10+1)}{2} = \frac{10 \times 11}{2} = 55$$

Therefore,

$$\text{First year depreciation charges} = \frac{10}{55} \times (60{,}000 - 5000) = 10{,}000$$

$$\text{Second year depreciation charges} = \frac{9}{55}(60{,}000 - 5000) = 9000$$

The depreciation charges for the remaining years continue to decrease by 1000 each year. Book value at the end of first year = 60,000 – 10,000 = 50,000 and continues to decrease at a fast rate till $s = 5000$.

A comparison of the above three methods at half the economic life shows:

(a) Straight line method   50%
(b) Sum of digits   73%
(c) Double declining   75%

ECONOMIC ANALYSIS

Most organisations use one or all of the following methods in their evaluation of equipment acquisition proposals:

(a) Pay-back period
(b) Return on investment (IRR—Interval rate of return)
(c) Net present value.

These methods are not perfect. Table 18.3 gives the comparative statement of the three methods. The two fundamental difficulties in all the methods are:

Table 18.3 Methods of Economic Analysis

Evaluation Method	Calculation	Applications	Limitations
Pay-back period	(a)	(Easy to calculate and easy to understand; can be used if it is short; used at best as a crude measure for arriving at difficult decisions.	Ignores time value of money; overlooks the total profitability of the project and the interest income potential.
Return on investment (ROI)	(b)	Easy to calculate and can be used at best as a crude measure for arriving at decisions.	Ignores times value of money; difficult to predict savings or income over many years in the future.
Net present value	(c)	Can be used to evaluate two or more investments opportunities (say two or more different machines for the same job); more dependable for arriving at decisions assuming the time value of money.	Difficult to deal with uncertainties; not easy to calculate precisely the savings and effects of technology over many years in the future.

(a) Whether an equipment should or should not be acquired depends upon what its cost of operation will be, compared to the savings. It is difficult to arrive at an absolute measure of profitability of investment.

(b) These methods of analysis are based by estimating the costs, savings, useful life, salvage values, and other factors used in evaluation. If they are under or overstated, often the decisions may prove to be very costly.

There are three methods of economic evaluation which help the decision maker in various areas. Each of these methods (a) focuses on the required length for an investment to pay for itself; (b) stresses not the actual amount received by investment but rather the per cent rate of return on the investment; and (c) considers the present value of the cash: inflows which will accrue due to an investment, relative to the cost of investment.

Briefly, the following guidelines can be given for facilities investment decisions:

(a) Consider cash flows only.
(b) Include the effects of taxes and depreciation.
(c) Account for the time value of money.
(d) Incorporate the economic life concept wherever possible.

Tables 18.4–18.6 are useful in calculating the economic analysis of investment decisions.

LIFE CYCLE COSTING (LCC)

Historically, most organisations in our country are known to have a tendency towards purchasing the least expensive equipment available in the market. This practice is being carried on even today. And this in effect has meant the choosing of the bidder (rather than the equipment of asset) who has quoted the lowest bid price in comparison to the different bid offers for a particular equipment. There are exceptions to this role as, at times, certain organisations have had the courage to choose a higher-priced product on the grounds of superior performance, or improved delivery schedules, or on the proven demonstrated capability of that equipment which is available with other users.

Some forward-looking organisations with an enlightened management have come to the conclusion that acquisition cost is only a part of the total cost of an equipment spread over its entire life. Sometimes the cost of operation and maintenance is far greater than the acquisition cost. As a result, they are becoming aware of the drawbacks of this system and have concluded, that the low acquisition cost of an equipment could eventually lead to high operating and maintenance costs. Thus, the most important criterion for selecting an equipment has become the total cost of the equipment which will be incurred on it till the end of its life, and therefore selecting equipment on the basis of lowest acquisition cost is progressively being perceived as a false economy.

Life cycle costing (LCC) is a concept which brings together, engineering, economic, financial and statistical techniques to quantify, by taking into account all costs in relation to an asset during the entire period of its ownership. It is concerned with the generation of alternatives which are quantified for scientifically arriving at an optimum choice of an asset.

Table 18.4 Single Payment Present Worth Factors (per cent)

Year	1	2	4	6	8	10	12	14	15	16	18	20
1	.990	.980	.962	.943	.926	.909	.893	.877	.870	.862	.847	.833
2	.980	.961	.915	.890	.857	.826	.797	.769	.756	.743	.718	.694
3	.971	.942	.889	.840	.794	.751	.712	.675	.658	.641	.609	.579
4	.961	.924	.855	.792	.735	.683	.636	.592	.572	.552	.516	.482
5	.951	.906	.822	.747	.681	.621	.567	.519	.497	.476	.437	.402
6	.942	.888	.790	.705	.630	.564	.507	.456	.432	.410	.370	.335
7	.933	.871	.760	.665	.583	.513	.452	.400	.376	.354	.314	.279
8	.923	.853	.731	.627	.540	.467	.404	.351	.327	.305	.266	.233
9	.914	.837	.703	.592	.500	.424	.361	.308	.284	.263	.225	.194
10	.905	.820	.676	.558	.463	.386	.322	.270	.247	.227	.191	.162
11	.896	.804	.650	.527	.429	.350	.287	.237	.215	.195	.162	.135
12	.887	.788	.625	.497	.397	.319	.257	.208	.187	.168	.137	.112
13	.879	.773	.601	.469	.368	.290	.229	.182	.163	.145	.116	.093
14	.870	.758	.577	.442	.340	.263	.205	.160	.141	.125	.099	.078
15	.861	.743	.555	.417	.315	.239	.183	.140	.123	.108	.084	.065
16	.853	.728	.534	.394	.292	.218	.163	.123	.107	.093	.071	.054
17	.844	.714	.513	.371	.270	.198	.146	.108	.093	.080	.060	.045
18	.836	.700	.494	.350	.250	.180	.130	.095	.081	.069	.051	.038
19	.828	.686	.475	.331	.232	.164	.116	.083	.070	.060	.043	.031
20	.820	.673	.456	.312	.215	.149	.104	.073	.061	.051	.037	.026

Table 18.5 Series Present Worth Factors (per cent)

Year	1	2	4	6	8	10	12	14	15	16	18	20
1	.990	.980	.962	.943	.926	.909	.893	.877	.870	.862	.847	.833
2	1.970	1.942	1.886	1.833	1.783	1.736	1.690	1.647	1.626	1.605	1.566	1.528
3	2.941	2.884	2.775	2.673	2.577	2.487	2.402	2.322	2.283	2.246	2.174	2.106
4	3.902	3.808	3.630	3.465	3.312	3.170	3.037	2.914	2.855	2.798	2.699	2.589
5	4.853	4.713	4.452	4.212	3.993	3.791	3.605	3.433	3.352	3.274	3.127	2.991
6	5.795	5.601	5.242	4.917	4.623	4.355	4.111	3.889	3.784	3.685	3.498	3.326
7	6.728	6.472	6.002	5.582	5.206	4.868	4.564	4.288	4.160	4.039	3.812	3.605
8	7.652	7.325	6.733	6.210	5.747	5.335	4.968	4.639	4.487	4.344	4.078	3.837
9	8.566	8.162	7.435	6.802	6.247	5.759	5.328	4.946	4.772	4.607	4.303	4.031
10	9.471	8.983	8.111	7.360	6.710	6.145	5.650	5.216	5.019	4.833	4.494	4.192
11	10.368	9.787	8.760	7.887	7.139	6.495	5.938	5.453	5.234	5.029	4.656	4.327
12	11.255	10.575	9.385	8.384	7.536	6.814	6.194	5.660	5.421	5.197	4.793	4.439
13	12.134	11.348	9.986	8.853	7.904	7.103	6.424	5.842	5.583	5.342	4.910	4.533
14	13.004	12.106	10.563	9.295	8.244	7.367	6.628	6.002	5.724	5.468	5.008	4.611
15	13.865	12.849	11.118	9.712	8.559	7.606	6.811	6.142	5.847	5.575	5.092	4.675
16	14.718	13.578	11.652	10.106	8.851	7.824	6.974	6.265	5.954	5.668	5.162	4.730
17	15.562	14.292	12.166	10.477	9.122	8.022	7.120	6.373	6.047	5.749	5.222	4.775
18	16.398	14.992	12.659	10.828	9.372	8.201	7.250	6.467	6.128	5.819	5.273	4.812
19	17.226	15.678	13.134	11.158	9.604	8.365	7.366	6.550	6.198	5.877	5.316	4.843
20	18.046	16.351	13.590	11.470	9.818	8.514	7.469	6.623	6.259	5.929	5.353	4.870

Table 18.6 Capital Recovery Factors (per cent)

Year	1	2	4	6	8	10	12	14	15	16	18	20
1	1.010	1.020	1.040	1.060	1.080	1.100	1.120	1.140	1.150	1.160	1.180	1.200
2	.508	.515	.530	.545	.561	.576	.592	.607	.615	.623	.639	.655
3	.340	.347	.360	.374	.388	.402	.415	.431	.438	.445	.460	.475
4	.256	.263	.275	.289	.302	.315	.329	.343	.350	.357	.372	.386
5	.206	.212	.225	.237	.250	.264	.277	.291	.298	.305	.320	.334
6	.173	.179	.191	.203	.216	.230	.243	.257	.264	.271	.286	.301
7	.149	.155	.167	.179	.192	.205	.219	.233	.240	.248	.262	.277
8	.131	.137	.149	.161	.174	.187	.201	.216	.223	.230	.245	.261
9	.117	.123	.134	.147	.160	.174	.188	.202	.210	.217	.232	.248
10	.106	.111	.123	.136	.149	.163	.177	.192	.199	.207	.223	.239
11	.096	.102	.114	.127	.140	.154	.168	.183	.191	.199	.215	.231
12	.089	.095	.107	.119	.133	.147	.161	.177	.184	.192	.209	.225
13	.082	.088	.100	.113	.127	.141	.155	.171	.179	.187	.204	.221
14	.077	.083	.095	.108	.121	.136	.151	.167	.175	.183	.200	.217
15	.072	.078	.090	.103	.117	.131	.147	.163	.171	.179	.196	.214
16	.070	.074	.086	.099	.113	.128	.143	.160	.168	.176	.194	.211
17	.064	.070	.082	.095	.110	.125	.140	.157	.165	.174	.191	.209
18	.061	.067	.079	.092	.107	.122	.138	.155	.163	.172	.190	.208
19	.058	.064	.076	.090	.104	.120	.136	.153	.161	.170	.188	.206
20	.055	.061	.074	.087	.102	.117	.134	.151	.160	.169	.187	.205

AIM OF LCC

The objective of LCC is to optimise the life cycle cost of owing and using a physical asset. Life cycle costing is the total cost of an asset accruing to it throughout its life. It includes the following costs:

(a) Specification cost
(b) Design costs
(c) Production costs
(d) Installation and commissioning costs
(e) Operating costs
(f) Maintenance costs
(g) Disposal costs

From the point of view of an equipment design, in order to be able to optimise the life cycle cost of any asset, it is necessary to have at first a forecast of the requirements of the asset and its costs and then to quantify the life cycle costs of each of the alternatives available. Thereafter, optimizing by making trade-offs amongst the production cost, commissioning cost, operating cost, and maintenance and disposal costs over the economic life periods of each of the alternatives available. The important point to note is that, unlike other concepts, life cycle costing focuses on engineering considerations and on the economies accruable from alternate engineering systems.

For most organisations, an asset is purchased, where the capital costs include the costs of (a) acquisition, (b) installation, (c) commissioning, (d) spares, and (e) recruitment and training of maintenance and operating personnel.

The revenue costs incurred during the operational life of the asset include the costs of the following:

(a) Operating of the equipment, including the costs of labour and material, tools, tackles, jigs, and fixtures and supporting services like ppc, quality control, material control, and a percentage of the overhead costs.
(b) Maintenance cost of labour, material, spares, etc.

However, certain costs under operating and maintenance costs may either be missed or be difficult to ascertain. These are:

(a) Cost of loss of production during a breakdown or during predictive maintenance.
(b) Lower capacity utilisation which could perhaps be due to the fact there is not enough market demand for the product manufactured. This could be due to a variety of reasons such as poor quality output and discouraging performance of required asset, which in turn may be due to poor maintenance.

The above factors will eventually be reflected in poor output quality, bad overall performance of the owner organisation, which will ultimately result in low profitability. Finally, the disposal costs include (a) removal and cartage cost, (b) dislocation caused to production, till the removal and installation of a new asset, and (c) actual physical disposal of the asset, which may be partially offset by the disposal value of the asset.

Each of these three basic cost groupings have an impact on each other. Disposal of an asset may bring in a fair amount of money and may not apparently cost anything to the owner organisation, because in an economy like ours, due to inflation, the resource crunch and scarcity conditions, the resale value of goods seems to keep pace with the spiralling prices. Fiat car in a good condition could fetch a substantial sum of money when disposed of. On the other hand, disposal of ash from thermal power station can be expensive, and the burden has to be borne for many years by the owner organisation. Perhaps, it is possible that in order to compensate for a higher acquisition cost, machinery may have to be genuinely designed to have a lower maintenance cost during its entire operating life period.

The LCC may result in the choice of an asset with lower operating maintenance costs and, therefore, a reduced total cost to achieve a targetted output in terms of quality and quantity. This reduction in cost would make it possible to release funds for other investment opportunities for the enterprise.

APPLICATION OF LCC

The following example shows the application of LCC.

A company is planning to acquire a truck. Two makes of trucks are available in the market. The cost of garaging and the driver's wages are the same for both. The other data on cost are provided in Table 18.7.

The difference in annual operating cost is calculated as follows for a period of 30,000 km operation:

$$\text{Truck A} = \left(30{,}000 \times \frac{4}{20}\right) + \left(30{,}000 \times \frac{2}{1000} \times 20\right)$$

$$= 6000 + 1200 = ₹7200$$

$$\text{Truck B} = \left(30{,}000 \times \frac{4}{24}\right) + \left(30{,}000 \times \frac{2}{1500} \times 20\right)$$

$$= 5000 + 800 = ₹5800$$

Advantage of truck B over A is $7200 - 5800 = ₹1400$

Facilities Investment Decisions (FID) and Life Cycle Costing (LCC)

Table 18.7   Cost Information

Parameters	Truck A	Truck B
Capital Cost	₹2 lakhs	₹3 lakhs
Annual Road Tax and Insurance		
Operating Costs		
(a) Fuel Consumption	20 km/litre	24 km/litre
(b) Oil Consumption	2 litres per 1000 km	2 litres per 1500 km
(c) Fuel Cost	₹4/litre	₹4/litre
(d) Oil Cost	₹20/litre	₹20/litre
Maintenance Costs		
(a) Service Interval	Every 5000 km	Every 6000 km
(b) Cost of Service	₹4000	₹2000
(c) Random Breakdown	Every 8000 km	Every 30,000 km
(d) Cost of Breakdown	₹6000	₹10,000
(e) Cost of Breakdown	₹6000	₹10,000
Expected Life	10 years	10 years

The difference in annual maintenance cost is calculated as follows:

$$\text{Truck A} = \left(30{,}000 \times \frac{4000}{5000}\right) + \left(\frac{30{,}000 \times 6000}{8000}\right)$$

$$= 24{,}000 + 22{,}500 = ₹46{,}500$$

$$\text{Truck B} = \left(30{,}000 \times \frac{2000}{6000}\right) + 30{,}000 \times \frac{10{,}000}{30{,}000}$$

$$= 10{,}000 + 10{,}000 = ₹20{,}000$$

Advantage of truck B over A = 46,500 – 20,000 = ₹26,500

Differential cash flow showing advantage of truck B over truck A is as shown in Table 18.8.

While calculating these details, the following assumptions have been made:

(a) The operating and maintenance costs have been assumed to be the same for the 10 years. The incidences of breakdown and fuel/oil consumption have also been taken to be the same for 10 years. In an actual situation this is not likely to happen.

(b) Annual run in kilometres, terrain conditions including road conditions, cost of servicing and breakdown, cost of oil/fuel are the same.

(c) Garaging and driver's wages are the same for both trucks A and B.

In the final analysis, it is seen that truck B has considerable advantage over truck A, mainly because of lower frequency of breakdowns during its operational life. For keeping the calculations simple, the effect of inflation, taxation, etc. have been ignored in this case.

Table 18.8 DCF Table

Year	Capital Cost	Annual Tax and Insurance (₹)	Operating Cost	Maintenance Cost (₹)	Total Cost (₹)	Discount at 100%	Net Present Value (₹)
0	1 lakh						1 lakh
1		1000	1400	26,500	26,900	.909	
2		1000	1400	26,500	26,900	.826	
3		1000	1400	26,500	26,900	.751	
4		1000	1400	26,500	26,900	.683	
5		1000	1400	26,500	26,900	.621	165,273
6		1000	1400	26,500	26,900	.564	
7		1000	1400	26,500	26,900	.513	
8		1000	1400	26,500	26,900	.467	
9		1000	1400	26,500	26,900	.424	
10		1000	1400	26,500	26,900	.386	
Total	1 lakh	10,000	14,000	265,000	269,000	6.144	

COSTING OF ALTERNATIVES

The best method to follow is to prepare a cash flow projection of expenditure and income for each of the alternatives under consideration. These are spread over the entire life period of that asset. These patterns can then be compared for different alternatives from the point of view of design use and maintenance or purchase. However, discounting the cash of the life time costs and incomes will indicate the present day value of that cash flow, which can be then compared for the choices to be made.

The main problem encountered in actual practice is the difficulty in getting the correct cost data to carry out the analysis even for a replacement decision. Take, for example, a company which is using the Tata and Ashok Leyland trucks. They have used these for a number of years and now wish to make a decision as to which make they would like to standardise. They would naturally carry out an LCC analysis. But it is quite likely that all the desired cost details may not either be available or are such that they create an undue bias due to internal organisational methods. This can be better understood from the following example, the data for which has been gathered from an actual transport operating organisation.

The following parameters are indicated for the two vehicles under consideration:

(a) Both vehicles A and T have the same load carrying capacity.
(b) Budgeted kilometres per vehicle for life is 7.5 lakh km of 7.5 years.

(c) Basic price (in lakhs)   A   T

    (i) Price of chasis   2.35   2.45

    (ii) Body-building cost, including taxes and excise   1.46   1.44

    (iii) Total cost to make the vehicle road worthy   3.81   3.89

(d) Manpower, material, fuel, and taxes. The costs of alternatives are given in Table 18.9.

Table 18.9 Costs of Alternatives

Description/ Calculation of Cost	Vehicle A		Vehicle T	
	Per 1 lakh km	Per 7.5 lakh km	Per 1 lakh km	Per 7.5 lakh km
Fuel	0.8125	6.093	0.795	5.969
Tyres	0.26	1.95	0.24	1.80
Stores	0.11	0.825	0.21	1.575
Taxes	0.47	3.525	0.33	2.475
Wages	0.22	1.65	0.35	2.625
Total	1.8725 or 1.87	14.043	1.915 or 1.92	14.444

Note: The price of HSD oil is taken as ₹3.90 per litre. Vehicle A gives 4.8 km per litre! and vehicle T gives 4.9 km per litre. Therefore,

$$\text{For vehicle A per 1 lakh km} = \frac{3.9}{4.8} = .8125$$

$$\text{For 7.5 lakh km} = .8125 \times 7.5 = 6.09375$$

$$\text{For vehicle T per 1 lakh km} = \frac{3.9}{4.9} = 0.795$$

$$\text{For 7.5 lakh km} = .795 \times 7.5 = 5.969$$

It should be noted that there is considerable difference in the wage structure costs between vehicle T and A, i.e. $.35 - .22 = .13$, which for 7.5 lakh km = $.13 \times 7.5 = .975$, i.e. ₹97,500.

This difference was created because the T vehicles fleet were used in metropolitan division whereas vehicle A fleet were used in a suburban division. The metropolitan division consisted of personnel with longer service and was made up of an older age group and hence drawing higher salary and allowances. But the disparity has created a bias due to the internal structure and the utilisation pattern of the fleet.

(e) Organisational policy with respect to repairs and maintenance, (i) During the life of vehicle when 7.5 lakh km or 7.5 years have been completed, it would have undergone one vehicle complete overhaul (VCO) and 1.5 sundry repairs. In addition, at every 2 to 3 lakh km, one full sundry repairs (full), and at 6.0 to 6.5 lakh km another half sundry repair would have been needed. (ii) A unit complete overhaul (UCO) is done during the life of the vehicle as follows:

1. Major units like engine rear axle, gear box, and front axle   3 times
2. Self-starter   7 times
3. Alternators   6 times
4. Fuel Injection pumps for A vehicle   6 times
5. Fuel Injection pumps for T vehicle   5 times

(f) Repair and maintenance costs (R&M costs) incurred during 7.5 lakh km of the life of vehicles is given in Table 18.10.

Table 18.10 Repair and Maintenance Costs

Activity	Vehicle A (₹ in lakh)	Vehicle T (₹ in lakh)
Vehicle Complete Overhaul (VCO)	0.70	0.65
Sundry Repairs (SR)	0.24	0.27
Engine	0.25	0.21
Rear Axle	0.12	0.14
Front Axle	0.08	0.06
Gear Box	0.06	0.09
Fuel Injection Pump	0.03	0.04
Self-Starter	0.06	0.04
Alternator	0.04	0.03
Total for Life of the Vehicle of 7.5 Years	1.58	1.53
Total for One Year	0.2106 or 0.211	0.204

(g) The cost of random breakdown (RB) is given in Table 18.11.

Table 18.11 Cost of Random Breakdown

Description	Vehicle A	Vehicle T
Frequency of Random Breakdown	Every 50,000 km	Every 50,000 km
Total Frequency during Life: 7.5 lakh km	15 times	15 times
Cost of Making Arrangements Every Time	₹2000	₹2000
Total Cost of Random Breakdowns	₹30,000	₹30,000

(h) The scrap value is as shown in the following tabular data:

Description	Vehicle A	Vehicle T
Scrap value received by disposing of the vehicle at the end of its life, i.e. 7.5 lakh km.	0.31	0.34

(i) Differential cash flow showing advantage of Vehicle A over Vehicle T is given in Table 18.12.

Table 18.12 Cash Flow Statement

Year	Capital Costs	OHM Cost	R&M Costs	Random Breakdown Costs	Scrap Value	Discount at 10%	Net Present Value
0	0.08 = 3.89 – 3.81						
1		0.05 = 1.92 – 1.87	(0.007) = .211 – .204			0.909	.039087
2		0.05	(0.007)			0.826	.035578
3		0.05	(0.007)			0.751	.032293
4		0.05	(0.007)			0.685	.029455
5		0.05	(0.007)			0.621	.026703
6		0.05	(0.007)			0.564	.024252
7		0.05	(0.007)			0.513	.22059
8		0.05/2 = 0.025	.007/2 = (0.0035)		(0.03)	0.467	.0240505
Total	0.08	0.375	(0.0525)		0.03		24,441

In the above calculations, certain assumptions have been made. These are:

(a) Garaging and overheads costs have been taken as the same for both vehicles.
(b) The levels of OMM, repair and maintenance and random breakdown costs have been assumed to be the same for both vehicles for their lives of 7.5 lakh km or 7.5 years.
(c) Annual run in kilometres, the terrain and road conditions, the cost of servicing, breakdown, etc. have been assumed to be the same.

However, this will not necessarily be so in practice. As an example, we shall now show as to how the DCF will change if we ignore the impact of wages on these calculations. Here the situation gets totally changed, and a positive advantage emerges for vehicle T over A, which is considerable.

Table 18.13 shows the differential cash flow indicating the advantages of vehicle T over vehicle A by ignoring the impact of wages.

A calculation of LCC can perhaps save one from making the purchase of an equipment with an attractive low acquisition cost but a high maintenance and repair cost. The cost of ownership of an equipment can be easily hidden because of the time scale, over which the costs are incurred, the accounting system prevalent for computing capital and revenue costs, and the correct perception and evaluation of down time costs.

LCC is basically concerned with planning for and achieving optimum life cycle costs for the assets. However, an optimum can never be achieved, because the knowledge of the present and more so of the future is never perfect. But LCC seeks to optimise the value for money in ownership of assets. Though dependable and accurate data is difficult to obtain, they are indispensable for reliable calculations. Nevertheless, even with all the departures from the ideal state that are normally prevalent, LCC is still an invaluable tool in the hands of the decision makers, though an ideal and fault-free assessment may be difficult to make.

Table 18.13 Differential Cash Flow

Year	Capital Cost	OHM Costs	R&M Costs	RB Scrap Costs Value	Total	Discount Factor at 10%	Total at Discount Rate of 10%
0	(0.08)						
1		0.08	0.007		0.087	0.909	0.079083
2		0.08	0.007		0.087	0.826	0.071862
3		0.08	0.007		0.087	0.751	0.065337
4		0.08	0.007		0.087	0.685	0.059595
5		0.08	0.007		0.087	0.621	0.054027
6		0.08	0.007		0.087	0.564	0.04512
7		0.08	0.007		0.087	0.573	0.044631
8		0.04	0.0035	0.03	.0435	0.467	0.020354
Total	(0.08)	0.60	.0525	0.03	0.6525		44,004 − 8000 = 36,004

A–Z OF LCC

Let us sum up the need and phases.

The conventional forms of capital investment appraisal methods take most of the cost variables as deterministic. As against this, life cycle costing pays a greater amount of attention to the probabilistic nature of the variables involved. This is particularly relevant because, the total cost of operation of any piece of equipment throughout its life depends upon a number of variables, a majority of which are probabilistic rather than deterministic. Particularly, items like failure of equipment, availability of spares for timely repair of breakdowns, repair times etc. are all matters that cannot be determined with certainty, and consequently probabilistic evaluation of these variables is likely to read to more meaningful results, than using deterministic values for cash flows. Life cycle costing seeks to achieve this by free resort to the use of probabilistic techniques, and thereby simulate the possible occurrences of failure, etc. and ultimately determining the net operating hours for which an equipment is likely to be available during its useful life of operation. In fact, this is one of the fundamentals of the approach in life cycle costing. Again it is this, that distinguishes life cycle costing from the conventional methods of capital expenditure proposal evaluation.

Though the concept of life cycle costing is recognized as a common and effective tool for use in equipment acquisition, there are different definitions that have been evolved. The Logistics Management Institute, defines life cycle cost as 'the total cost of ownership of a system during its operational life'. It embraces all costs associated with the feasibility studies, research and development, design and production, all support training and operating costs generated by the acquisition of the equipment.

Another definition states 'Life Cycle Cost as the cost of usage of plant, equipment, facilities and systems, with particular emphasis being placed on the cost of usage, rather than on the original cost of the plant. Total life cycle cost are all the money spent for design, construction, operation and maintenance of plants, equipment and facilities, from the moment, the decision is made to acquire them until the moment they are retired down from service.

Comprehensive applications of LCC are today made possible because of the availability of techniques like simulations and forecasting, that can help taking care of the elements of uncertainty on forecasts. In fact, the comprehensive applicability of the technique is such that it makes possible a forecasting of the relationship between design and development costs and the operating maintaining costs throughout the equipment's life.

In other words, the technique is capable of providing guidance even from the stage of determining the initial cost of an equipment with design cost as an element and also facilitating the measurement of probable maintenance costs during its life so that a beneficial trade-off can be arrived at. The life cycle phases and tasks are given in Table 18.14.

Table 18.14 LCC Phases and Tasks

Concept Formulation	Contract Definition	Development	Production	Operation
Identify need for equipment and accomplish feasibility study	Accomplish equipment functional analysis, optimization, synthesis and definition	Accomplish design liaison and support services	Fabricate, assemble, test and deliver support items	Operate and support equipment and associated support elements in the field
Define initial equipment operational requirements	Accomplish allocation of R&M, human factors and support requirements	Accomplish LSA	Prepare operational and maintenance sites for receipt of support items	Provide interim support capability until full operational capability is achieved
Define initial maintenance concept	Establish support design criteria	Prepare data for the acquisition of supply support items	Plan, implement and participate in tests and evaluations	Initiate and maintain field data collection system
Define initial support targets and constraints	Accomplish initial LSA	Plan, implement and participate in tests and evaluations	Update LSA to evaluate equipment and verify support factors	Evaluate impact of plan and carry out modifications
Accomplish initial support planning	Accomplish design liaison and support services	Prepare forward support plan	Initiate transition from supplier to user operation	Update LSA and accomplish resupply of support items
Participate in conceptual programme reviews	Prepare initial support plan			Plan and implement equipment phase-out and disposal
	Participate in equipment design reviews			

Facilities Investment Decisions (FID) and Life Cycle Costing (LCC)   221

LCC being a summary of number of elements of costs, these are identified by phases, forecast and aggregate. The following is a list of some typical elements by phases. The capital costs include the following elements.

   (a) Specification
   (b) Design
   (c) Development
   (d) Manufacture
   (e) Installation and commissioning
   (f) Manuals and training for operations
   (g) Manuals and training for maintenance
   (h) Provision for spares, inventory, space and tools

The revenue costs include:

   (i) Direct materials
   (j) Direct labour
   (k) Direct expenses and overheads
   (l) Indirect materials
   (m) Indirect labour
   (n) Establishment overheads

Maintenance costs cover the following:

   (o) Spares
   (p) Labour
   (q) Facilities and equipment
   (r) Establishment overheads
   (s) Down time
   (t) Disposal costs consisting of demolition; dislocation; disposal

The following eleven steps are recognized in the life cycle cost formulation for capital equipment acquisition.

   (u) Establish the cost structure.
   (v) Identify the cost phases in the cost structure and the conceivable cost elements in each cost phase including the stage of design.
   (w) Identify the cost variables.
   (x) Establish the behavior pattern of the cost variables.
   (y) Simulate the values for the stochastic cost variables of failure times, diagnostic times and repair times using appropriate methods.
   (z) Identify the part that failed and determine the status of availability of the failed part in stock by using simulation techniques.
      (z1) Compute preventive maintenance time, corrective maintenance down time and the operating time of equipment.

(z2) Account for the money value of each of the variables, either in the form of throughputs or estimated values, using appropriate cost estimating relationships.
(z3) Provide for cost escalation due to inflation.
(z4) Choose the appropriate cost of capital to be used as the discount rate.
(z5) Compute the individual costs, aggregate, discount and account for the life cycle cost of the equipment. Thus life cycle cost formulation consist of identifying the cost elements, during its life, computing the individual costs, aggregating discounting, and accounting for the life cycle cost of the equipment. Simulation takes care of the uncertainties associated with the equipment breakdowns leading to maintenance costs and facilitates a more reliable operating cost forecast for the entire life of the equipment.

Chapter 19

Evaluation of Maintenance Function

NEED FOR EVALUATION

India, in order to catch up with advanced countries, is investing heavily in sophisticated plants and equipment in all the sectors of the economy. A substantial portion of the capital is likely to be invested in maintenance-oriented infrastructure sectors like energy, oil exploration, coal, petrochemicals, transport, steel, mining, defence, fertilisers and process industries. This phenomenon is not only a challenge but also a great opportunity to the plant engineers.

Technology is developing fast, and the well known principles of management are constantly being stretched beyond boundaries drawn earlier. Such advanced technology brings severe conditions of temperature, pressure, corrosion, etc., which are big challenges to the plant engineer. In the challenging tasks, evaluation of plant engineering function is of great significance to the professionals.

Sound corporate planning involves the evaluation, refinement and integration of various functional objectives into overall corporate goals. Hence, the need for an objective evaluation of plant engineering function cannot be overemphasised. It is well known that the primary responsibility of the top management is to identify and recognise competence and to ruthlessly weed out incompetence. The rewards and punishments should be commensurate with the results achieved so as to infuse motivation and fix responsibilities. Such responsibilities will be both a challenge and an opportunity for the plant engineer since the need for savings, cost reduction and professionalisation is crucial.

The major difference between successful and unsuccessful organisation lies in the results achieved, which can be measured in terms of effective utilisation of available resources. Since the maintenance management plays a crucial role in contributing to the corporate profits by effective asset utilisation, evaluation of maintenance function assumes great importance. The objective should be to utilise each employee potential so that both the individual and the organisation are benefitted. The organisation of evaluation process of the maintenance function should be so geared as to motivate the right persons towards the corporate objectives. This chapter must be read with Chapter 12 on Training and HR.

A–Z MAINTENANCE CHALLENGES

It is desirable to identify the challenges in maintenance function before proceeding to the evaluation process. The plant engineering group is called upon to possess adequate knowledge, skill-mix, practical expertise, and technical know-how to tackle the maintenance related aspects in the following diverse fields, in spite of low priority, least glamorous, without universities denying justice to maintenance subjects.

(a) Mechanical jobs relating to machinery and equipment, and systems oriented towards better productivity, higher production levels, capacity utilisation, efficient capital employment and reduced losses
(b) Civil and environmental jobs relating to maintenance, upkeep and construction of new facilities, engineering of environmental control programmes, comfort engineering, and illumination engineering, etc.
(c) Electrical jobs relating to all electric prime movers, power control and optimum efficiency by minimising losses and planning load distribution
(d) Electronic jobs relating to solid-state devices and microprocessor control
(e) Air-conditioning, ventilation and refrigeration
(f) Fluid power and mechanical power transmission
(g) Hydraulics, pneumatics and structural engineering
(h) Piping, valves and fittings
(i) Pumps and compressors
(j) Construction, erection and project management
(k) Material handling
(l) Value engineering, standardisation, codification and spares management
(m) Energy management
(n) Anti-corrosion method
(o) Paints, protective coatings and application of preservatives
(p) Fire protection and safety engineering
(q) Equipment maintenance and house keeping

(r) Reliability failure analysis, MTBF and MTTR
(s) Overhauling and network analysis
(t) Tools and welding
(u) Pollution control and environmental engineering
(v) Instrumentation and controls
(w) Cost estimation control and cost reduction
(x) Maintenance budgets and variances
(y) Workshop and in-house manufacturing/repairing for replacement
(z) Constantly updating technical and other maintenance-related areas to become a true professional with ideas on LCC.

EXPECTATIONS: A–Z PARAMETERS

Before proceeding to the evaluation of maintenance function, we may identify the parameters expected of the maintenance function viewed by the top management and other departments. Some of these parameters are:

(a) Reduction in unscheduled breakdowns
(b) Elimination of time-scheduled overhauls
(c) Increase in operating life of the assets
(d) Reduction in down time
(e) Reduction in maintenance manpower
(f) Reduction of spares inventory
(g) Maximising equipment effectiveness
(h) Establishing a total system of preventive maintenance covering the whole life of equipment
(i) Failure modes and effect analysis
(j) Criticality analysis
(k) Preliminary hazard analysis
(l) Fault tree analysis
(m) Event tree analysis
(n) Hazards and operatability studies
(o) Cause-consequence-cost analysis
(p) Integrating different maintenance systems such as breakdown, predictive, corrective, preventive and productive
(q) Installing information systems for control, corrective, productive, and predictive maintenance systems
(r) Determining areas for condition monitoring
(s) Installing documentation systems and procedures
(t) Bridging the gap between knowledge and practice
(u) Identifying down time losses due to failures and adjustment for speed losses and minor stoppages

(v) Locating defect losses which result in reduced yield
(w) Having a lubrication system which leads to energy savings
(x) Pursuing economical cycle costing in terms of products, systems, programmes, equipment and technology
(y) Periodic reporting and updating knowledge
(z) Taking steps to improve business results and creation of pleasant workplace.

BACKGROUND OF MAINTENANCE

Before proceeding to the evaluation process, it is essential to acquaint oneself with the background conditions in which the plant engineer operates in an organisation. In spite of the challenges, the maintenance function is still one of the most neglected areas in many developing countries. Many organisations do not have systematic organised approach to the plant engineering function. The maintenance executives in a large number of organisations report only to the production executives. Hence, planned maintenance, in the form of predictive and preventive maintenance, is often neglected because of the priority given to the stipulated production targets and is given importance only during breakdowns.

Usually, the plant engineer is not consulted in the procurement of capital equipment even though he is responsible for maintaining the same. The maintenance engineer has to be an expert in maintaining such diverse items as sophisticated plant machinery, civil constructions, buildings, electrical accessories, electronic device, transport equipment, and service utilities.

EQUIPMENT DISCARD POLICY

In developing economies, the equipment is stretched beyond endurance beyond its economic life in view of the scarce capital, thereby posing great challenges to the maintenance engineer. Most of the machinery is over twenty five years old in India. Hence, the cost of maintenance is more than the original value of the equipment. Digboi oil refinery and the Indian Railways transportation system are about 125 years old. We have also procured second hand plants such as Scooters India Limited. Uncertainty of delivery, non-availability of good quality spares for old imported machinery, spurious indigenous spares, poor reconditioning infrastructure and inadequate testing facilities are faced by the Maintenance Manager every day, posing a great challenge to him.

The total life cycle cost involved in installation to discarding of equipment for plant engineering is not considered in many organisations while formulating equipment cost estimates, maintenance budget, and operating costs. Hence there appears to be no scientific policy for discarding or phasing out the equipment in many firms. Procurement on lowest lender basis, without considering reliability and availability could perhaps be a problem in government organisations. The non-uniformity of equipment due to collaboration from different countries is causing serious problems with regard to interchangeability.

MBO/SWOT AND MANAGERIAL STYLES

The organisational culture, corporate style of decision-making process, managerial styles, role of training and top management attitude to maintenance function adopted play a crucial role in the evaluation process of the maintenance function. The role of plant engineering function is such that it is more often blamed than praised and its performance is measured on the basis of breakdowns or negative points as in long distance motor rallies. In this context, the various styles of management practices adopted in India are:

(a) Management by objective or participative style
(b) Management by exception where routines are delegated to lower levels
(c) Management by crisis or everybody works only when the house is on fire
(d) Management by precedence or previous occurrences
(e) Management by committees
(f) Management by procedures
(g) Management by results
(h) Management by delays or postponement of issues
(i) Management by facts or analytical approach
(j) Management by peace (साम), money/gift (दाम), division (भेद) and punishment (दंड).

Management by objectives involves the identification of key result areas by the mechanical engineer and set targets by himself and compare the same with actuals at the end of a month or year.

The evaluation process normally involves the following steps:

(a) Set acceptable standards of performance which accurately reflect the corporate objectives by scientific methods.
(b) Check operations against the standards.
(c) Identify deviation of actual against the standards.
(d) Quickly identify the reason for the variations.
(e) Evaluate the causes.
(f) Set indicated corrective measures in motion.

Any performance appraisal, by comparison, will be of no value unless the norms used are valid. This factor is very important while fixing norms in maintenance function because various factors influence its behaviour. Moreover, it is difficult to generalise these effects for all industries. Through SWOT (strength, weakness, opportunity and threat) analysis, the maintenance function can identify its strengths and weaknesses in the face of external opportunities and threats. Before the evaluation process, the course of action is to increase the strength of Maintenance Department and reduce the weakness so that the opportunities can be capitalised and the threats minimised. Such SWOT analysis has to be done for the Maintenance Department, organisation, industry, country and even in international field so that all relevant influencing actors can be identified. Sometimes, it may not be possible to review all aspects of performance, and the top management must select certain strategic check points which provide an indication of what is happening in the area of plant engineering.

EVALUATION PROCESS OF PLANT ENGINEERING

The evaluation of maintenance function can be done by comparing one organisation with similar organisations in the same year, known as *inter-firm comparisons*. If data on identical organisations are not available, the performance of the same organisation over a time horizon, known as *intra-firm comparison*, for certain plant engineering characteristics can be compared.

This evaluation process can be done by external agencies like Institution of Plant Engineers, other professional agencies, consultants, government departments, suppliers, foreign collaborators, etc. Within the organisation itself, various activities connected with plant engineering such as production, operations, design, maintenance, stores, purchase, costing, finance, sales, servicing, and top management can contribute to the evaluation process. The evaluation process is usually based on criteria to be developed by reports prepared by individual work centres of plant engineering, coordinated by the Chief Plant Engineer and approved by the top management.

The evaluation of the plant engineering function can be done by reports. No organisation can function effectively without regular reports on the performance of the constituents to the control centre. The frequencies and maintenance activities depend upon the level for which it is intended. Quarterly and annual reports with specific recommendations, on future long range impacts are intended for the top management. Thus, a report to the Chief Executive may have to be sent once in a month while the finance executive may need a weekly report. Monthly reports containing detailed information may be sent to all work centres in the Plant Engineering Department as well as to Operations/Costing Departments. It cannot be overemphasised that the reports should be accurate, uptodate, brief, clear and to the point, supported by graphs, charts, tables, footnotes, suggested action plan, comparative previous performance, and appropriate appendices. These reports should aim at enhancing the objectives of plant engineering function in the organisation, namely, maximum equipment availability and reliability with minimum maintenance costs. The purpose of a report is not a mere presentation of facts and data but a summary of the total picture of the maintenance function. Unless corrective measures are taken on the basis of reports, these tend to become only papers in the file. For this purpose, the periodic reports must reach the concerned persons in time, and not at the end, for by then, they could become useless. It may also be noted that the performance of any organisation depends to a great extent on the interrelations between the persons at various levels, and this should be borne in mind while actions are taken on the reports.

EVALUATION A–Z SUBJECTIVE METHODS

The evaluation of plant engineering function cannot be done in isolation. It will have to fit in with the corporate culture, organisation objectives, and must be consistent with the socio-political economical-technological environment. Any performance appraisal of the function will be abortive unless the norms used are valid. It should be noted in this context that it will not be possible to set quantitative norms for subjective characteristics in the field of plant engineering function. Some of these characteristics have been covered under the section Parameters on Expectation.

The subjective characteristics include:

(a) Attitude to latest scientific techniques
(b) Training subordinates
(c) Manpower development
(d) Morale of maintenance staff
(e) Good inter-departmental relations
(f) Satisfaction to the user
(g) Development of drawings
(h) Preparation of manual
(i) Frequent breakdown
(j) Equipment availability
(k) Effectiveness of maintenance system
(l) Maintenance documentation
(m) Accuracy of plant history cards
(n) Adequacy of maintenance organisation
(o) Professionalisation in plant engineering

(p) Computerised maintenance control
(q) Lubrication effectiveness
(r) Cost-conscious culture
(s) Attitude to safety
(t) Plant engineering information system
(u) Technical literature
(v) Identification of critical equipment and vital spare parts
(w) Control of obsolete inventory
(x) Use of energy saving devices
(y) Maximising the availability of all machinery by minimising wear and tear
(z) Ensuring the safety of all equipment and employees.

Admittedly, numerical values can be given to some of the above characteristics, but in many organisations, these are usually expressed by subjective estimates as *good, large, adequate* and *fair*.

A–Z OBJECTIVE CRITERIA

Ratios, when expressed objectively, form the major characteristics in the evaluation process. These ratios use meantime between shutdown, meantime to repair meantime between failure, meantime down for the equipment, maintenance expenses, equipment availability, etc. Some of these ratios, usually computed at the end of a month or year, are presented below. The values of these ratios vary from one industry to another.

(a) Breakdown repair hands: $\dfrac{\text{Total hours spent on breakdown}}{\text{Total hours available for repairs}}$

(b) Maintenance cost: $\dfrac{\text{Total maintenance cost}}{\text{Total annual sales}}$

(c) Repair intensity ratio: $\dfrac{\text{Down time due to repair work}}{\text{Total operating time}}$

(d) Planned maintenance work: $\dfrac{\text{Planned maintenance cost}}{\text{Total maintenance expenses}}$

(e) Emergencyfailure intensity: $\dfrac{\text{Time slot due to emergency breakdown}}{\text{Total operating time}}$

(f) Maintenance breakdown frequency:

$\dfrac{\text{No. of breakdowns by poor maintenance}}{\text{Total number of breakdowns}}$

(g) Equipment availability: $\dfrac{\text{Equipment running time}}{\text{Total available time}}$

(h) Craft workers activity: $\dfrac{\text{Total craft hours clocked}}{\text{Total maintenance hours}}$

(i) Work order turnover: $\dfrac{\text{No. of workorders processed}}{\text{Total number of work orders on hand}}$

(j) Work order turnover: $\dfrac{\text{Total man-hours on incentive}}{\text{Total man-hours available for maintenance}}$

(k) Manpower efficiency: $\dfrac{\text{Total man-hours spend on the job}}{\text{Total man-hours allowed on the job}}$

(l) Jobs under control: $\dfrac{\text{Total man-hours spent on schedules}}{\text{Total man-hours available}}$

(m) Spare parts ratio: $\dfrac{\text{Value of spares consumed}}{\text{Total maintenance expenditure}}$

(n) Overhaul expenditure: $\dfrac{\text{Overhaul expenses}}{\text{Value of equipment}}$

(o) Spare parts ratio: $\dfrac{\text{Value of spares consumed}}{\text{Value of equipment}}$

(p) Overdue jobs: $\dfrac{\text{No. of jobs of one week backlog}}{\text{No. of jobs completed at the same time}}$

(q) Degree of scheduling: $\dfrac{\text{Total direct hours on scheduled service}}{\text{Total direct hours available}}$

(r) Scheduled service cost: $\dfrac{\text{Annual cost of scheduled service}}{\text{Total production cost per year}}$

(s) Breakdown cost: $\dfrac{\text{Cost of total breakdown repair}}{\text{Total number of breakdowns}}$

(t) Machine utilisation: $\dfrac{\text{Total running time in hours}}{\text{Total shift-hours of working}}$

Besides the above ratios, a questionnaire on the following lines is also found to be useful in the evaluation process:

(u) Are machines and the plant covered by scheduled inspection including lubrication checks, and the specification of inspection routes?

(v) Have optimum inspection frequencies been determined and reviewed?
(w) What is the follow-up system to detect failures to complete inspections on schedules?
(x) Are historical records kept of major repairs and what steps are taken to improve the records?
(y) Other possible indicators are maintenance expenses per man-hour, maintenance cost per unit of output, maintenance expenses per kilowatt of power consumed, work measurement, work load identification, incentive determination trend in total maintenance expenditure, trend of down time ease of maintenance, ease of repair, replacement policies, trend of maintenance expenses per unit rupee of sales, etc.
(z) Trend analysis of spares inventory/obsolete item values, design imperfections, cost control, incentives, overtime, control on inventories, tool stores, and use of mathematical techniques like inventory models, queueing and network analysis. Are the plants insured against sabotage, earthquake, disasters, floods, etc.

An evaluation of the maintenance function can also be done by asking questions on structure, strategies, systems, skills, staffing, styles and techniques used.

It is desirable that each organisation, based on the above indicators, should develop its own ratios and norms for evaluating the maintenance function, bearing in mind that the major objective of evaluation should be oriented towards equipment reliability, availability and maintainability at lowest operating costs.

MAINTENANCE: FUTURISTIC SCENARIO

It is an accepted fact that with increasing automation, the design of the modem plant and machinery is becoming more and more complex and sophisticated each day. And with it the maintenance function is becoming increasingly crucial, and its responsibilities getting more expanded.

The types of changes and developments which are likely to take place in the area of maintenance are both interesting and innovative. And some of them are revolutionary while others are evolutionary in concept. Increasingly more and more of the preventive maintenance and predictive maintenance tasks will be carried out by sophisticated inspection, testing and surveillance equipment using real time computer controls to try at first to eliminate and, if that is not really possible, then predict machine failure and breakdown well in time and thereby not allow a critical part to even reach the breakdown point, by activating a control mechanism which will stop the plant before it can reach the breakdown stage. The majority of the equipment will be so designed as to require only little maintenance or no maintenance at all. This design effort will be made possible with the increased knowledge and application of *reliability* and *design out maintenance concepts*. The entire design will be made up of modular components built into the machines, to facilitate quick replacements to be made modulewise. The removed part can then be repaired separately as a rotable or discarded if it is deemed to have gone beyond repair.

Computers will be increasingly used in a variety of operations in the coming years. Maintenance, spare parts and materials management offer no exception to this trend. In future, even the indenting of materials spares will be done by computer.

All these changes involving higher technology will obviously make it necessary for the maintenance supervisors to be engineering graduates, and for the maintenance workers to be either well trained multiskilled craftsmen or to be highly competent specialists in their individual spheres of work.

Technology upgradation has increasingly led to the use of high technology maintenance systems, principles, practices and concepts. For instance, *System Effectiveness* is an important area for consideration. It is defined as the *probability that a system will operate a full capacity over a defined period of time*. Its effectiveness depends on spares availability. Therefore, spare parts availability and the supporting infrastructure are important factors which have a marked effect on maintainability. We shall discuss these in the following section.

Due to research and development all over the world, today the Maintenance Manager has a large variety of techniques that help him in facing the challenges. Some of these are:

(a) Criticality analysis
(b) Condition-based maintenance
(c) Computerised maintenance and information system
(d) Fault tree analysis
(e) Hazard analysis
(f) Failure analysis
(g) System effectiveness concept.

Despite all these developments, there is a big gap between what is known and what is actually applied in practice. Therefore, the main challenge facing us is to bridge this gap as soon and as effectively as possible.

Section III
Core Spares Issues

20. Indian Spares Scenario
21. Spares Practices Survey
22. Cost Reduction in Spares
23. Music-3D-beyond Cost Criticality Method
24. Inventory Control of Spares
25. Maintenance Spares
26. Simulation for Spares Control
27. Insurance Spare Parts
28. Rotable Spares
29. Overhauling and PERT (Programme Evaluation and Review Technique)

Chapter **20**

Indian Spares Scenario

CAPACITY UTILISATION

Today, we find that a lot of machines are lying idle for want of spare parts. This certainly does not mean shortage of spare parts. Our experience shows that for every spare used, five or more parts are held in stores, which perhaps may not be used. This point has been highlighted in several seminars, courses, bureau of public enterprise studies, reports of the Committee on Public Undertakings (40th COPU report, p. 10), and Public Accounts Committee proceedings (53rd report, Seventh Lok Sabha, p. 2) and several conferences.

The role of effective spares management cannot be overemphasised in minimising the down time of equipments and in making optimal use of investments already made in different sectors of the economy. The Indian Institution of Plant Engineers estimates that 54% of machine breakdowns are repetitive. The studies of the National Council of Applied Economic Research indicate that the capacity utilisation in engineering industries is only about 65%. The plant load factor in the energy sector is only about 65%, and the capacity utilisation, according to several other studies in other key sectors, is around these levels, necessitating the optimum utilisation of existing capacity to be achieved by proper spares management, before thinking of further investments. In infrastructure industries for petroleum products, there is slope for improved utilisation instead of increasing the consumer price.

A–Z FINANCIAL ASPECTS OF SPARES

(a) The public enterprises survey indicates that the investment in the public sector has risen from a mere ₹29 crores in 1950 to nearly ₹1,50,000 crores in 2010, and this trend is continuing with each five year plan. A substantial portion of the plan investment is likely to be invested in equipments in industry, mining, transport, energy sectors and infrastructure industries.

(b) The return on the above investments depends upon the condition of the equipments, which relies on the maintenance policies, which in turn depends upon the spares management.

(c) Two to five per cent of the total project cost is usually invested in the spares inventory at the time of starting a project itself.

(d) About 10% of the cost of an equipment is invested in the recommended spares at the time of procurcment/import of the plant. Many of these recommended mandatory spares are found to be non-moving, becoming obsolete later on.

(e) The original equipment manufacturer usually exploits the user's ignorance by dumping slow moving parts, resulting in obsolete and project surplus items at the user's stores.

(f) The former Union Finance Minister, Mr. C. Subramaniam, said on 10 December 1974 that over ₹2500 crores worth obsolete spares inventory were lying idle. This figure is likely to be around ₹99,000 crores today with industrialisation and inflation.

(g) About 40% of the working capital of spares intensive process, mining and transport industries is tied in the spare parts inventories, out of which 25% usually forms non-moving and obsolete items.

(h) The rate of obsolescence in spare parts is higher in technologically advanced industries like electronic communication and instrumentation-oriented industries.

(i) The Reserve Bank of India—through the Tandon Committee and Chore Committee reports of the study group to frame guidelines for follow-up of credit—has stipulated that the "banker should keep a watchful eye if the spares inventory exceeds five per cent of the total inventory". Spares intensive industries like process, mining, transport, etc. will find it impossible to adhere to these norms.

(j) The bank finance for working capital will also not be available for hypothecation of non-moving spare parts. The inventory carrying charges, including storage charges and cost of capital, come to 30% per year while the Central Government claims that the rate of inflation is maintained at about 22%.

(k) The ordering charges or acquisition cost per order or order processing charges vary considerably and are about ₹1000 per order.

(1) The stock of spare parts is guided not merely by the prices or carrying cost but by the loss of income without a spare part. This understock cost is only talked about and is found to vary widely with the criticality

of an item and is yet to be estimated correctly by many organisations. But it is usually found to be greater than the price of spare parts.

(m) 60–80% of the maintenance expenditure is accounted by spare parts consumption in industries.

(n) The average spare parts consumption for the machineries is found to vary between 1–3% of the annual sales value of the output. Due to non-availability of power in summer months between March to July, the consumption of generator spares is very high in all parts of India.

(o) The spare part sales for replacement market account for nearly ₹5 crores per year in the country. Leading organisations selling spares include BHEL, TELCO, BEML, BEL, L&T, TAFE, HMT, Greaves Cotton, MAMC, Voltas, Kirloskar, Ashok Leyland, Hyderabad Engineering, Allwyn, Escorts, auto sector, textile sector, CMC, Eicher, Mico, TVS Group, Maruti, Jyoti, Josh Engineering, Associated Ball Bearing, HEC, and Texmaco. Some of these firms sell more than ₹1000 crores worth of spare parts per year handling over 10,000 assemblies and subassemblies. Due to power shortages during March–July in most parts of the country, generator spares for generator sets constitute an important item of sales of these organisations.

(p) The profit margin of spares varies from 50%–300% in the replacement market. This can be easily proved by comparing the cost of any equipment, say a diesel engine, and assembling the same after buying the items individually.

(q) In some cases like aircraft—with higher technological sophistications—the price of some spare parts may even be more than the "original aircraft price".

(r) In spite of the huge margin in spare parts, many original equipment manufacturers have not set apart separate capacity to manufacture spare parts for replacement market—resulting in the use of spurious spares or spares made in 'USA' (Ulhasnagar Sindhi Association).

(s) This phenomenon is observed more when the buyer prolongs the life of the equipment due to lack of capital and the seller has changed his models several times in order to meet the technological challenges in competition.

(t) The companies selling spare parts aim at the sales-to-inventory turnover ratio of about 6, indicating an average of two months stock.

(u) A very high inventory turnover ratio indicates that those spares with high manufacturing lead time will be frequently out of stock with the dealer, thereby inviting spurious spares suppliers to jump into the market.

(v) On the other hand, a low inventory turnover results in locking up of scarce working capital and more often leading to accumulation of obsolete spares inventory.

(w) The recent concept of modular approach of replacing only assemblies modular replacement—recommended by the original equipment manufacturers in a few cases—renders obsolete the other used parts in the assembly, besides involving high cost of replacing by the module.

(x) India imports spares worth about ₹5 lakhs to ₹50,000 crores per year from different countries to service and maintain the imported equipments.

(y) While importing capital equipments, a scientific analysis of the total life cycle approach—involving original equipment costs, operating costs, maintenance costs, fuel expenses, spares cost, depreciation, salvage value, etc.—is not usually carried out. This affects reliability, availability and maintainability of the equipments.

(z) According to the equipment users, the role of after-sales service—even though costly—needs to be improved, particularly in the area of ready availability of replacements. No wonder many organisations have switched to 'AMC' or annual maintenance contract.

SPARES MANAGEMENT A–Z ISSUES

(a) The financial problems discussed above which are relevant to spare parts management pose not only challenges to the supply chain manager, parts executive/mechanical engineer, maintenance engineer, and the materials manager, but also offer opportunities to the professional executives as the problems in this field are indeed unique. In spite of the great challenges, the available literature is not only scanty, but also highly theoretical and hence deters the initiative of the professional manager.

(b) As a national policy, the developing nations have gone in a big way for State-owned public sector enterprises. Since the managers of these companies are accountable to the nation at large, the tendency is to play safe, resulting in the overstock of spare parts. This safe policy of overstocking is adopted because the frequencies of stockout with consequent machine down time is known to every one in the organisation, while there is no incentive or pressure to reduce the high spare parts inventories. High spare parts inventories imply insurance against machine down time and hence the tendency to overstock except on occasional audit query.

(c) Due to paucity of capital, the developing countries use the equipments for longer duration, sometimes to the extent of overworking the machinery beyond the operational/economic life. This makes the plant engineer's job unenviable and challenging, forcing the organisations carry increased stock of spare parts.

(d) The user continues the outmoded machine, but the supplier changes his models in order to keep abreast of the technological changes and competition. In such cases, the user is forced to estimate and buy lifetime requirement of spare parts, resulting in overstock, most of which is likely to become obsolete in course of time.

(e) The developing countries normally buy equipment from advanced nations (imported equipments and spares) who have their own systems, specifications and standards. This makes interchangeability of spares a difficult problem in practice. For instance, a state electricity board has 10 generating stations from 15 countries. The Indian Petrochemicals Limited has about 50,000 types of one group of items known as *bearings,* since it has entered into technical collaboration with more than 20 countries. India possesses about 40 types of aircrafts depending upon the usage, training, light transport, heavy transport and passenger traffic. The number of components in each aircraft may go up to even 40,000, Obviously, it is difficult to manufacture spare parts of economic batch size for these aircrafts.

(f) In case of imported equipments in sensitive sectors like defence, space, nuclear sectors, it may not be easy to get the emergency related replacement spares due to vested interests and political pressures. A case in point is the dispute in the supply of critical spares to Tarapur Nuclear Reactor, where the Indian contention was that understanding on the supply of critical spares had been reached between the former Prime Minister, Mrs. Indira Gandhi and the former President of the United States, Mr. Ronald Reagan. Similar high level contacts have been necessary to procure critical spares in the sensitive sectors. Russian inability to supply spares for fighter aircraft was common in opting for global tender.

(g) The annual import policies of the Central Government—even though they are supposed to be stable for spare parts—fluctuate widely due to socio-political, financial, and national and international pressures.

(h) In order to get renewal of the import licences, many firms buy spare parts not needed by them just to show fuller utilisation of the import licences.

(i) The procedures and systems at customs/clearance need to be simplified and improved for easy clearance.
(j) The estimates of total lead time—recognition of the need of the spare parts, processing, ordering, manufacturing, transporting, inspecting etc.—vary from 24 months to 30 months. This makes planning, forecasting, ordering, follow-up and procuring the item a difficult process.
(k) The original equipment manufacturer in foreign countries constantly improves the models in order to keep pace with technological innovation and is unable to supply spares for the old equipment already supplied, which are obsolete models for the seller.
(l) In view of the high obsolescence, in terms of spare parts availability for imported equipments, producers and users must start indigenisation at the design stage itself.

NECESSARY INFORMATION

(m) Progress in import substitution, however, has not yielded the desired results in the spare parts field because of the aspects like increased price, inferior quality, non-availability of drawings, less reliability, limited required quantity, lack of infrastructure facilities, non-availability of crucial input quality material, inability to take risk with indigenised parts in high precision jobs, etc.
(n) Maintenance management, in terms of staff, quantity, quality, managerial attitude skill, knowledge, documentation, technical know-how, and recognising the importance, needs to be improved in most organisations.
(o) Provision of spare parts is a service function to maintenance which in turn serves production interest. Because of service-to-service job, maintenance and after sales service have not been able to attract highly skilled staff in many firms.
(p) The most important question concerning the responsibility to control spare parts—measured in terms of stockouts, obsolescence, satisfaction of the user—is being discussed even today.
(q) The role of maintenance, inventory control, stores, marketing, materials and production in controlling spare parts is not clear in a large number of organisations.
(r) In this context the plant engineers and users sarcastically define spares as those available in plenty when not needed and not available when needed!
(s) Information on such items as total number, categories, inventory value, failure, consumption, sales rate, lead time, stockout cost, order

processing charges, carrying cost, stock levels, minimum, maximum, reorder point, reorder quantity, reliability, drawing features, preservatives, maintenance aspects etc. is not readily available for each spare part.

(t) The cost-conscious culture of asking relevant questions on specifications and adoption of cost control techniques is absent in many organisations.

(u) In many firms, policy manuals relating to maintenance, reconditioning, replacement, inventory, after sales service, preservatives, make-buy-lease are not usually prepared and not readily available. The practice of insisting on three quotations for monopoly items must be discontinued.

(v) The powers of delegation for procurement are outmoded and have not been updated in many organisations to match the ever increasing inflationary spiral.

(w) Reconditioning facilities, particularly for each industry group, must be available in the form of regional workshops.

(x) In view of the large number and variety of spare parts—occasionally running into millions—identification of spare part is the biggest problem in the industry. Sometimes the parts are identified only by the mechanic. Spare parts, in user organisations, are often identified by supplier's part numbers and not by internal codification, resulting in duplication, in terms of storage location and documentation.

(y) Even though India has a large number of computer mainframe systems with terminals as well as table top personal computers, the computer has rarely been used in the field of spare parts. The advantages of using computer for spare parts management are immense and a beginning must be made at the earliest.

(z) Communication gap between different parties—sales, after sales service, finance, buyer, maintenance, stores, design, production, plant engineer etc. each having his own expertise—needs to be bridged. There are a few cases when spare parts are ordered on the basis of previous year's experience, for machines which have been phased out. The plant engineer interested in total safety resorts to categorising many items as insurance spare parts and inflates the requirement. Sometimes the maintenance department, at different levels, resorts to informal/multiple storage of the parts in the workshop, thereby creating a difference between issue and consumption. In view of the above communication gap between different departments, sometimes standby machines are ordered in order to cannibalise a machine and use its parts in times of emergency. The problem of communication is more acute in multiplant organisations, located in far-flung areas with centralised system of spare parts purchasing.

The spare parts management is comparatively easy in industries if there are a large number of batch process identical machines, like textiles, engineering, pharmaceuticals, etc., as compared to the high technology single stream spares intensive process industries. It is necessary to have an integrated and innovative approach covering all aspects of the spare parts field in order to meet the complexities and challenges. In this context, government organisations, public sector undertakings, private sector firms, industry associations, professional associations, and academic institutions can play a crucial role in improving spares management and increasing the capacity utilisation in our country.

INITIAL SPARES PROVISIONING

A survey on spares conducted recently by us has pointed out the area of concern which is the initial provisioning of spares recommended by the machinery suppliers. The recommendatory becomes mandatory and some suppliers even mention that the warranty and guarantee of performance is subject to acceptance of recommended spares. In spite of test performance and visits to the suppliers plant by operating and maintenance executives, the supplier's recommended list invariably lands as obsolete items. None of the suppliers provide the drawings of critical spare parts. Usually 5–10% value of machinery is dumped in this process.

The supplier also introduces major modifications in machinery, and even though parts cost more than 100%, Original Equipment Manufacturer is unable to supply spare part after ten years.

The list of initial provisioning, at the time of introduction of the equipment must be based on a critical study of technical literature, visits to manufacturer by maintenance R&D technological experts. The initial provisioning depends on (1) whether the equipment is indigenous or imported, (2) the lead time for supply of spares, after sales service facilities, (3) practices prevailing in similar industries, (4) annual maintenance contract details etc. In some cases the suppliers even vouchsafe that they have not even supplied the machines and washoff their responsibility.

Chapter 21

Spares Practices Survey

RESEARCH METHODOLOGY

It is desirable to assess the principles, practices, policies, problems and procedures—in all their aspects—followed in respect of spare parts from the user and selling organisations, since the available literature is scanty and is treated as confidential.

The author has been gathering information from 500 organisations belonging to various sectors of the Indian Industry during the last three years regarding the spare parts and maintenance problems. This has been done with the help of carefully designed and tested questionnaires based on the a–z issues in all chapters on maintenance and spares and discussion points in Chapter 48. In addition to this, the author has conducted indepth interviews with about 1000 senior executives, from different sectors of the economy, who participated in materials management, spare parts management, after sales service management, maintenance management, financial management, production management, general management, and development programmes at the Administrative Staff College of India, Hyderabad. Follow-up visits were made to a few organisations, wherever necessary. The list of respondents also includes a few organisations in the Middle East and Africa, where the author was associated as a United Nations Advisor on Inventory Control. The identity of individual organisations has not been revealed and only broad features have been outlined to ensure the confidential nature of the information from the organisations. Only a summary of high points are presented.

In many organisations spare parts are defined as parts of machinery, which are kept standby to be substituted, when a part of machinery breaks down or is worn out. However, the maintenance department, not directly controlling spares, ironically defines spare as those "available in plenty when not needed and not available when needed!" The questionnaire is given in Chapter 48 and the results relevant to maintenance have been discussed earlier.

DIAGNOSTIC STUDY

Many organisations say that they conduct a diagnostic study in the following manner to understand the spare part problems. The equipments in the company are classified as VEIN vital, essential, important and normal according to the criticality of performance. For instance, power generator sets have been classified as vital equipment in these organisations. The cost of spare parts required for each machine is then identified by the maintenance group. In this context, some firms find that the Equipment History Card is the most useful document for it details (a) the date of installation, (b) supplier's profile, (c) operational significance, (d) lubrication data, (e) spares replaced, (f) servicing history, (g) failure data, (h) false detection analysis, (i) fault repair, (j) expected span of life, (k) information life cycle costing, etc. The diagnostic survey in some organisations includes conditions of working environment, such as dusty, corrosive, excessively hot/humid conditions, working machines beyond specification, overloading, negligence, lack of safety appliances, etc. This has enabled the organisations to decrease spare parts consumption by resorting to preventive maintenance.

Some organisations have conducted a survey to locate all areas of storage conditions of formal and informal warehouses. This survey has revealed that lockers, cupboards, and corners of the workshop have been converted into informal area of storage, due to lack of confidence at different levels of maintenance staff on the capabilities of stores and purchase department. A list of parts salvaged during the survey has been submitted to scrap the useless items and the remaining spares are identified with the requirements of the machineries. The materials that need to be retained are then placed in pre-assigned storage space after attaching tags or slips to them. It is claimed by the responding organisations that the above exercise has been used for confidence building and to reduce the spare parts stock in unauthorised places.

ORGANISATIONAL ASPECTS

The research survey has revealed that for a multiplant, responsibilities in terms of delegation of powers and reporting between materials department and others significantly affect the internal lead time. In the organisations surveyed, the spares inventory function is under the control of maintenance department, whereas the materials department is responsible for processing the order. Only 10% of the organisations have written policy manuals for various categories of spare parts. However, the users admit that the spare part function is the most neglected function in the organisation with maintenance and purchase trying to put the blame on the other as maintenance is the least glamorous jobs.

The materials department controls the spares function in Bokaro Steel Plant while the Chief Mechanical Engineer is incharge of spare parts function in the Port Trust. In government departments such as irrigation and public works the Chief Engineer controls all aspects of spare parts. The finance department has a crucial role in capitalising insurance spare parts in organisations such as Madras Fertilisers. While maintenance and materials planning play a critical role, the other departments which have a marginal interest in spares planning include design, stores, operations, purchase, costing and computer. The maintenance engineers, however, complain that they are not consulted at the time of capital equipment buying which they have to maintain.

There are a large number of industries based entirely on spares, like bearings, V-bells, hardware items, motors, spark plugs, etc. Organisations like MICO, Jyoti, HMT Bearings, Associated Ball Bearing, Lucas TVS, TELCO, Hindustan Motors, Voltas, Bharat Electronics, Bharat Earth Movers Limited, Maruti Udyog, Bajaj Auto, Greaves Cotton etc. attach great importance to spare parts sales as original equipment manufacturers. They have well established spares departments attached to the marketing department dealing with capital equipment sales and after sales service. Some of these organisations report more than ₹1000 crores in spare parts sales per year and claim to have reserved 10% manufacturing capacity only for replacement market and after sales service operations. These organisations claim to use computer for order processing, invoicing, billing and after sales service operations. The pricing policy is determined by the marketing executive in consultation with operations and finance after ascertaining market intelligence.

TECHNOLOGY AND CAPITAL EQUIPMENT

As witnessed in any developing economy, India also has sought technical expertise from many advanced countries. In this process, the transfer of technology is taking place in all sectors from many countries, as a result of which we see the major steel plants in operation with the assistance of USSR, Germany, Britain and South Korea. A state electricity board is operating nine generating stations with equipments from UK, USA, USSR, Yugoslavia, Japan, India, Sweden, Germany, and France. The same phenomenon is observed in process industries like Indian Petro Chemical Complex, defence and oil sectors. What is meant by this is that interchangeability of vital and costly spares is rendered impossible, leading to wasteful accumulation of spares inventory. This, some feel, is unavoidable in high technology and sensitive defence and nuclear sectors.

Like many other items, spares of electrical equipments are not available beyond 10 years of their original introduction in the market. Same is the case with electronic devices, instruments and computers, where the model in the original country becomes obsolete as soon as it is supplied to India. Railways are using steel-made steam locomotives supplied by the Canadian firm M/s Fair Banh Morse, who are not in a position to supply spares. Similar is the case with all imported engines from Japan/UK/USA. In such changing technological scenario, the managers now feel that they should constantly keep abreast of developments so as to balance the introduction of new technology, keeping in mind the interchangeability of spares within the same industry. The survey has revealed that in most capital equipment acquisition decisions, the Maintenance Managers have never been consulted and the concepts like life cycle costing have rarely been used by the organisations. British oil major BP has completed acquisition of 30% stake in oil and gas production—sharing contracts operated. Reliance Industries with an investment of $7.2 billion, while Reliance will receive the latest technology in deep sea oil drilling from BP.

OVERHAULING AND DISCARD POLICY

The respondents to the survey unanimously agree that they adopt overhauling the equipments in order to give these a new lease of life. Since the overhauling date is known in advance, spares planning, replacement of oversize bearings, removal of rubber items, etc. are done meticulously by a committee. Procurement is initiated well in advance after assessing the lead time and, wherever necessary, stocks are drawn from the maintenance stores. Many organisations admit that despite best efforts, there is considerable amount of errors in quantity estimation in the case of First overhauling. Transport organisations have clear-cut overhauling practices after running specified number of kilometres. The shutdown/turnaround in process plants is indicated to all employees three months in advance of the overhaul. Except for an occasional delay in imported spares, overhauling does not pose a major problem. Few organisations claim to practise "just in time and zero inventory" in case of overhauling the spare parts. We will discuss overhauling in Chapter 29.

Reconditioning assemblies and part of the equipment is done in the central workshop, particularly when the original supplier fails to deliver the spare part assemblies. Organisations such as Century Enka, TISCO, Indian Railways, Lucas TVS, etc. have well-defined reconditioning policies, which depend upon criticality of the machine, maintenance practices, price of spare parts, cost of reconditioning, life expectancy, workshop capacity, etc. For instance, one electrical equipment manufacturing unit resorts to reconditioning only if five years life will be available for 25% cost.

Refurbishing is done essentially in the aviation industry in order to introduce technological upgradation after complete overhauling of aircraft, by attaching the latest gadgets, in order to match the most recent model but with competitive price. Many organisations have technical committees, for make-buy-lease overhaul, reconditioning, refurbishing, and overhauling, In view of the scarcity of capital and high cost of new equipments, all the organisations prolong the life of the equipment by making alterations, as far as possible. For instance, Digboi Oil Refinery is more than 100 years old and Scooters India Limited has only second-hand machinery. In some cases, the equipment is discarded only if maintenance cost is more than the original capital cost. Reconditioning is discussed in detail in Chapter 37.

OBSOLESCENCE

Due to the rapid technological changes in the advanced countries, spare parts are no longer produced by the original equipment manufacturers, as the earlier models have become obsolete. Due to capital scarcity, the Indian Industry has to keep the plant running for a longer time and hence is forced to play safe in the initial stages; otherwise, it faces the situation of non-availability of the original manufacturer's spare parts or the unreasonably high lead time or inflated price. According to some respondents, the original equipment manufacturer even has asked the photo of the machine supplied by him, due to frequent updating of the models. Such heavy accumulation and overbuying results in obsolescence. Other causes are rationalisation, technological upgradation, application of standardisation, use of modular approach of replacing composite assembly, making other parts as obsolete as per the survey.

Every organisation has admitted that it has obsolete spares which are not likely to be consumed. The accumulation of obsolete spares, according to the respondents, is due to overbuying at initial stages, project surplus, increased buying for fear of non-availability, and categorising many items as insurance emergency spare parts. The rate of obsolescence ranges from 10% of the stores inventory value in engineering and new refineries to 40% for on-shore oil exploration with an overall average of 20% in most cases, including organisations selling spare parts. The obsolescence rate is found to be high in spare parts intensive sectors like mining, processing and transport. Chapter 36 discusses Management of Obsolescence in great detail.

In many organisations, movement analysis is carried out once in two years and in others XYZ analysis or inventory value analysis to find out slow moving/non-moving high value spares. The user departments rarely declare the non-moving as obsolete items for fear of repercussions of stockout. Even if declared obsolete in procedure-oriented government organisations, the top management does not dispose of these items because of the difference between book value and expected realisation value resulting in CAG audit and ministry questions.

INITIAL PROVISIONING

All the respondent organisations admit that initial provisioning is the most error prone area and they feel that a decision at this crucial stage has a direct bearing on the non-moving stock. The initial provisioning is done by most organisations at the time of procurement of capital equipment based on suppliers' recommended list. In a few cases this value may go up to 20% of the value of equipment. The safe policy of buying more than the requirement is done, according to the survey, to avail the facility of capitalising on the initial spares because of fear of non-availability at a future date, unwillingness to take the risk of stockout if there is a demand for the spares, and ignorance of the technical knowledge relating to future requirements. All the users opined that, for these reasons the original manufacturers exploit the technical ignorance, and dump the non-moving expensive spares on them at an enormous price, thereby making huge profits. On the contrary, the selling organisations vehemently deny this and claim that they only recommend the list of initial spares. The users point out to the large surpluses from incorrect initial procurement. A few organisations appoint technical committees to shortlist the recommended list of spares based on past experience or getting machine history from other users and segregate proprietary spares separately. In view of problems faced in initial provisioning, we will discuss this topic in marketing segments.

CATEGORISATION AND CODIFICATION

In view of the large number, sometimes going up to a million spare parts as in the case of the Oil and Natural Gas Commission (ONGC), the Indian Army and the Air Force, identification of spare parts is one of the key problem areas in many organisations. According to the survey of the Association of Indian Engineering Units, one unit responded to the question of insurance spares saying that it had insured all its plants and stores! This is the state of ignorance in the field of spare parts. Categorisation of spare parts into consumable, insurance, overhauling and project surplus have been done in many cases. Organisations equipped with repair shops for floats or assemblies like defence, transport and mining sectors, use the concept of floats or rotable spare parts.

All the selling companies have an excellent system of cataloguing the spare parts, which is used for identification at all levels. They also ensure that the users try to adopt these part numbers so that interdependence is assured. On the contrary, only 50% of organisations have developed scientific codes with check digit, the total number varying from 9 to 16 digits. Several multiunit organisations like the Railways, Bharat Heavy Electricals Ltd., etc. have developed common codes for the same part in all locations. Many users opine that common codes should be developed at the industry level and national level, for easy identification and availability of spare parts.

The importance of developing drawings, particularly for critical components, have been appreciated by all firms. But, due to patents and secrecy considerations, detailed drawings are not given by the suppliers to the users. A slow start has been made by the design and maintenance section of a few organisations to develop drawings as it is the best way to identify and understand spare parts. But many of the new organisations and oil refineries have developed drawings for the critical spare parts. Many user departments, however, complain about lack of infrastructure facilities and wake up only when the original equipment manufacturer stops supplying them.

STANDARDISATION

Lack of progress in standardisation has been attributed by the responding organisations to the import of equipments from different countries, for considerations such as global tenders, only supplier tied credit, well known manufacturer, etc. For instance, Indian Petrochemicals Complex has more than 20,000 bearings. India has more than 40 types of aircrafts, the various steel plants have equipments from different nations, the oil exploration equipments and power generation units are from various countries, etc. In spite of such constraints, many organisations report initial progress on standardisation of hardware items, fasteners, oil seal, hoses, gaskets, filler elements, V-belts, formats, etc. The coal-based fertiliser plants at Ramagundam, Thalcher, and Korba have formed a central spares planning cell to introduce uniform classification. There is an awakening at all levels with regard to standardisation of mother equipment, subassemblies and spare parts, even though only a limited success has been achieved in the field of standardisation.

LEAD TIME ANALYSIS

Some public sector organisations that responded to the survey point out that public accountability forces them to adopt policies which are not always optimal, thus necessitating in increased stock. Admittedly, comparison of the policies between public and private sectors is difficult in view of policy differences in objectives, product lines, capacities, etc. Some public sector organisations are inviting open tenders through newspaper advertisements, resulting in inordinate delays. The lead time in most organisations has not been estimated but only guestimated. It is only very rarely that the four components of the total lead time ordering, making, transporting, inspecting—are available. The total lead lime for spare parts is found to vary from 6 to 18 months for indigenous spare parts and from 24 to 36 months for imported items. The organisations admit that the lead times are more in the case of spare parts compared to other items. The coal sector, for instance, states that the lead time goes up to three years for heavy earth moving machinery spares.

REQUIREMENT PLANNING

The area of planning and provisioning has been considered to be critical by many companies due to unpredictable failures and uncertain lead times. They opine that reliable data on failures, consumption and lead time can be fixed only after using them for a period of four years. The factors affecting the planning process, according to the respondents, include operating conditions, maintenance policy, lead time, availability of spare parts, reliability, estimated life, categories of spare parts, failure data, price of spare parts, cost of carrying, obsolescence rate, type of industry and organisation culture. Since it is difficult to correlate the effect of the above factors, many firms set apart at least one standby set per machine for insurance items and about three to five years as average stock levels. Suppliers advice weighs heavily in determining the requirement up to 10 years in many organisations.

Some defence establishments try to use sophisticated formulae in determining the requirement of spare assemblies. For example,

$$Q = N \left(\frac{L+R}{T} \right)$$

where N is total number of equipment, T the life of equipment, L the lead time, and R is the review time. Many organisations claim to use ABC analysis, criticality analysis, and availability information to develop criteria for requirement planning.

SELECTIVE CONTROL AND STOCK LEVELS

The need for exercising adequate inventory control and easy availability of spare parts to the user has been expressed by all the respondents. Many organisations claim that the conventional ABC analysis has been applied with advantage, particularly because the number of spare parts is very large. The selling companies use ABC analysis as a basis for pricing the spare part, with a lower margin of profit for the fast moving spare parts and a higher margin for the slow moving items. VED or criticality analysis has been done by the petrochemical industries and oil refineries. But combining the cost criticality and availability into a single analysis has not been done by any organisation. Reliability analysis, failure analysis, Poisson distribution, bath tub curve are only being talked about.

· Some organisations have worked out economic order quantities purely as of academic interest in the context of fixing inventory levels, but no uniform pattern of fixing stock level has been established even in firms belonging to the same industry. Cost information such as ordering charges, order processing cost, inventory carrying charges, stockout cost are not available in most organisations. There has been some criticism that scientific formulae are too theoretical to be applied to real world problems. The typical response is that for indigenous items, minimum stock is fixed as 6 months average monthly consumption, reorder level as 12 months, and maximum stock as 24 months. For the imported items, the above figures are doubled. The stock levels in utility government departments are higher than in public sector undertakings, which in turn are higher than those in the corresponding private sector units. Units belonging to irrigation, power, mines, chemicals and transport hold consumption stocks up to 60 months as the average stock, including slow moving and obsolete items.

SPARES BANK

The respondents to the survey point out that the concept of spares bank is ideal for a capital scarce economy like India, particularly for storing non-moving, costly insurance spare parts. However, this concept is applicable only if mother equipments or the subassemblies are standardised. At present only the supplier's warehouse acts as a spares bank in most cases. Similarly, local agents of imported equipment may be encouraged to import spares under their own licence and supply to consumers from their warehouses. It is also observed that multiunit organisations resort to central stocking of slow moving items.

The fertiliser plants for Barauni, Durgapur, and Namrup are of similar design. Similarly, the coal-based fertiliser plants at Ramagundam, Thalcher and Korba have a common spare parts bank for slow moving costly spare parts. The service and spares division of Bharat Heavy Electricals Limited is applying the spares bank concept for the state electricity boards by planning and ordering them sufficiently in advance. The road transport corporations, port trusts, and the mining sector feel that they have to resort to the concept of spares bank in the near future to reduce the working capital commitments.

IMPORT SUBSTITUTION

Large public sector organisations have set up import substitution cells. The petroleum refineries have set up a technical development committee to give a fillip to import substitution. The responding organisations attribute the slow progress in import substitution due to the following reasons:

(a) Technical people unwilling to take the risk of indigenous items
(b) Drawings not readily available
(c) Lack of indigenous production facilities
(d) Lack of local inspection facilities
(e) Easy availability of foreign exchange for importing spare parts
(f) The quantity required for spares uneconomical for manufacture or low demand
(g) Lack of drawing and detailed specification
(h) Higher price for indigenous items
(i) Inferior quality of local items
(j) Non-availability of key raw material like special steels
(k) Power cut and voltage fluctuations
(l) Lack of detailed information on all aspects of spare parts
(m) The suppliers increasing the price after approval
(n) Foreign collaborator discouraging indigenous items.

However, in view of the foreign exchange crisis, many organisations claim that they are trying to overcome the above difficulties.

SPARES INFORMATION SYSTEM

The information system is better organised in the selling organisations, where the data on actuarial equipment population, their working condition, and the probable spare parts requirement of the spares are readily available. Some of these organisations even advise the users about the date of overhauling and replacements to be effected. However, the information system at the user end needs to be considerably improved. Some report that they do not even have the list and location of equipment population or their history cards. Many firms report that they do not have a proper system of logging machinery down time due to spare parts stockout and the resultant monetary loss. The Shipping Corporation has calculated the stockout cost as ₹800,000 per ship of average d.w.t. per day. The Railways have estimated the stockout cost as ₹50,000 per diesel loco per day, while the inventory charges are estimated around 30%, the ordering cost is found to vary from ₹100 to ₹6000 per order. The total purchase cost or selling cost is about 1% of the total value purchased, or sold. The inspection cost, inclusive of salary and depreciation of equipment and facilities, for spares has been estimated as 1% of the value inspected.

The spare parts consumption value is about 70% of maintenance budget in refineries and steel plants, while it is 50% in textile mills. The refineries have estimated the consumption value around 2% of the value of output. The Engineering Projects India allocates up to 2% of the project expenses as initial spare parts. A budgetary provision of 10% of value of machine is made in some firms. Many organisations estimate, the spare stock value between 10% and 15% of the total value of machinery at a given point of time. The average inventory carried is 24 months for indigenous and 60 months for import spares. The transport, power, process industries and mining sectors carry much more spare parts than given by the Reserve Bank Tandon Committee guideline of 5% of total inventory value. The spare parts inventory value is about 25%–65% of total stores inventory. The estimates of slow moving/non-moving inventory value vary from 15 to 40% in different industries.

HIGH INVENTORIES: A–Z CAUSES

There are a large number of reasons for holding high spare parts inventory; these are discussed below:

(a) Many organisations have pointed out that the inventory management function is evaluated on the basis of the number and duration of stockouts.
(b) They adopt a conservative policy of avoiding stockout at any cost, leading to increased stock level.
(c) According to the research survey, for instance, the castings industry faces non-availability of good quality scrap.
(d) Alloy steel shortage is a major problem in some heavy industries.
(e) Raw materials and components with varying lead time and availability characteristics have to be planned to balance the inventory levels of different categories of spares, which is not an easy job.
(f) Protracted manufacturing cycle time is responsible for high work-in-progress and spare parts should be readily available to reduce the frequency of machinery breakdown.
(g) The respondents involved in marketing spare parts are anxious to meet the unpredictable demands of the consumer, thereby forcing him to keep higher spare parts inventory.
(h) The distribution chains at different channels such as warehouses, depots, stockists, retail outlets and dealer network are controlled by different organisations, leading to higher spare parts inventory.

Other reasons for higher inventory include the following:

(i) Random pattern of failures with difficulty in predicting failure rate.
(j) Long lead time for imported items.
(k) Impractical difficulties in operating the spare parts bank.

(1) Lack of standardisation and interchangeability.
(m) Lack of adequate transportation facilities and logistics.
(n) Inadequate movement and handling facilities in the plant.
(o) Too much time in transit.
(p) Project surplus transferred as regular spares.
(q) Non-moving and obsolete spares not disposed of.
(r) Tendency to declare many parts as insurance and increasing the stock level.
(s) Aversion to application of scientific techniques.
(t) Lack of motivation to reduce the stock levels of spare parts.
(u) Non-availability of up-to-date and accurate information system.
(v) Exaggerated forecasts of future requirements of spares.
(w) Many items on temporary codes, resulting in duplication.
(x) Standardisation not feasible due to equipments from different countries and makes.
(y) Maintenance practices not properly codified, resulting in breakdowns requiring immediate spare parts.
(z) Lack of professional scientific cost-conscious approach to spare parts management.
(z1) Complicated impracticable mathematical models cannot be applied in Indian culture.

RESEARCH SURVEY SUMMARY

Almost all the organisations feel that the spares field is the most neglected area in the organisation and attempts are being made to use scientific techniques in order to improve the situation. The spare parts intensive industries like refineries, fertiliser units, steel plants, power sector corporations, port trusts, mines, oil exploration companies, process industries and transport sector, perceive that the spares problem in their organisations is more serious than in traditional sectors such as textiles and engineering. Hence these capital intensive units are taking steps with adequate investment in spares. Highly conservative policies are adopted in these areas as they feel any bottleneck in these infrastructure industries may lead to a syndrome of repercussions in the entire economy. Recently started organisations and those with foreign collaboration claim to have better systems of spares control. Their plant engineers also claim to have learnt from the mistake of the earlier plants. The defence services claim to have applied scientific concepts with success, whereas other government departments feel that they are still on the lookout for optimum policies. The spare parts problems have been highlighted in various reports of the Committee on Public Undertakings and the Public Accounts Committee. The respondents are optimistic about reaping better benefits of spares management in the future by introducing a cost-conscious culture at all levels.

Chapter 22

Cost Reduction in Spares

DEFINITION OF SPARE PARTS

In this interdependent world, spare parts management is influenced by a variety of factors. In order to understand the problems of spares, let us first define spare parts. In Chamber's Twentieth Century Dictionary, a *spare part* is defined as "a part of a machine ready to replace an identical part of it, if it becomes faulty". The import trade control defines *spare parts* as follows: "Spare parts are those parts of machines which because of wear and tear, use or breakage, need replacement". We can define *spare parts* as parts identical to the part of a machine which need replacement due to wear and tear during the operating life of the equipment. Spare parts may look small and appear cheaper than the machine or raw material, but they play a vital role in maintaining, ensuring and reinforcing the reliability of any equipment.

Spare parts include materials such as (a) pipes, tubes, springs, electrical cables, knobs, wires, hoses, beltings, (b) sub-assemblies of essential parts of the machines like engines, motors, compressors, and alternators, and (c) complete units which are to be fitted with a machine, e.g. water circulating pumps, motors, panels, controls, bolts and gears. Items such as plates, sheets, rods, strips, etc. which have to be fabricated for manufacturing complete units are not treated as spare parts.

SPARES: A–Z FEATURES

The features peculiar to spare parts which distinguish them from other materials are:

(a) Their requirement is very small.
(b) There is excessive stock in all positions of the distribution channels.

(c) Their requirements are uncertain.
(d) They are uneconomical to manufacture as their demand is uncertain and small.
(e) They have a high tendency for obsolescence.
(f) They have a large variety.
(g) They are difficult to standardise.
(h) There are problems in identification.
(i) Decision making is delegated to lower levels.
(j) A small range of item is able to meet a large percentage of requirement.
(k) Stockout cost is greater than the spare part price.
(l) Price includes large margin of profits.
(m) It is difficult to forecast future requirements.
(n) Lead time is long.
(o) It is difficult to get failure data.
(p) Issue from stores may not reflect realistic consumption due to repairing of old spare part.
(q) There is lack of information system.
(r) Spares are critical from operational point of view.
(s) Spares are increasingly used with age of machine.
(t) Number of suppliers is smaller.
(u) There are difficulties in import substitution or development of new sources, or development of drawings.
(v) It is not possible to control them by the usual inventory control techniques as several departments are involved in controlling.
(w) Use of supplier's part number is common instead of internal codification.
(x) Adherence to part number results in duplication in storage as the same part can be supplied by more than one supplier with his part number.
(y) Inspection is not always easy as testing facilities are not readily available.
(z) Incomplete specifications are known by supplier's part number.

CATEGORISATION OF SPARES

Organisations classify spares in a variety of ways like regular, fast moving, emergency, consumables, major, minor, moving, non-moving, electrical, obsolete, mechanical, instruments, proprietary, permissible, and project spares. But spare parts must be classified as maintenance, overhauling, commissioning, rotable, insurance, and capital spares for introducing scientific controls.

Maintenance spares are those which are fast moving like bearings, belts, and hardware items. Normally, these are available in plenty and the spare parts can be stocked after building a database on the consumption pattern. Overhauling spares are those which are specially needed during regular overhauls in order to give a new lease of life to the equipment. Hence, these need not be stocked and ordered just in time before overhauling.

Commissioning spares are needed to start a project or commission an equipment and these parts are declared as project surplus after the machine starts its operation. Costly assemblies like motors, engines, and pumps are repaired and stored for use. For example, if the engine of a bus breaks down, it can be removed for repairs and the bus can be refuted with another good engine from the stores. This rotation process gives the spares assembly the name of *rotable spares* or *floats*, and these have to be tackled separately. Insurance spares are those vital parts of a machine, which have life nearly equal to that of a machine itself and are held as a standby against any breakdowns. These standby units have a high reliability of performance and can be capitalised.

FINANCIAL CONSIDERATIONS

There are a large number of factors that affect the spare parts costs in any organisation. Location of a plant is a key influencing parameter on spare part policies. If a factory or the using point is located in a backward area or far away from industrial centres, airports, national highways, railway lines and seaports, then it is bound to carry more stocks to account for transportation, and communication bottlenecks. Another aspect of unavoidable delay is the government regulations on imports, foreign currency, customs, octroi, and other taxation policies. The sophisticated nature of industry, degree of automation and high-tech nature of the firm are important influencing factors as a highly mechanised automated plant is sensitive to breakdown of components in the production system. Process power, transport, single stream continuous plants, imported equipments, and other spares intensive industries have to carry more stock. The availability of drawings and the existence of infrastructure facilities for reconditioning of major components will facilitate production of spare parts in the firm itself. Availability of working capital which can be blocked in spare parts inventory may be a major constraint in a few situations. Proper budgeting of spare parts, by anticipating failures and costing spares is a difficult job.

CHOICE OF EQUIPMENTS

In choosing the equipment, the user must get adequate guidance from the manufacturer regarding maintenance problems and spares support. The supplier should guarantee adequate spares for a specified time, and sufficient notice must be given if the supply has to be discontinued. He should provide all technical details covering maintenance manuals, spares catalogue, failure data, reliability information, drawing for critical items, etc. Before purchasing new equipments, compatibility with the existing machineries must be considered from the spares point of view. At the time of initial provisioning, the user should ensure that he is not dumped with unwanted spares. The recommended list of the supplier must be scrutinised after assessing factors like stress, strain, and wear and tear. Based on the number of equipments sold, the supplier must anticipate the user's spares requirement by means of a good feedback information system. A detailed analysis on techno-economic studies of equipments as well as cost criticality availability of spares and reliability quality, criticality, and maintainability of machineries must be carefully done while make/buy/lease decisions are made for the equipments in the organisation.

COST OF ORDERING

Customer order processing in a selling organisation, or ordering a spare part in a consuming unit, involves a large number of activities which cost money. This cost includes the following elements:

  (a) Salary and statutory payments of sales/purchase staff
  (b) Office space
  (c) Depreciation of office equipment
  (d) Stationery and typing charges
  (e) Advertisement, tender forms, tender opening formalities, lender committee time
  (f) Follow-up, travel, telephone, telex, telegrams, postal
  (g) Costs of source development
  (h) Cost of entertaining
  (i) Computer usage costs.

The total costs incurred on all these heads during a year divided by the number of orders in that year will give the average figure. The major elements are advertising and follow-up costs. The estimates are found to vary considerably depending upon imports, government department procedures, number of items procured, but an average of ₹1000 per order can be taken as a reasonable estimate. Corresponding to the ordering cost, we have the set-up cost in case of production, which considers changes in the job, tooling, jigs, fixtures, etc. to determine optimum manufacturing size.

INVENTORY CHARGES

The motivating factor to control inventory is the cost incurred by carrying it. The elements involved in the calculation are as follows:

  (a) Cost of capital invested in inventory or the opportunity costs of money (which amount to about 17.5% today)

(b) Costs of storage due to rents, depreciation charges incurred on the space, and other equipments used in stores
(c) Costs due to deterioration of the part
(d) Costs of preservatives used
(e) Cost of obsolescence (which is a major component in the case of spare parts)
(f) Salaries and statutory payments to stores and stock verification personnel
(g) Losses due to pilferage, theft, wastage, and breakages while handling
(h) Stationery and documentation charges
(i) Insurance costs incurred to protect against fire and related risks
(j) Computer charges pertaining to the usage in stores (in Indian conditions it has been found that this cost is about 30% on the average inventory value).

The overstocking cost is the same as inventory carrying charges except for the period for which excess stock is carried. This cost is basically the opportunity lost due to the investment in inventory for a longer period than necessary. For spare parts which cannot be used after a certain time period, this cost will be the difference between the cost of an item and its salvage value,

COST OF STOCKOUT

Stockout cost or understocking cost arises due to non-stocking of the spare part. This is usually measured in terms of opportunity lost due to loss of production by the idling cost of a line. If the stockout results in an expedited order, then the extra charges incurred will have to be added to this cost. There are other intangible elements like loss of customer goodwill, loss of image, reduction in future sales, loss of morale of workers, efforts in restarting the equipment, etc. But many organisations consider only the profit lost due to loss of production for want of a spare part as stockout cost. This cost is particularly useful for grading the spares into vital, essential and desirable categories depending upon the degree of damage. Similar stockout cost on machineries enables one to do the equipment classification on vital, essential, important and normal categories of equipment and the classification in turn enables one to decide on carrying standby equipment.

A–Z COST REDUCTION OF SPARES

There are a large number of cost reduction techniques which can be applied for spare parts. These include:
(a) Lead time analysis
(b) Consumption control

(c) Budgetary control
(d) Codification
(e) Standardisation
(f) Simplification
(g) Variety reduction
(h) Value engineering
(i) Timely disposal of obsolete items
(j) Spares bank
(k) Minimum-maximum levels
(l) Inventory control
(m) Selective control
(n) Cost criticality availability analysis
(o) Negotiations
(p) Reconditioning
(q) Preventive maintenance policies
(r) Maintenance documentation
(s) Vendor development
(t) Vendor rating
(u) Use of learning curve for pricing
(v) Price forecasting
(w) Computer applications
(x) Development of drawing
(y) Reliability, availability, maintainability and condition monitoring
(z) Transportation and optimum service level.

We shall now discuss some of the important cost reduction techniques and the remaining in subsequent chapters.

LEAD TIME REDUCTION

Lead time can be defined as *the period that elapses between the recognition of the need of a spare part and its fulfilment.* Spare parts are available from diverse sources such as the shop floor, substores, factory main stores, supplier's warehouse, company's own workshop, local dealers, supplier's factory and foreign sources. In the above cases the average lead time will vary considerably, resulting in increased stock. The average lead time of a spare part may also fluctuate from time to time. The total lead time can be broken into four components:

(a) Internal administration lead time of converting an intent to a purchase order
(b) Manufacturing time

(c) Transportation period
(d) Inspection lead time.

The first component, which is very high in procedure-oriented government departments, can be reduced by reviewing the systems, procedures, delegation of authority, and restricting to purchase by limited tender instead of advertised tender. The inspection lead time can be reduced by coordinating and communicating with the inspection department and by providing adequate test facilities. The manufacturing lead time can be reduced by follow-up and supplier plant visits. The transportation lead time depends upon the choice of transportation method and completion of documentation relevant in transit.

IDENTIFICATION BY CODIFICATION

It is very common to have the number of spare parts running into over 50,000 items as the number and range are very large. Hence, identification of spare parts in the stores is a major problem. Misleading nomenclature, faulty numbering in the use of temporary code, adherence to supplier part number, identifying by names are reasons which make identification difficult. The classic example is from an electrical firm in UK where a 3/8 inch diameter, 6 inch long screw was known by 111 names, e.g. plunger, drive pin, dowel pin, locating plug, and so on, resulting in 111 stocking points and sets of records. Rationalised codification, similar to pin-postal index number, telephone number, and permanent account number of income tax, helps the parts executive to avoid the above mentioned problems and enables easy identification. Here one item is referred uniquely by a code.

Codes can be numbers, alphabets, or both. The number of places varies from seven to fourteen, depending upon the type of information. The total number of places is split into groups and subgroups, each subgroup signifying a classification based on some characteristic. Normally the major group consists of important categories such as tools, hardware, mechanical spare, electrical spare, and electronic item. The other digits can be used to depict (a) metals such as steel, copper, aluminium, brass, etc. (b) dimensional characteristics, (c) supplier's name, (d) user department or location, (e) equipment category, and (f) the check digit indicating the veracity of proper codification. The process of codification takes a very long time in order to ensure uniqueness, flexibility, understand-ability and utility value. In multiplant organisations, it is desirable to have the same codes, facilitating inter-plant transfer of spare parts. Different units belonging to an industry may adopt the same code which can be developed into the national codification where all industries using the same part will identify each item by a single code. Such a step will pave the way for a universal codification. You may also add a check digit in codes so that you will know whether right codification has been given.

VALUE ANALYSIS

Value analysis is an important approach which is synonymous with cost reduction and is relevant to spares planning as it reduces the cost and enables one to substitute imported spares with local ones. It is defined as an organised creative approach to identify and eliminate unnecessary costs without affecting the functional utility, performance, guarantee, safety and quality. The basic framework for value analysis approach is to ask a series of questions like what, why, how, where, when and what else. The various phases involved in value analysis are:

   (a) Identification of the function
   (b) Evaluation of the function by comparison
   (c) Development of alternative strategies
   (d) Choice of the best strategy
   (e) Implementation of the strategy.

For this purpose, value analysis teams comprising executives from design, marketing, finance, purchase, stores and maintenance are formed. They undergo brainstorming exercises for improving their intelligence and analytical ability.

Spare parts accounting for maximum annual sales or consumption offer the best results when value analysis can be carried out. Highly critical and production holding spares are subjected to value analysis in order to examine the rigidity of the specifications. Parts requiring high reliability or right specifications can be considered. The products with least contribution per unit are also selected for value analysis. Non-critical spares accounting for high consumption value are the ideal ones to start the approach. In all the above cases, drawings are obtained and the specifications examined thoroughly by putting rupee value on the tolerances.

VARIETY REDUCTION BY STANDARDISATION

Since the number of spare parts are too many, standardisation is one of the tools available to management to optimise the number and improve the quality of service to the user. A standard is a model or general agreement of a rule established by authority, consensus or custom, created and used by various levels and interest. With standardisation the number of parts purchased, manufactured/sold will be reduced, facilitating easy identification. Interchangeability of parts reduces the inventory level. Quantities can be manufactured to economies of scale and the efforts can be routinised, resulting in cost reduction. Besides standardising spare parts, one can think of standardising procedures systems, formats, reports, measurements and nomenclatures.

A large number of techniques like frequency distribution, preferred number series, market research, profit analysis, and quality control are used to standardise the item. Besides the standard departments, design, maintenance, production, purchase, stores, costing, finance, marketing and industrial engineering are also associated in standardisation efforts. The department standard can be extended to the company, industry and nation as a whole. The Indian Standards Institution (now called the Bureau of Indian Standards) has developed over 15,000 standards which are used by all, but in the case of standardising spare parts, we have to go a long way as this process depends upon the difficult task of standardising various equipments.

SPARES BANK

Every organisation is interested in reducing the working capital commitment, particularly in locking up of funds in non-moving spare parts. The concept of spares bank is a positive step in this direction, particularly for very slow moving, non-moving, and insurance spares. The procedure consists of pooling, by organisations of the same industry, their resources and having a common bank of spares instead of each one carrying its own stock of non-moving insurance spare parts. To start with, the slow moving insurance, high cost spares may be kept in the bank. The supplier himself can be thought of as a source for spares bank provided assured supply of spare parts is ensured. Admittedly, the spare parts bank concept will be applicable only if the equipments are standardised and inter-changeability of spares is assured. The operational details as to who should be in charge of banks, how much spares to be stocked and when to replenish them, can be dealt with through industry associations or professional agencies. It should be noted in this context that the fertiliser industry has been very successful in operating the catalyst bank. BHEL has achieved some success in operating the spares bank for imported/indigenous power equipments for the state electricity boards. Applications of the cost reduction methods, we have discussed, will enable the organisations to provide the right quality of spares at the right time with optimum investment in working capital. This is more applicable to chemicals, fertilisers, oil, petroleum refineries and at heat process industries. Industry Associations like FICCI, ASSOCHEM and CII were a well for stocking the items, only little progress has been made by the users in this direction. In view of more relevance to spares intensive industries, same concept has been emphasized again and again in many chapters.

Chapter 23

Music-3D-beyond Cost Criticality Method

LIMITATIONS OF ABC ANALYSIS

The traditional 'ABC' approach—Pareto's law or 80/20 rule—vital few, trivial many, witnessed in every walk of life—is the beginning only (one has to go beyond ABC—annual usage, annual consumption) annual sales, annual stores issue—is the basic first dimension and cannot be an end itself. You have to consider other cost control parameters like criticality and availability. ABC is the starting point of the three-dimensional cost-criticality-availability called MUSIC-3D multi-union spares inventory control. This provides a combined decision-making on cost-criticality availability or finance operations–materials. The same item is viewed from these angles of finance maintenance–purchase. Unfortunately people are accustomed with the first step and do not want to go for better accurate control of the three dimensions called ABC-VED-SDE, providing a total $3 \times 3 \times 3 = 27$ groups. Since this is complex for usage, we suggest each dimension is viewed only in two classes of high and low. High cost/low cost, highly critical/low functional utility, very long lead time or short lead time. This enables us to divide the total items problems/activities/or any other issue into $2 \times 2 \times 2$ or 8 groups for better control from the three-dimensional view of MUSIC-3D. You will agree that 8 groups (which is minimum) is better to control than 27 groups. Hence we will discuss MUSIC-3D in greater depth in this chapter for forecasting, planning, cost reduction and other forms of control purposes. The top management must ensure that for better control of inventions, the three departments should meet and evolve MUSIC-3D.

ABC ANALYSIS

The spare parts are grouped into three classes known as A, B, and C, respectively, based on the annual usage value or sales or consumption value.

For this purpose, the quantity issued/sold from the warehouse of each spare part in a year is multiplied by the average price or standard price. The items are arranged in a descending order of the consumption rupee value. Experience indicates that the top 10% of the items account for a vital portion of about 60% of annual consumption/sales value and are therefore classified as A category items. The middle 20% account for about 20% of annual sales usage value and are known as B category items. The remaining 70% of items account for a trivial annual consumption sales value of about 20% and are classified as C category. Computer programs are available to demarcate these items on a scientific basis. However, the spare parts can be grouped on an ad hoc basis as A items, accounting for more than ₹1 lakh annual consumption value and C items below ₹10,000 consumption, with the B category falling in between. The exact percentages will vary from one organisation to the other according to the consumption pattern industry process, corporate culture, top managing interest, commitment of maintenance/spare parts personnel, public–private sector, imported, etc. and hence are reviewed periodically. It is advisable to carry out this analysis separately for each group such as imported, indigenous, insurance and maintenance items, if the items are too large.

The items accounting for the bulk of annual spares sales value (which are usually very few) should be closely controlled and watched strictly with regard to management information, planning, control, follow-up, monitoring, deliveries, records, auditing, stock levels, replenishment, forecasts, and application of cost reduction techniques. They must be reviewed and ordered frequently to reduce the working capital commitment for inventories. The records for the management information system should be accurate, and uptodate for the A category items. Since they are fast moving, high sales value, fast rotation stocks, the percentage of profit margin could be lower as compared to other categories of spares.

The C category spares, on the other hand, are slow moving in nature and account for only a small percentage of annual total sales/usage/investment. For these items, there can be high safety stocks, annual records, high margin of profit, etc. It is possible that a large number of spares have little or no sales/usage over a period of time and may be non-moving in nature and categorised as C item, Care should be taken to avoid obsolescence in such cases. It stands to reason that the middle consumption value B items should be in between these two extremes: Using ABC analysis alone would not suffice to control the spare parts as it does not take into account "criticality and availability" aspects. When will organisations wake up from ABC analysis, which is below basement since Pareto found this and use MUSIC-3D for better control?

VEIN-VED ANALYSIS

The ABC analysis focusses only on annual sales or consumption rupee value and ignores the criticality of spare parts. VED denoting vital, essential, desirable-analysis, on the other hand, looks at the criticality of the spare part from the utility technological/production/design/safety/pollution-point of view, and hence the decision is based on technological considerations to ensure smooth functioning of the production system (refer Chapter 5). Vital items are those items without which the plant system will come to a grinding halt, a typical example being the spark plug in a car. Vital spare parts include all items which if not in stock, would result in huge losses and complete closure of the plant, for a considerable period of time. Criticality is discussed at length in Chapter 5.

In a spare part selling organisation, non-availability will result in loss of image of the firm. Hence a very heavy cost has to be incurred to procure the items on an emergency basis. However, attempts to classify all items as *vital* by the technical people must be resisted. Here again the equipments can be classified as VEIN—vital, essential, important, normal—if a machine is the starting point feeding to a large number, then it is more vital than the downstream equipment. Vital equipment is also known as mother equipment at the origin of the total process. Stoppage of vital machinery rests in closure of downstream machines resulting in closure of the plant. In a textile mill the blow room may be the vital equipment. The process industry classifies the equipment as upstream and downstream machines. A vital part getting into a vital equipment is more critical than the same part getting into a normal machine and maintenance forces must devote full attention.

Essential spares are those in which stockout would result in moderate losses while the non-availability of desirable spares will cause only minor disruptions for a short duration. A desirable item is one without which the system can effectively function, usually an extra fitting, like the seat cover in a car, or items used for extra comfort or decoration.

SPARES INTELLIGENCE

The third dimension is the availability of an item, based on lead time considerations and materials intelligence. Spares could be available from diverse sources such as company's own workshop, local dealers, ancillaries, other firms, imports, and monopoly suppliers. Obviously, the manufacturing/transporting time will vary in the above cases. The total lead time includes (a) internal administration lead time of converting an indent to a purchase order, (b) manufacturing time, (c) transporting time, and (d) inspection time. Depending upon procedures, systems and powers of delegation, the internal administration lead time may go up to 36 months in the case of imported items with advertised global tenders. Similarly, the manufacturing time for high precision spares will be large. The delays in inspection may be due to communication difficulties, lack of inspection staff, testing and facilities of location inspected. Thus, the total lead time, in case of imported spares for procedure-oriented organisations, may be up to four years. This implies that apart from keeping a high inventory, the organisation has to plan four years in advance. Organisations can control the lead time by critically examining the procedures, updating life powers of delegation, and reducing the advertised and open tenders by resorting to limited tenders. Usually the lead time is considered on the basis of SDE (scarce, difficult to get, easy to obtain), GOLF (government controlled, ordinarily available, locally available and foreign purchase), and SOS (seasonal and offseasonal). If there is only one monopoly supplier, the item becomes scarce to obtain and is a long lead time item. The above three classifications of cost, criticality and availability cannot operate in vacuum and must be compared for managerial control.

MUSIC-3D CONCEPT

MUSIC-3D tries to integrate all the above three types of analysis, viz. ABC/ VED/SDE, and each dimension is taken at two levels. For instance, spare parts are classified into two categories: (a) high sales value (HSV), if it is beyond ₹50,000 per annum—and this may go up to 80% of annual sales—, usage value accounts for 20% of the number of items, and (b) low sales value or low consumption value of about 20%; this accounts for the remaining 80% of the items. Instead of three categories as vital, essential, desirable, the spares are separately classified as critical and non-critical by the technical personnel. The lead time is also classified as long lead time or LLT with more than 12 months and short lead time (SLT) of less than one year. Combining the three dimensions we obtain cost-criticality availability as shown in Table 23.1, classifying the items into eight categories after using all the three dimensions.

Table 23.1 Cost Criticality Availability

		ABC Base		
		High Consumption Sales Value (₹)		Low Consumption Sales Value (₹)	
VED base		LLT SLT		LLT	SLT
		Lead Time Base			
Critical		1	2	3	4
Non-critical		5	6	7	8
		Number 20% (small) Sales value ₹80% (large)		Number (80%) large Sales value (20%) low	

INTERPRETATION OF MUS1C-3D

In integrating, we have taken only two categories in each of the three dimensions, leading to eight categories. Otherwise, by conventional methods we get three categories in each dimension (ABC × VED × SDE), leading to 27 categories. Thus the spare part will fall into any of the eight categories. For easy understanding of operators, the bin cards may be in eight different colours. We also know that items falling in cells 3, 4, 7 and 8 account for 80% in terms of number but with a low sales value of 20% only. For items falling in (low cost cell 3), the inventory level should be high—about two years stock, there should be no stockout at any cost, storage should be always nearest to the user, purchase should be highly decentralised, with maximum possible service level of 100% and minimum audit efforts as it is low value hint critical. The control will gradually be tightened to items falling in cells 4, 7 and 8. For items falling in cell 1, with LLT, HSV and high criticality, the planning should be perfect with maximum possible market intelligence system, minimum margin of profit per item due to increased turnover, low inventory but minimum stockout by application of mathematical formulae, scientific application of norms, and maximum follow-up and records. Items falling in cell 2 will have treatment similar to that of items in 1, but to a lesser degree. Items falling in cell 6 (high value items) will receive treatment opposite to that in cell 3, namely low/zero inventory, storage for the entire industry in the form of a bank, purchase decisions taken at the highest level with the unavoidable delays, very low service levels delivered at the last moment, etc. It is not useful to visualise cost reduction techniques like value engineering, value analysis, etc. for items in cells 3, 4, 7, 8 as the cost of application of these techniques will be greater than the cost of the item and initially it is dangerous to apply for items in cells 1 and 2 as they are critical. The starting point of application of cost reduction techniques is for items in 5 and 6. It is unfortunate that still professionals talk only about ABC analysis and unable to comprehend classification into the other dimensions of criticality and availability.

A–Z ADVANTAGES OF MUSIC-3D

Thus MUSIC-3D helps the spare parts executive to selectively control, in a better manner, the following aspects:

    (a) Planning—Immediate/Short-term
    (b) Forecasting—Short-term/Medium/Long-term
    (c) Records/MIS computerisation
    (d) Materials/Market intelligence system

(e) Inventory levels—minimum/maximum safety stock levels
(f) Storage points—Depots-divisions-wholesalers/Channels
(g) Purchase authority/Sales priority/Powers of delecation
(h) Consumption/Sales norms
(i) Service level/ASS/AMC contracts
(j) Pricing strategies and margin of profits
(k) Application of cost reduction techniques like value engineering
(l) Obsolescence analysis/Movement/Review/Write off
(m) Audit of all types
(n) Stock reviews—Physical, periodical
(o) Reduction in lead time
(p) Movement analysis
(q) Use of mathematical models—Simulation, inventory
(r) Stock verification—Sample basis, periodicity
(s) Development of drawings
(t) Ancillary development
(u) Follow-up efforts
(v) Computerisation/Communication
(w) Priority allocation/VIP customers with repeat orders
(x) Writing off/Charging off discrepancies
(y) Better managerial control
(z) Efficient spare parts control by marketing and users.

OTHER SELECTIVE APPROACHES

Table 23.2 lists all selective controls used in spares management even though all of them culminate in the MUSIC-3D model:

Table 23.2 Selective Approaches

Approach	Basis of Classification	Major Application
ABC analysis	Annual sales value (₹)	Stock control
HML (high, medium, low value)	Spares unit price	Delegation of powers
FSN: fast, slow, non-moving	Issues from stores	Movement control
XYZ analysis	Inventory value	Obsolescence control
VED approach	Part criticality	Storage point
VEIN (vital, essential, important, normal)	Equipment criticality	Standby decisions
RAM (reliability, availability, maintainability)	Equipment parts	Capacity utilisation
FAN (failure analysis)	Design of spares	Reliability engineering

(Contd.)

Table 23.2 Selective Approaches (*Contd.*)

Approach	Basis of Classification	Major Application
SDE (scarce, difficult, easy to get)	Availability of items	Source development
GOLF (government controlled, ordinarily and locally available, foreign purchase)	Source of origin of spare parts	Purchase follow up strategy
SOS (seasonal, offseasonal)	Nature of supply and transportation	Stock levels
MUSIC-3D combing all	Cost criticality availability	All above uses

SERVICE LEVEL

We have noted that the service level or issue from warehouse to the user (customer) gradually decreases from 100% for items in cell 3 to very low level for items in cell 6. This can be expressed as $K_u/(K_u + K_o)$, where K_u is the stockout cost of a spare part and K_o is the overstock cost (or inventory carrying charges till the item is used) of the same item. It is noticed that the service level has a bearing on inventory of a spare part. This can be derived mathematically.

Let the demand for a spare part be distributed as per the probability density function, $F(x)$. Let S be the stock on hand, K_u the understocking cost, and K_o overstocking cost. The expected cost of understocking is

$$\int_S^\infty K_u(x-S)F(x)dx$$

expected cost of overstocking is

$$\int_0^S K_o(S-x)F(x)dx$$

For optimum results, the first derivative should be zero-deriving with respect to 'S' + simplifying we get

$$K_o(S) = K_o - K_o[F(S)]$$

$$\therefore \quad F(S) = \frac{K_u}{K_u + K_o}$$

Usually when this formula is applied to stockout is very high for items in cell 3, 4 and 1 and 2 resulting to the service level close to 100 as the items are all critical.

On the contrary 7, 8, 5, 6 cells contain items which are non-critical and senile level could be near to zero. As a matter of fact items in cell 8, low value, immediately available and non-critical can be delegated at the lowest level and can be eliminated. For items in 5, 6, 7, 8, one can ask all questions and value analysis can be started with those falling in items under cell 5. Cell 1 is high value, high critical and long lead time—must be monitored by higher authorities every minute or day. Thus, cost–criticality–availability has to be done by a team of finance, operations and materials for better control, while ABC is based only on one dimension of consumption values.

Chapter **24**

Inventory Control of Spares

OBJECTIVES OF CONTROL

Inventory is defined as an idle resource of any kind having an economic value. Inventory control aims at developing procedures to optimise the working capital and satisfy the user. As long as the spare part is lying idle in the warehouse, the working capital is locked up till the part is sold to the consumer. The Finance Manager is wary of servicing the idle working capital at 30% per annum and considers it as part of the current asset of the company. On the other hand, if the item is not kept in the store, there will be a stockout if the demand arises. The Warehousing Manager sees inventory control as physical storage, location, age of the item, security from theft, etc. for servicing the consumer. The purchase personnel are concerned with the ability of inventory to satisfy the supply and demand process with optimum spares stock. The maintenance men using spare parts normally complain of low stock as the major cause of maintenance holdups. The Marketing Manager is unable to satisfy the consumer even though he has plenty of non-moving stock. The best inventory control is one that allows for maximum availability of service with minimum cost; this is, however, a difficult exercise.

Primarily, the inventory is held for transaction purposes, as in any organisation there is invariably a time lag between recognition of the need of a spare part to the satisfaction of the need and during this lead time, the spare part in the warehouse becomes handy. Inventory is also held as a precaution or contingency for increase in the lead time or consumption rate. There is a speculative element that the spare part may not be available or prices may increase. In selling organisations spare parts inventories are held in the distribution channels to decouple the consumer from the manufacturing plant. Thus spares inventory is held for adequate customer service, to take advantage of price discounts, to make possible economies in transportation, for batching in the manufacture, to serve as an insurance in case of delays and to even out the workloads on the shops in case of fluctuating demand.

STOCKOUT COST

Though the stockout or understocking cost is the most difficult to calculate, it is very relevant in the spare parts field as the inventory level will be higher if stockout charges are higher and vice versa. It is obvious that the stockout cost in most cases will be higher than even the price of the relevant spare parts which in turn will be higher than the carrying charges, leading to adoption of conservative policies of overstocking in many cases. The stockout cost is measured in terms of opportunity costs due to loss of image/sales/production/goodwill by the idling capacity of a machine, due to non-availability of a specific spare part for the duration till the machine is commissioned. The components of the stockout cost are as follows:

(a) Profit lost due to lost sales which in turn is due to production loss as the part is not available.
(b) Loss of goodwill of the customer and future sales due to his switching over to another supplier wherever possible.
(c) Penalties for late deliveries.
(d) The down time of the equipment and low morale of workers.
(e) Waste of products preceding, during, and following condition after repair.
(f) Cost of bringing back equipment to working condition after repair.
(g) Additional price/transportation charges to get the spares.
(h) Interest burden on investment on equipment which is lying idle due to want of spares.

While the components (a) and (b) are more serious and may affect the buyer–seller relations, components (c)–(b) will be considerable in case of single stream spares intensive continuous process industries.

As against the stockout cost, in most practical situations, there is overstocking cost or stocking beyond the required quantity. This cost is the same as the inventory holding cost except that the increased quantity is carried for a specific period of time to reduce down time. Duration of the down time usually consists of (a) time for reporting the failure and transit to the place of repair, (b) initial inspection to locate the fault and to diagnose its costs, (c) the physical process of repair including the procurement of spare parts, (d) trial test to specify performance, and (e) transit time for return to the user.

INVENTORY CARRYING COST

Spare parts stored in the warehouses cost the organisation in terms of men, material, time, space, insurance, etc. and the total stocking charges are estimated at 30% per annum of the average value of the stores material. The major components consist of the following elements:

(a) The cost of the money invested in the stocks that can be equated to the present lending rate which is 17.5%. In order to borrow the working capital, Financial Managers have to spend their time and fill up relevant forms, and taking this figure into account, the cost of working capital is 18% in India. In the absence of borrowed funds, we can assume the weighted averages of the cost of capital or an opportunity cost of the funds invested in lieu of the bank rate. Even public utilities like state electricity boards borrow money through debentures with 12% interest,

(b) The cost incurred on physical facilities like space, taken on rental value or opportunity cost basis.

(c) The cost of racks, bins and handling equipments taken at the depreciated value.

(d) The salary and statutory payments of stores staff.

(e) Documentation used in the stores.

(f) The data processing cost connected with stores.

(g) The cost added by deterioration of terms with low shelf life.

(h) Losses due to evaporation and natural spoilage.

(i) Cost of application of preservatives to prevent deterioration.

(j) Losses due to pilferage, theft, wastage, breakage and mishandling.

(k) Accommodation of obsolete items that may have to be sold at throw away prices. (It should be observed that spare parts tend to become more obsolete compared to other materials.)

(l) Insurance cost incurred as a protection against fire, theft, burglary and other risks.

The total inventory cost is sizable, about 30% per year—and lays a great burden on the Warehouse Manager, whose duty is not only to reduce the cost but also to manage his stores efficiently without diluting the service to the customers.

ORDERING CHARGES

Other than the price that is paid for the materials purchased, the very act of ordering costs the purchases. This is called *ordering charges* or *acquisition cost* or *order processing cost* in a selling organisation. In addition, one has to take into account freight, insurance and inspection. The acquisition cost is measured in terms of money spent in organising the purchasing department, while the order processing cost reflects the cost incurred on organising the sales department. This cost is found to vary from one organisation to another depending upon imported spares, indigenously manufactured items, powers of delegation, systems, procedures, documentation, control and quality as well as number of staff, it is around ₹1000 per order.

The following components constitute the ordering cost: (a) The salary and statutory payment to purchase (or sales) staff—the time spent by finance/user in tender committees should be included in this; (b) stationery/documentation used in ordering; (c) rental charges for office accommodation and depreciation of furniture; (d) advertisement expenses in trade journal and other papers; (e) communication cost in the form of telegram, telex, postage, telephone, courier charges, etc.; (f) follow-up cost, including travel and messenger cost; (g) computer charges relevant to purchasing; and (h) source development expenses. The total amount spent by the purchasing department is divided by the total number of orders to get the acquisition cost per order. Similarly, in selling organisations, order processing charges may have to be taken into account. For staggered deliveries, costs of additional documents, additional follow-up, and additional efforts in stores are considered. The total purchase cost expressed as a per cent value of items procured should be about 1% as the Directorate General of Supplies and Disposal charges 0.5–1% on the indebtors as procurement service charges. Similarly, norms for order processing and selling expenses can be fixed by spare parts sellers.

SERVICE LEVEL

The efficiency of a warehouse is measured on a negative scale, i.e. in terms of stockouts. On the other hand, the achievement of the Utopian goal of zero stock-out will require a very high idle investment and, even then, such a service cannot be assured. Depending on the cost and criticality, we can fix these levels. Thus, for a vital spare part, we can fix the level as 99.5%, which implies that if there are 100 demand requisitions, they will be met at least on 99 occasions. As the criticality of the spare part decreases, the service level can be decreased.

The service level is usually fixed on the basis of two opposing costs, namely, overstocking cost and understocking costs. It is noted that understocking cost may be only an objective estimate as it is difficult to quantify such concepts like loss of image or low morale. But, it should be possible to grade them for different spare parts. The service level is usually fixed by the formula $K_u/(K_u + K_o)$ where K_u is the understocking cost and K_o the overstocking cost. It can be deduced at once that as the understocking cost increases, the service level has to be increased, i.e. the criticality of the item is greater. On the other hand, if the understocking cost is comparable to the overstocking cost, then it would be advisable to have a much lower service level. Thus, depending upon the cost estimates, appropriate service levels can be fixed. It may also be noted that whenever service level is fixed on other considerations, the understocking cost can be estimated since the overstocking cost is fairly easy to compute. However, many customers and maintenance engineers always complain about the low service level of about 70% from the warehouse or service after sales, or suppliers of spare parts. Selling organisations must endeavour to increase the service level, particularly of low value critical spare parts. While determining the service level, it should be remembered "uncontrolled inventory is industry's cancer"!

ECONOMIC ORDER QUANTITY

Any inventory system tries to answer the question, how much to order and when to order. We shall now discuss the economic order quantity formula to answer the question how much to order, which is applicable only for fast moving spares. Let us illustrate this. In a firm, the inventory carrying cost is 30% per annum and ordering charges per order are ₹600. Let the annual consumption (or sales) be 10,000 units priced at ₹10 each. It is assumed that the part is freely available in any quantity so that stockout cost does not arise. Let us now arrive at the optimum by means of trial and error by ordering any quantity and measuring the consequences. Then we arrive at Table 24.1.

Table 24.1 Cost and Consequences Table

Order Quantity	Orders Per Year	Ordering Cost	Inventory Average Quantity	Value	Carrying Cost	Total Cost
1000	10	6000	500	5000	1500	7500
2000	5	3000	1000	10,000	3000	6000
5000	2	1200	2500	25,000	7500	8700
10,000	1	600	5000	50,000	15,000	15,600

Thus, in this given case, the optimum ordering quantity that minimises the total cost is arrived as 2000 spare parts, and at this quantity the carrying charges are equal to the ordering charges. The quantity which optimises the total cost is known as *economic ordering quantity* (EOQ). In the manufacturing context, it is known as economic lost size with set-up cost or cost incurred due to job changes replacing the ordering cost.

We shall now derive the EOQ formula in a general case. Let C_a be the order cost and C_c be the carrying charges of 30% per year, M the annual consumption, p the unit price, and q be the quantity to be ordered. The number of orders is M/q and ordering charges are MC_o/q. The average inventory—when q units are ordered—is $q/2$. The inventory carrying charges are $(q/2)pC_c$. The total cost is

$$(M/q)C_o + (q/2)pC_c$$

Differentiating with regard to q and equating to zero, the optimum q or EOQ is obtained as $\sqrt{2\,MC_o/(pC_c)}$. Substituting the original values of the illustration, we obtain

$$\text{EOQ} = \sqrt{2\frac{(10,000)(600)(100)}{(10)(30)}} = 2000 \text{ units}$$

Figure 24.1 illustrates how EOQ is obtained.

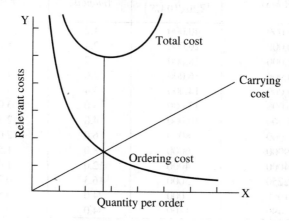

Figure 24.1 EOQ curve.

Calculation of EOQ Formula

The Warehouse Manager handling a large number of spare parts will find it difficult to calculate the costs and tries to devise a simplified system. The EOQ formula is

$$\text{EOQ} = \sqrt{\frac{2MC_o}{pC_c}}$$

The EOQ value (₹) is

$$\text{EOQ} = \sqrt{\frac{2MC_o}{pC_c}}\, p \times p$$

$$= \sqrt{\frac{2C_o}{C_c}\, MS} = \sqrt{\frac{2 \times 600 \times 100}{30}}\, Mp = 20\sqrt{10Mp}$$

In the previous example, $20\sqrt{10(10,000)10} = ₹20,000$. By using this formula, we compile Table 24.2 as MS is the annual consumption value or annual sales value. The ordering and carrying costs have been taken as ₹600 and 30%, respectively. Thus EOQ is very high for high consumption value items and is hence ordered individually. Low consumption value items can be ordered once a year and the items can be launched, particularly since the EOQ curve is flat at the bottom, allowing for deviation according to convenience, as can be seen in Table 24.2.

Table 24.2 Ready Reckoner Table

Annual Consumption (₹ MS)	EOQ ($₹20\sqrt{10MS}$)	EOQ Duration (months)	Orders Per Year
400,000	40,000	1.2	10.0
100,000	20,000	2.4	5.0
81,000	18,000	2.7	4.4
64,000	16,000	3.0	4.0
49,000	14,000	3.4	3.5
36,000	13,000	4.0	3.0
25,000	10,000	4.8	2.5
16,000	8000	6.0	2.0
9000	6000	8.0	1.5
4000	4000	12.0	1.0
2250	3000	16.3	0.7
1000	2000	24.0	0.5
490	1400	34.0	0.3

REVIEW PERIOD

The EOQ time (months) is usually known as review time. For high value maintenance items, the review period is lower, resulting in frequent reviews. Conversely, the low value items will be reviewed less frequently. A common review period can be fixed for maintenance items with similar consumption values, so that they can be grouped and ordered to take advantage of market forces. It is enough if the spares are reviewed cyclically in the above periodic review time cycle. The ordering quantity will depend upon the total lead time, review time, fluctuations in demand, and lead time. Then the maximum physical stock that must be carried is dependent upon the service level, and the above parameters are known as desired inventory level or DIL. This consists of average consumption during lead time and review period plus the minimum stock to cater to fluctuations in lead time and consumption. During every review, as per the cyclical review time, orders are placed to make the stock reach the DIL Level.

There are many mathematical models for inventory control of spare parts, but the simplest formula that the Warehouse Manager should have under his control is $KLTC$. Here LTC denotes lead time consumption and K a factor depending upon cost-criticality availability. For instance, K may be 2 for low cost long lead time and highly critical item, and will be half for high cost, short lead time non-critical items. If the consumption is zero, then $KLTC$ becomes zero, resulting in nil stock. This is subject to at least one item will be in the stock if consumption is zero. This formula can be easily applied in actual practice to determine the physical stock with the Warehousing Manager for various spare parts and reviews done during review periods.

Chapter 25

Maintenance Spares

CONCEPT OF MAINTENANCE SPARES

The spare parts which are consumed regularly are called *maintenance spare parts*. These spare parts, which are required for replacements of old parts due to frequent breakdowns caused by wear and tear, can be treated as maintenance spare parts. They are generally fast moving and repetitive spares. It is possible to build adequate data base on failures and consumption, according to actual withdrawals from the stores. Hence, it is possible to work out stocking policies and consumption norms for withdrawal from the stores. Typical examples include belts, bearings, oil seal, wire ropes, fasteners, hardware items, couplings, gears and shafts.

Any equipment consists of numerous individual parts, with each part having its own in-built reliability and failure rate. The equipment becomes inoperative because of the failure of a component due to the inferior quality of a part which results from the conditions in which the equipment is operated. The failure of a part is usually an exceptional phenomenon. A substantial portion of the part that has failed may be adjusted repaired/retained/reconditioned suitably; the proportion of replacing the spare parts is often less than the failure rate.

In the context of failure rate, it should be noted that usually the design of an equipment is such that the parts which fail frequently are easily accessible to the mechanic from the repairing point of view. Usually the parts that fail often are also cheaper compared to other categories of spare parts. On the other hand, the critical spares in the equipment are generally made of strong materials with higher reliability and lower failure rate.

POISSON PATTERN

From practical experience as well as theoretical considerations, the frequency distribution of data on maintenance spare parts consumption figures conforms to the discrete statistical distribution known as Poisson probability distribution. The Poisson distribution shape is skewed to the right like a fish, and is analytically tractable. The Poisson probability distribution is expressed by the mathematical formula

$$P(x) = \frac{\left(e^{-m}\right)\left(\frac{x}{m}\right)}{\lfloor x}$$

where m is the average failure rate, e is the Naperian logarithm with a constant value of 2.77, and x is the value for which the probability of failure is designed.

It is interesting to know that one of the earliest attempts to fit the Poisson distribution to random events was for the number of cavalry men killed due to horse kicks during the course of a year. The Poisson pattern describes only those failures where the probability remains constant from one period to another. This implies that it is likely that a failure will occur just after several other failures, i.e. when a considerable time has elapsed since the occurrence of the last failure.

The physical assumptions corresponding to the Poisson distribution are as follows:

(a) The number of demands for a spare part occurring in any interval of time—say today, does not give any information about the number of demands for tomorrow.
(b) The process is stable over time.
(c) If the time duration is divided into very small intervals, the probability of two or more demands in the same interval is negligible.

It is interesting to note that in practice the above assumptions are generally satisfied for most maintenance spares and it is possible to check the assumptions by statistical tests. The distribution is uniquely determined by the average failure rate or sales pattern for selling organisations in a given duration.

STOCK LEVELS

The average issues from the stores consumption/failure/sales data should be compiled at least for a period of two years in order to represent a realistic representative picture for future projections. If the average consumption is very low or a small fraction, then this tends to become an insurance item for which the stock level should be at least one item. In a few special cases the Poisson pattern may not give a proper fit. This may be due to unexpected failures when some part of one equipment leads to preventive replacements of other items or the overhauling requirements are also combined with the routine maintenance demand. These discrepancies should be investigated before applying Poisson pattern as the Poisson distribution provides answer to the basic inventory questions as to when to order and how much to order in the area of maintenance spares.

We usually calculate the average consumption/sales of a specific spare part. Let this average value range from 1 to 12 in sequence. The corresponding maximum stock level relating to a service level of 95%—or a risk level of 5%—using Poisson distribution formula, would be 3, 5, 7, 8, 10, 11, 13, 14, 16, 17, 18 and 20, respectively. The maximum stock level obviously will decrease if the service level decreases e.g. for a service level of 75%, the maximum stock levels are 2, 3, 4, 5, 6, 8, 9, 10, 11, 12, 13 and 14, respectively. Thus, depending upon service level and average consumption rate in a given period, the maximum stock can be easily determined for maintenance spares using Poisson distribution. It is beneficial to calculate the consumption during lead time as the norm and obtain the maximum stock by using Poisson distribution. The review time based on the periodic review system or economic order quantity converted to the time scale may be used to replenish the stock level for each maintenance spare.

SERVICE LEVEL

The formula for getting the maximum stock is $M + K\sqrt{M}$, where M is the representative average consumption during the lead time and K is a factor depending upon the service level with values 0.7, 1.3, 1.7 and 2.3 for service levels of 75%, 90%, 95% and 99%. The service level can be obtained as the ratio of stockout cost/stockout cost + overstocking charges. The safety stock or minimum stock is $K\sqrt{M}$ units, The service level depends upon a large number of factors such as (a) maintenance policy, (b) cost of spare parts, (c) criticality of the item, (d) reliability, (e) reconditioning nature, (f) lead time, (g) purchase policies, (h) purchase delegation, (i) procedure for work authorisation, (j) equipment design, warranty, (k) guarantee, (l) accuracy of records, (m) calibre of maintenance staff, and (n) availability of working capital.

Let us calculate the maximum stock for different values of M, the average consumption/issues/sales/failures in a given duration. Here the average value M has been taken as 16, 9, 4, 1, .1 and .01. If the K factor corresponding to service level 75% service level, the factor is .7, then we get $M + K\sqrt{M}$

as $16 + (.7)\sqrt{16}$, i.e. $16 + 2.8$ or 18.8. Since we cannot have the physical stock as 18.8 items, we round off to the higher integral level of 19 in order to ensure the desired service level. Similarly, when $M = 1$, we have $M + K\sqrt{M}$ as $1 + (.7)(\sqrt{1})$ or $1 + .7 = 1.7$ which is rounded off to the higher integer of 2 as the number of spares to be stocked must be in integral values. Due to rounding off errors, we add large percentage for lower values of average consumption, in order to ensure the desired level, for obtaining the maximum stock level. If the value of M is very low or near zero, the relevant spare part could be an insurance item where one number of stock has to be carried. Table 25.1 gives the maximum physical stocks to be maintained in the warehouse, corresponding to different consumption/sales values rounded off to the nearest higher integral value. For other values of M, the formula $M + K\sqrt{M}$, can be used to obtain the physical maximum value of Stock, as shown in Table 25.1.

Table 25.1 Physical Stock Levels

Service Levels K Factor	75% 0.7	90% 1.3	95% 1.7	99% 2.3
M	$M + K\sqrt{M}$	$M + K\sqrt{M}$	$M + K\sqrt{M}$	$M + K\sqrt{M}$
16	19	22	23	26
9	12	13	15	16
4	6	7	8	9
1	2	3	3	4
.1	1	1	1	1
.01	1	1	1	1

CENTRAL STOCKING

A common question relating to maintenance spares is which items should be centrally stocked and which spare part must be stocked in each substore. For this purpose we consider an item with an average failure rate of .2 and a stock level of 1 at each of the six stations leading to total stock of 6 items. On the other hand, if the six stores can be merged, without causing any difficulty to the user, then we get an average of 1.2 and the maximum stock level of three spare parts.

Let us take a road transport undertaking with 10,000 buses consuming on an average 1000 spark plugs in a given duration, say one year. The maximum consumption will be $1000 + 2.3\sqrt{(1000)}$ or about 1100. If the same number of buses operate at different locations, each with 1000 buses, then the spare part consumption will not be exactly 100 per year in all locations. This will vary considerably with die average of 100 and the maximum of 300. If this fleet is distributed into 100 depots with 100 buses each, the average consumption will be 100 with a maximum of 17, leading to a possible maximum of 1700 in all depots. The stock levels in the decentralised locations have to cater to the above fluctuations. Hence, if other factors like service, number of machines, user satisfaction etc. are equal, it is advantageous to centrally stock the high consumption value spare parts as the benefits in the form of decreased working capital will be substantial.

ILLUSTRATION OF STOCK LEVELS

Let us consider a pan failing at the rate of 0.001 failure per hour, mean time between failure as 1400, and the total lead time in one year. In a continuous process plant for three shifts working, after allowing for a turnaround period of 15 days, the total number of working hours is 350 × 24 = 8400 hr. During the lead time of one year, the lead time between order placing and receiving the stock, the average expected failure is 8400/1400 = 6 failures. With an average failure of 6, from Poisson distribution we get a stock level of 10 units for a service level of 95% and this could be the physical stock. We can work out the probability distribution, indicating service level and stock corresponding to each stocking level by using Poisson distribution; this is given in Table 25.2.

Table 25.2 Poisson Probability: Mean 6

Possible/ Stock Values	Probability of Occurrence	Cumulative Service Level
0	.0025	.0025
1	.0149	.0174
2	.0446	.0622
3	.0892	.0522
4	.1330	.2851
5	.1606	.4457
6	.1606	.6063
7	.1377	.7440
8	.1033	.8470
9	.0688	.9161
10	.0413	.9574
11	.0225	.9799
12	.0153	.9912
13	.052	.9964
14	.0022	.9986
15	.0069	.9995
16	.0003	.9998
17	.0001	.9999
18 and above	.0001	1.000

It may be noted that the stockout can be obtained by subtracting service level from one. It will be seen from this that the service level corresponding to the stocking level of 11 is 98% and 99.95% for 15 spare parts as shown in Table 25.2. For the locking up of working capital in four additional spares, the improvement in service level is only 1.85% due to the tapering nature of Poisson distribution at higher values. The statement—the higher the stock, the larger the assurance or safety—is not always true since in stocking beyond certain levels, the increase in assurance is only marginal and may not be worth locking up of working capital.

The information on costs and consequences is available in order to enable the decision-maker fix the service level for individual maintenance spares after weighing the costs and benefits in the system. We have used the discrete Poisson pattern, while we can use normal curve if the number of observation is large.

Chapter 26

Simulation for Spares Control

PLANNING AND SIMULATION

There are several approaches and formulae to determine spare parts stock either for procurement or for sales, for optimum results they may be combined in order to crosscheck the usage of each. These methods include:

(a) Application of Poisson distribution to maintenance spares
(b) Use of queuing principles for rotable category
(c) Role of cost-benefit analysis for insurance spare
(d) Use of failure rate, bath-tub curve, and Weibull distribution for maintenance items
(e) Resorting to selective inventory principles such as ABC-VED
(f) Use of the simple formula of $KLTC$, where LTC is the anticipated lead time consumption and K the varying factor, depending upon cost-criticality availability of the spare part and is at least one.
(g) Forecasting methods such as moving average and exponential smoothing
(h) Simulation approach—either manual or computer—to determine optimum levels
(i) Multiechelon models to determine the stocking levels at each warehousing point.

Each of these methods has its own advantages and disadvantages and areas of applications. Let us now discuss the simulation method in detail.

SIMULATION METHOD

The first cave drawing was a simulation as are all maps, charts, graphs, robots TV plays, photographs, models, aeroplane simulators, etc. In fact, a simulation is any scaled down or symbolic representation. What is new is that it is rapidly becoming possible to create working models of complex dynamic systems such as spare parts market situations.

A simulation is also an analogy because it is only analogy and not really the objective it represents. It enables experimentation which cannot be done in real-life situations. Because an architect's drawing simulates materials and space, he can build and rebuild the same structure over and over again, in order to see which plan is best. Airline and train schedules simulate time relationships and facilitate the travellers to decide on the best route and the best carrier available without taking all routes and carriers themselves. A missile expert's formula simulates an object's flight through space and he can predict precisely where a rocket will land. Similar is the case of simulation of war games, where one can afford to take risks for various alternatives and measure the costs and consequences.

Simulation is particularly useful when the spare parts executive has to deal with a number of problems with inadequate knowledge on aspects such as failure rate, lead time, consumption pattern, availability, buffer stocks, reorder levels, and forecasts. In such cases, simulation creates reality in advance on paper or computer or scale model and test various strategies to meet the reality in order to choose the best.

TWO-BIN SYSTEM

The two-bin system—S, s inventory policy—is the simplest—age old easy to follow—system and can be operated by even clerical staff and is particularly suited for spare parts. Start with an initial stock S which represents the quantities of both the bins and is gradually depleted. As the stock falls to s, which is equivalent to the second bin's capacity, place an order for the quantity consumed, which is $(S-s)$, and further depletion of spare part continues. The quantity ordered at each time is $(S-s)$ and the maximum stock is S with the ordering point as s. Thus s stands for one bin and $(S-s)$ for the second bin. The time difference, when the stock is at S and $(S-s)$, is the lead time. For convenience of working of this system, it is necessary that s be less than half of S. Suppose S is 100, then s should be less than 50 to avoid stockout and periods of shortages, when stock reaches zero before fresh stock arrives. We shall work out an inventory simulation with S, s policy through the following illustration. Even though it is called 2 bin system, one bin of size 's' is enough for lab item and 's' is marked inside the bin.

ILLUSTRATION OF SIMULATION

The simulated demand for a spare part in a factory for 35 days is given in Table 26.1. The number of spare part units demanded in the past has been

Table 26.1 Scoring Sheet for Spare Part Simulation for 35 Rounds

Period (1)	Cycle (2)	Lead Time (3)	Initial Stock (4)	Demand (5)	Final Stock (6)	Shortage Units (7)
1				0		
2				1		
3				2		
4				0		
5				0		
6				1		
7				0		
8				2		
9				1		
10				0		
11				1		
12				0		
13				0		
14				1		
15				1		
16				1		
17				0		
18				2		
19				1		
20				0		
21				0		
22				1		
23				2		
24				0		
25				0		
26				0		
27				1		
28				0		
29				2		
30				0		
31				1		
32				0		
33				0		
34				1		
35				0		

Number of orders – Total stock units;
Number of stockout – Average stock per day;
Inventory carrying charges ₹; – Ordering charges ₹–;
Stockout cost ₹ –; Total cost ₹ –.

varying as 0, 1, 2, 3 and 4 with the percentage of occurrence of demand being 54.9, 32.9, 9.9, 2.0, 0.3, respectively. The total demand is worked out as

$$(54.9 \times 0) + (32.9 \times 1) + (9.9 \times 2) + (2.0 \times 3) + (0.3 \times 4)$$

and the average demand is obtained by dividing the above total by 100 as

$$(32.9 + 19.8 + 6.0 + 1.2) = 59.9/100 = .6$$

The total lead time has been fluctuating in the past as 3, 4, 5, 6 days with frequency percentages as 20, 40, 30, 10, respectively, yielding the weighted average lead time of 4.3. The cost of carrying the spare part is one rupee per day and the ordering charges are ₹900 per order, with the stockout cost being ₹10,000 per unit of stockout per day. The simulated data for future lead time for various reordering cycles are 6, 3, 4, 5, 4, 3, 4, 5, 4, 6. Given the above information, one can choose a set of S and s and run the manual simulated working for the total duration and compare the costs and consequences for each (S, s) in order to decide the optimum set-up.

This example can also be illustrated analytically as follows. The average lead time has been estimated as 4.3 and the average demand as 0.6, with the resultant average lead time consumption as $(4.3 \times 0.6) = 2.58$ or 3 numbers, which constitutes the reorder point s. The order quantity given by the familiar Economic Ordering Quantity or EOQ formula is

$$(S - s) = \sqrt{\frac{2 \text{ (Annual consumption) (Order cost)}}{\text{(Unit price) (Carrying charges)}}}$$

$$= \sqrt{\frac{2 \,(0.6)\,(365)\,(900)}{(1)\,(366)}} = 33$$

Hence,

$$S = (S - s) + s = 33 + 3 \text{ or } 36$$

In order to reach the real optimum, a search is made around (14, 3) by examining combinations such as (14, 2), (14, 4), (15, 3), (15, 4), (15, 2), (13, 3), and 03.4).

The above simulation can be extended to 100 or more rounds in order to further determine the maximum stock and reorder point in the two-bin inventory system (S, s).

ILLUSTRATION ON INITIAL ORDERING

The spares for a certain type of turbine disc used in an atomic reactor costs ₹1 million each and that a loss of ₹100 million is likely to occur due to nonavailability of each turbine, including the stoppage of electricity production. This turbine disc is a special purpose spare part which is made only to order. Although it is very unlikely that any of these spare parts will be needed, they will have to be procured as a safeguard against the consequences and rather serious loss which will occur if a spare part was not available when needed. It may be mentioned that this spare part cannot be used in any other type of turbine except the one for which it is specifically manufactured against the order. Analysis of past data indicates the spares will be needed as replacement during the lifetime of the type turbine is as follows: How many spares should be procured, given that the stockout cost is 100 times the overstocking cost?

Number of spares required	0	1	2	3	4	5	6 or more
Probability of requirement (Per cent)	77.8	4.6	6.5	10.5	0.5	0.1	0

The optimum service level given by $[K_u/(K_u + K_o)] = 1/101 = .99$. This service level is arrived at by maintaining three spares as the level is 99.4%, which is more than the required 99%. Even if the stockout is more by 50%, then also the above formula holds good. We can generalise this approach by using the following formula to determine the optimum stock (N):

$$f(N-1) \leq \left(\frac{K_u}{K_o + K_u}\right) \leq f(N)$$

USE OF POISSON PATTERN

A steel mill uses five important spare parts—P, Q, R, S and T and the plant will come to a halt if the parts are not available. But of these parts Q, R, and S can be procured on a rush basis in a month's time and the rush order costs ₹5000 more than the normal order cost. Q, R and S can be procured in two months, whereas P and T need six months. However, if P and T are out of stock, the defective part can be reused after repairs, pending the arrival of spares, but only with 50% efficiency. The rush lead time for A and E is two months and a rush order costs ₹20,000 more than a normal order. The present prices of the spares are ₹60,000, 60,000, 12,000, 30,000 and 35,000, respectively. The frequency distribution of consumption of the five spare parts in the preceding four years, converted into percentages is given in Table 26.2.

The average consumption of spare part P can be obtained as

$(74 \times 0) + (23 \times 1) + (3 \times 2)/100 = .29$ per month.

Simulation for Spares Control   293

Table 26.2 Frequency Distribution of Consumption (per cent)

Spare Part	0	1	2	3	4	5	6	7	8	9	10	11	12	13
P	74	23	3											
Q	0	1	4	8	12	15	16	14	12	8	5	3	1	1
R	4	13	21	22	14	16	6	3	1					
S	33	38	20	6	3									
T	54	34	11	1										

Thus the average expected demand P, Q, R, S, T can be obtained as 0.29, 6.30, 3.22, 1.08, and 0.59. With these as averages, we can check with Poisson probabilities (Table 26.3).

Table 26.3 Poisson Probability Table

Spare Average	P 0.29	Q 6.30	R 3.22	S 1.08	T 0.59
Demand					
0	.748	.002	.041	.333	.549
1	.217	.012	.130	.366	.329
2	.032	.036	.209	.201	.099
3	.003	.077	.223	.074	.020
4		.121	.178	.020	.003
5		.152	.114	.005	
6		.159	.061	.001	
7		.144	.028		
8		.113	.011		
9		.079	.004		
10		.050	.001		
11		.029			
12		.015			
13		.007			
14		.003			
15		.001			

We thus see that the Poisson pattern fits the observed frequencies of P, Q, R, S and T. We can cumulate the individual probabilities to get the desired service level for determining the stocking policies and investment decisions as in Table 26.4. Further sophistications can be done by simulation if necessary.

Table 26.4 Service Level with Investment in Spares

Spares	Stock Quantity	Service Level (per cent)	Inventory Investment (₹)
P	2	99.7	1,20,000
Q	12	99.6	78,000
R	8	99.5	96,000
S	4	99.4	1,20,000
T	3	99.6	1,05,000

Chapter 27

Insurance Spare Parts

TYPICAL INSURANCE PARTS

Insurance parts, has not bring to do with life insurance or general insurance, but emergency high value parts.

In spare parts management, maintenance engineers use the term *insurance spare part* as if it is a holy cow. These spares, also called emergency spares, critical spares, or standby spares, have a very high in-built reliability of performance. Hence the probability of the spares functioning satisfactorily during the lifetime of the equipment is very high. These spares are usually brought when the original equipment is imported/purchased, after making a lifetime assessment and are usually capitalised by depreciating over a period of time. Since they are capitalised, they are also called *capital spare parts*. These spare parts may not be immediately available and kept in stock ready to cater to unprecedented breakdowns, resulting in very high stockout and severe bottlenecks in the production process.

Typical examples of the insurance spares are: generator in a factory, propeller in a ship, compressor unit in a fertiliser plant, and are usually referred to as *standby units*. Many of these parts may not be used at all even throughout the lifetime of the equipment. Hence, if not used throughout the operating lifetime of the equipment, these parts and assemblies get accumulated in the stores, when the equipment itself is finally discarded. Sometimes insurance are classified as slow moving, non-moving and obsolete.

In view of the high reliability of insurance spare parts, they can be compared with having air insurance or life insurance for a passenger travelling by air. Unlike the compensation paid in air disaster, the compensation is small in case of rail/road/river travel accidents. While it is true that the relatives benefit (monetarily) from such mishaps, at the same time, the insurance cost could have been saved if someone could guarantee that there would be no accident. In industry, it will be very difficult to forecast which insurance part of which equipment is going to breakdown and when exactly the breakdown will occur and how long it will take to get another standby.

STOCKING POLICIES

The reliability of the insurance part is very high, almost as near as the life of the equipment, and hence the part may not be needed at all. But, if an unprecedented catastrophic failure occurs, its consequences will be very high. Hence the dilemma before the spare part executive is whether to carry these insurance spares (with stock level being one or two items) or not to carry them, with stock remaining at zero. Since the previous consumption will be zero in most cases, the inventory policy of insurance spare is different from other categories of spare parts.

It is also observed in many organisations that a substantial portion of non-moving spare part is accounted by the category of insurance spare parts. The insurance spare part is often called a *standby part,* and the concept can be extended to every single item. Perhaps this concept can be extended to standby machine or standby factory, and standby for the standby, and so on. An ideal situation from the user's point of view would be to have a spare part for each piece of equipment in operation throughout the entire plant, but the cost of such a proposition will definitely be prohibitive.

The parts executive can carry out an ABC analysis on all the insurance spare part based on value of the item and ask relevant questions on A category of spare part with a view to reducing the inventory value on the insurance spare part.

The authority to declare these A items as insurance should rest at the highest level. In this process, factors such as price of spare part, reliability, stockout cost, availability, lead time, obsolescence, etc. influence the categorisation. It may be noted that 100% service level or zero risk policy is an impossible proposition. Stock level of one item is recommended to only those parts which fall in the real insurance category and as soon as the stock is issued, efforts are made to procure the item on an emergency basis in order to maintain the stock level of one. In practice, however, stocking of insurance spares represents insurance against stoppages. Hence the system encourages the parts executive to adopt conservative stocking policies. It should be noted that it is not the parts executive who pays the premium, but it is he who will be blamed for the stockouts. In this context of grading the insurance spare parts, it is observed that the compensation is ₹200,000 in case of air tragedy, whereas it is far less for the same human being in case of rail/road/river tragedy. We can also compare the reliability of insurance spare part with the human anatomy where man has been endowed only with one sturdy part like heart, without any standby. Only in exceptional cases like the kidney, ear, hand, leg, etc., we have the standby or do have another part. The above discussions show that the parts executive should have a thorough look at the insurance spare part categorisation and plan for one standby insurance item.

FINANCIAL ASPECT

We have earlier defined insurance spares as spare parts or complete assemblies, usually of high value, which are not normally required for routine maintenance but will be needed in the event of an unforeseen breakdown to avoid lengthy shutdowns of a vital equipment. At the time of ordering, it is desirable to carry out a cost-benefit analysis of the following aspects:

(a) the scope of repairing the damaged part locally or at the factory without holding any stock;
(b) ordering a new part and awaiting its arrival before repairs are undertaken to make it immediately available when required.

From financial considerations, it is difficult to estimate the stockout cost. As emphasised earlier, the stockout cost consists of down time, production loss, idle time cost incurred due to labour and machine, loss of image, and the costs associated with the emergency purchase of the part which will obviously be higher than if the part is purchased with the machine. Organisations in the transport sector like railways, shipping, air, road transport etc. consider this as loss of revenue per shift, without considering the loss of image in the customers' eyes.

The parts executive must consult the financial executive in determining the order processing cost, ordering charges, stockout costs, and inventory carrying charges to derive optimum stocking policies. It is necessary to consult the finance executive about the depreciation aspects while considering the capital and insurance spares. As per the Income Tax Act and finance bills, the depreciation rates vary from year to year. The rates are 10% to 100%, thereby indicating a range of technological changes from 1 to 10 years. For instance, any energy saving spare part device is entitled for 100% depreciation. If the insurance spare part is depreciated, it would be treated as a capital asset and will not be included in the spares inventory. In order to be entitled for depreciation rates, the price of an insurance spare part should be more than a specific value; since April 1993 this figure is ₹10,000. The Finance Ministry also stipulates that the benefits derived from the usage of the spare part should be spread over a longer duration and the insurance part should not be issued from stores frequently, and most of the insurance spares satisfy these stipulations.

SPARES BANK

Though we discussed spares bank earlier, it has more relevance to insurance spares. One way to minimise the costs is by applying the concept of spares bank. In India, the fertiliser units have been successfully operating, since the last six decades, catalyst banks to which every member has access; the member can withdraw the planned quantity from the central catalyst bank or pool, thereby reducing the total quantity of stock and satisfying the individual units. Similarly, the concept of spares bank should be introduced for the high value capital intensive insurance spare parts of similar industries provided the equipments are interchangeable. Basically, the process consists of similar organisations cooperating and pooling their resources in the common spare parts bank rather than the individual unit carrying the stock. The original equipment manufacturer may also keep such stock in major consuming centres to reduce the individual unit's inventory. In such a case the seller himself becomes the spares bank.

Admittedly, the bank concept will be applicable only if the equipments are standardised and the interchangeability of parts ensured. The operational details as to who should be in-charge of the bank, the quantity to be stocked, how to service the individual units, etc. can be dealt with, after accepting the concept of spares bank. The different industry associations as well as the various ministries of the Central Government can give an impetus in the formation of spare parts bank in the initial stages. In this context, a good beginning has been made by Bharat Heavy Electricals Ltd. (BHEL) in consultation with the state electricity boards. The steel industry, particularly the Steel Authority of India, should evolve common codes for items like gears, couplings, rollers, bearings, conveyor belts, etc. as the first step towards setting up spare parts bank for the industry. Industry associations, professional organisations, FICCI, ASSOCHEM, CII may take initiate in this context.

Example: Let us evolve an approach to determine the optimum stocking strategies with the help of a simple example. Let the price of an insurance spare part be ₹20,000 with the reliability stipulated by the manufacturer as 0.98. Since the reliability is 0.98, the chance of using the spare part is (1 − 0.98) or 0.02. So the expected cost of not stocking the spare part is ₹20,000. The understock cost, at which a company will be indifferent to the above two stocking policies, i.e. whether to stock a spare part or not to stock, is obtained by equating $0.02\,K_u = 20{,}000$ or $K_u = ₹1$ million. Thus the optimum policy can be scientifically determined as not to carry out any stock if the stockout cost is estimated beyond ₹1 million according to the optimum strategies! But stockout cost alone may not be the criteria as the implications are complex.

COST-BENEFIT ANALYSIS OF STOCKS

In order to determine the costs and consequences of various stocking strategies, let us work out an example and find the optimum stocking strategy. A factory has 50 central lathes working for the last 10 years. During this period, failure data on a costly spare part has been compiled; it has been noted that one spare part was used to five lathes during this period, two spare parts were used on two lathes, and three spare parts used on one lathe. The probability of spare part consumption, based on past data can be expressed as follows:

The probability of requiring one spare per lathe is 5 out of 50 lathes or 0.1. The probability of requirement of one or 2 spares per lathe is 5 out of 50 lathes plus 2 out of 50 lathes = 0.1 + 0.04 = 0.14. The probability of requiring 1 or 2 or 3 spares per lathe is 5 out of 50 lathes plus 2 out of 50 lathes and 1 out of 50 lathes = 0.1 + 0.04 + 0.02 = 0.16. With the above probability data, we can explore the cost effectiveness of the various alternative stocking strategies, policy of zero stock, 1 stock, 2 stock and 3 stock items. Since the probability of requiring 1, 2 and 3 are 0.10, 0.04, 0.02, respectively, the probability of spare not required is obtained by subtracting the total of 0.16 from the total probability of 1 or 0.84. Let us now study the effect of each policy by assuming the price of spare part, including carrying charges as ₹1000 and the stockout cost as ₹20,000. We get the following pay-off table (Table 27.1) in this context. The total cost is obtained by multiplying the cost figures with the respective probabilities.

Table 27.1 Pay-off Table

Spare Stock	Possible Requirement	Cost (₹)	Probability Requirements	Total Cost (₹)
0	0	0	0.84	
	1	20,000	0.10	
	2	40,000	0.04	4800
	3	60,000	0.02	
1	0	1000	0.84	
	1	1000	0.10	
	2	21,000	0.04	2600
	3	41,000	0.02	
2	0	2000	0.84	
	1	2000	0.10	
	2	2000	0.04	2400
	3	22,000	0.02	
3	0	3000	0.84	
	1	3000	0.10	
	2	3000	0.04	3000
	3	3000	0.02	

The total cost is optimum for two spare parts and hence the stocking policy for this insurance spare part is 2.

We can illustrate the above example with the same probabilities for a general understocking cost. Let this general understocking cost be represented by ₹10,000 of A and let us analyse the equations for stocking policies of 0, 1 or 2. The expected cost in this case with zero stock policy will be as follows:

$$(0.84)(0) + (0.1)A + (0.04)(2A) + (0.02)3A = 0.1A + 0.08A + 0.06A = 0.24A$$

Since the expected cost per spare part will be $1000 + 0.03A$, if we consider the cost effectiveness of the policies of 0 and 1 stock, they are equal from the indifference point of view. For optimum A, we get

$$0.24A = 1000 + .08A, \quad 0.16A = 1000, \quad A = \frac{1000}{0.16} \text{ or } 6250$$

Up to an understocking cost of ₹6250, it is not necessary to carry any stock in the stores in this example. Similarly, if we consider the cost effectiveness of 1 spare and 2 spares stock as equal, then we get the optimum

$$.02A + 2000 = .08A + 1000, \quad 0.06A = 1000$$

$$A = 1000/0.06 \text{ or } ₹16,666.$$

If understocking cost is more than ₹16,666 then we carry items. Similarly we consider the cost effectiveness of stocking 2 and 3 spares as equal to get optimum stocking policies. We have $0.02A + 2000 = 3000$, $A = 1000/.02$ or 50,000. Only for understocking cost of more than ₹50,000 we get three sets of capital spares. Thus, the stocking policies of 0, 1, 2 and 3 spare parts for insurance categories can be determined with the given probabilities of requirement and estimating the stockout cost. But the mechanical engineer, who is in-charge of the productivity of equipment, will not be interested at all in the logical mathematical explanation, until and unless the top management takes the hard decision. Its repercussion of the loss of production due to nonavailability of the critical insurance spare part can be seen on the market, competitiveness and customers.

Chapter 28

Rotable Spares

CONCEPT OF ROTABLES

Rotables and floats are associated with the transport sector, namely air, rail, road and shipping as well as the Army, Navy and Air Force. The concept of rotables is also applicable in mining sector and processing industries. The part has to be removed from assemblies for overhauling and repairs should be done after a specified number of hours or after its failure. The repaired spare assembly, after overhauling and testing, is fitted back in the main equipment. Spare part assemblies such as rotors, pumps, engines, compressors, motors and clutch assemblies are typical examples of rotable spares in the industry. These rotable spare parts are not always scrapped, but must be repaired and used again. Factors pertaining to these rotables are considered while developing stocking policies for such spare parts. Rotables are also called "Floats".

In order to reduce the down time of the equipment to the barest minimum, replacement should be done in such a way that a stock of rotable spare part has to be maintained in the stores or in the base repair shop. In the event of a failure of rotable, it is repaired from the reserve and the failed item is repaired and brought to the stores.

FINANCIAL CONSIDERATIONS

In practice, the rotable spares are expensive and hence the inventory carrying charges of these parts are quite high. Similarly, the stockout cost or cost due to non-availability of the rotable spare part is also very high. Some of these spares are even made to order, with resultant increase in lead time, and not carried by the supplier on the shelf. The choice before the executive is either increase the equipment down time while the rotable part is repaired or alternatively fix another readily available rotable spare on the equipment, to make it immediately functional while the part is repaired. In the first alternative, since the equipment is idle when the rotable part is repaired, it is more costly in actual practice. Hence let us consider only the second alternative. The rotable spares usually have a high unit cost and share a major portion of total investment in spares and hence optimum allocation of the working capital becomes important in organisations where budgets are allocated separately for spares.

QUEUING APPROACH

In order to solve the stocking policy of rotable spares, several approaches are available; we shall now discuss one of these, viz. the queuing approach. Let us consider a rotable part which has an average monthly demand rate for the units. Let us assume the total repair cycle time, including time for removal, transit to and from the repair base workshop, waiting time at the base shop, actual repair time, etc. to be two months. This implies that a total stock of 20 units has to be maintained and reserves for the uncertainties in the above estimate have to be added. Queuing theory builds rigorousness to this simplistic approach.

Queues or waiting lines are commonly encountered in everyday life in the consumer/industrial scene. Typical illustrations include people waiting in the hospital, ticket booking counters, super market, petrol pump and telephones. In all the above cases, a flow of customers towards the service facility is turned into a queue as it is not possible to serve all customers immediately on arrival simultaneously. If the service facility can serve customers immediately on arrival, there will be no queue at all. A queue is formed in the absence of a perfect balance between the arrival of customers and the service facility.

The principles of queuing approach reduce the waiting time to a minimum after considering the service cost and waiting cost.

Let us now consider a machine shop having a large number of identical machines in operation. A demand for rotable spares arises due to the breakdown of the part from time to time. In the queuing theory, these breakdowns, necessitating rotable spares replacement, are treated like customers. If adequate spares are not available to meet the demand, then a queue of customers or breakdowns or services occurs according to the queuing theory. The point where service is provided to the customer is the service station. If we decide to hold too many rotable spares inventory in the workshop, then the inventory carrying cost is very high. On the other hand, if only a few stocks are held, then we face the risk of formation of a queue of customers or a breakdown, resulting in production losses and loss of image of the company.

The theory of queues helps to determine the optimal balance between the servers and customers which will keep the total cost minimum. The economy involved is best understood by an example of airline repair shop. Here, the customer is an aircraft such as Boeing which is ground for want of a spare part and the server is engine of the aircraft. The application of queueing theory has got a special significance in such cases.

APPLICATION OF QUEUING MODEL

Several approaches are available, but the queueing is handy. In order to apply the queueing theory to spare parts control, it is important to have a knowledge of certain basic principles involved in the theory. It can be seen that in a spare parts system, the 'customers' or breakdowns leading to a demand for spare parts occur at random intervals with no specific time pattern. For applying queueing theory, the following data are needed:

(a) What is the pattern of arrival of customers? Does it follow any distribution? In the context of spare parts management, this would mean certain failure history data resulting in a demand for spare part.

(b) What is the service time from the commencement to the completion of service? Again with reference to spare parts this would mean location of spare parts stores, lead times involved in procurement if the spare is not available, etc.

(c) The discipline of the queue system—this will again depend upon the urgency of a particular spare part demand for a critical machine. In some cases the 'first come first served' policy is adopted.

(d) Demand for equipment, number of equipments, cost of replacement when spare is available, repairing cost, cost of repairing when replacement (spare) is not available and purchase cost of spare part.

If the data mentioned above is collected, the queueing theory can be applied with certain simplifying assumptions to answer inter-alia the following important aspects:

(a) the average number of customers waiting for service—this may be the number of spare parts demanded not 'served'

(b) the average waiting time of the customer, and

(c) the average idle time of facility. This may refer to the idle investment in spare parts.

SOLUTIONS TO QUEUING QUESTIONS

There are two basic approaches to solve a queueing problem, namely, the *mathematical approach* and the *simulation approach*. In this chapter we will, however, focus our attention to the mathematical approach only.

The structure of the queueing problem can be construed as customers/breakdowns arriving to use the service facility at the repair shop and the customers are served in the sense that the equipments are fitted with new spare part or repaired spare part, depending upon the spare parts maintained in the service facility. As the spare parts consumption follow the Poisson distribution in many industries, the arrival rate also follows Poisson distribution in the queueing theory. Here, the probability of an arrival of a customer/breakdown at any time is independent of the previous arrivals. The service time is assumed to be an exponential in the queueing models.

Let 'A' represent the mean arrival rate of customers and 'S' represent the mean service race. It is clear that when 'A' is greater than 'S', there will be ever-lengthening queue of customers. This is not a practical situation and hence will not be discussed further. The ratio A/S represents the measure of the utilisation of the service facility and it actually expresses the probability of a customer having to wait for service.

The relevant formulae have to be suitably modified and used depending on the one service facility or parallel service facilities. If there are 'm' parallel service facilities then, the service rate will depend upon whether all the service facilities are busy serving the customers or some are idle. Thus, when 'n' customers are present, i.e. when 'n' breakdown occurs—the service rate will be 'n, S', where the value of 'n' is less than 'm', i.e. the number of service facilities. However, if the number of customers present exceeds 'm', the service rate will be 'm, S'. These modifications can be used to solve the queueing problems with multiple parallel service facilities.

QUEUEING AND ROTABLES

In India, the application of queueing theory to rotable spares is being practised in Indian Railways, Airlines, Shipping and Road Transport as well as in some reconditioning shops. In all the above cases, the defective engine is removed in the workshop and a spare engine is fitted.

Rotable spares are very expensive in nature and the cost of stockout of such spares are indeed very high. Some of the spares are even made to order by the supplier and not carried by the supplier on the shelf. Typical examples are large electric motors, compressors, special jigs, selected assemblies and tools. Hence, for such spares the carrying costs are very high and the out of stock or the shortage cost is also very high. However, the demand for such spare parts is low for any single machine. Let us consider the following example:

Cost of the spare part	= ₹40,000
Delivery lead time	= 4 months
Estimated understocking cost	= ₹5,00,000
Estimated carrying cost	= 30% per year

Past consumption data for this spare part was analysed and found to be 0.5 units per month. It is desired to fix the optimum stocking policy. The analysis of the delivery lead time shows an exponential distribution and the demand was found to follow Poisson distribution.

The example above is a typical spare parts problem faced in industry.

$$\text{Average demand (A)} = 0.5 \text{ per month}$$

$$\text{Average delivery lead time } \frac{1}{S} = 4 \text{ months}$$

$$\therefore \quad \text{Average demand during lead time} = \frac{A}{S} = 2 \text{ units}$$

Let us assume that a stock of 'k' units of spares be carried for minimum total cost. The cost of carrying one spare part with 30% carrying charges is obtained as follows:

$$40,000 \times 0.30 \times \frac{4}{12} = ₹4000$$

Hence, the cost of carrying k spare parts = 4000 k

The other cost is that of stockout. This is where queueing theory is useful to us. The probability of a stockout when k units are carried in-stock can be proved to be, according to queueing theory:

Probability of stockout, i.e. no service when k units are available for $\frac{A}{S} = 2$ is given by the following expression.

$$\frac{1}{\left[1 + \dfrac{\underline{k-1}(k-2)}{2^y}\right]\left[1 + 2 + 2^2 \dfrac{\cdots + 2^{k-1}}{\underline{2} \quad \underline{k-1}}\right]k}$$

POLICY FOR ROTABLES

Using Poisson tables the value of k (stockout) can be easily determined. They are shown below.

Y (Stockout)	Probability (Stockout)
3	0.444
4	0.174
5	0.060
6	0.018
7	0.005
8	0.002

Total cost of alternative stocking policies is carrying cost plus understocking cost. For any stocking policy in the service repair shop, the understocking cost is given by 5,00,000 multiplied by probability of stockout.

The total cost of stocking 3 units will be:

$$(4000 \times 3) + 5{,}00{,}000 \,(0.444) = ₹ \, 412{,}000$$

Similarly for the alternative policies costs are shown:

Stocking policy	Carrying cost	Understocking cost	Total cost
3	12,000	2,22,000	2,34,000
4	16,000	87,000	1,03,000
5	20,000	30,000	50,000
6	24,000	9,000	33,000
7	28,000	2,000	30,000
8	32,000	1,000	33,000

Therefore it is optimal to stock six units of the spare parts and order as and when consumption takes place. Therefore, it is clear that queueing theory helps in arriving at optimal stocking policies for spare parts.

Let us now consider another example where overhauling repair is also taken into account. An airline company has determined based on past records that the engine of the aircrafts are demanded at the rate of 1 per month. Demand pattern was found to be closed approximated to Poisson distribution. The repairing shop is in a position to replace the engine in 1 month, when there is no stock of fresh/repaired engines in stock. The management of the company wishes that the stock-out probability should not be more than 2%. Let us now determine the optimum stock level to be carried.

$$\text{Average demand (arrivals) A} = 1 \text{ engine per month}$$

$$\text{Replacement lead time} = 1 \text{ month}$$

Hence the average lead time demand is $1 \times 1 = 1$ engine

If we carry h engines in stock plus in order, then we know that the probability of stockout is given by (for h service channels) the following equation.

$$\text{Probability (Stockout)} = \cfrac{1}{\left[1 + \cfrac{\lfloor h-2\,(h-1)}{1^h}\right]\left[\cfrac{1+1+1^2}{\lfloor 1 \lfloor 2} + \cdots + \cfrac{1^{h-1}}{\lfloor h-1}\right]k}$$

For different stocking policies the probabilities of stockouts are worked out and shown in the table below:

h	Probability of stockout (%)
1	100
2	33
3	17
4	6
5	1

Hence, it is clear that the stocking policy to ensure that stockout is less than 2% is to carry 5 spare parts in stock.

A–Z ISSUES IN QUEUING

It is necessary to know the arrival of customers or breakdowns, leading to demand for spare parts at random intervals with a specific pattern. Answers to the following questions will lovable spare parts executive to understand the application of queueing to rotables.

(a) What is the pattern of arrival of customers or breakdowns?
(b) Does this pattern follow any recognised statistical probability distribution?
(c) How does the arrival distribution look like?
(d) What is the inter-arrival—time between two arrivals—distribution?
(e) Is the time between two arrivals equal or unequal?
(f) What is the average number of customers waiting per unit of time?
(g) What is the mean time between arrivals?
(h) What is life service time from the commencement to the completion of service?
(i) What is the service mechanism employed in the base workshop?
(j) What is the availability and capacity of the infrastructure service facility and duration of service time in the base workshop?
(k) Where are the locations of part stores?
(l) Discuss the estimated lead time for procurement?
(m) What is the policy of service in the base repair shop?
(n) What is the queue discipline at the service station?
(o) Is the policy strictly based on first come first served or first in first out?
(p) Is there any priority of jumping the queue?
(q) If yes, what is the basis of priority?
(r) What is the servicing policy of catering to emergencies or criticalities?

(s) When is selective service in a random order (SIRD) adopted?
(t) Is a policy of last in first out (LIFO) ever adopted?
(u) If so, when and why?
(v) How is the data on history cards and failure data compiled?
(w) What is the demand pattern for the equipment needing the rotable spare part?
(x) Do you have data on reliability of reconditioned part?
(y) What is the total number of such identical machines?
(z) Has the cost of replacement been compared with the cost of repair?

It is advisable to get answers to the above influencing parameters on the rotable spare part before attempting to apply the queuing approach. The queuing analysis can help the management to study the average number of rotable parts arriving for service, the average waiting time of the rotable part before repair, repair time, average working capital locked up in the inventory of rotable parts, etc. Based on the answers to the questions and data generated, it is possible to expand the repair facility, adding balancing equipment, optimum number of repair etc. in order to improve the service.

When there are parallel service facilities which would be utilised by customers, the customer may join a queue and finding some other queue moving faster may change over to that queue, Queuing approach also answers other complex questions like the customer not entering the lengthy queue or entering the queue but leaving it because it is too lengthy. In such cases, e.g. in telephone industry, sophisticated techniques like Erlang distribution, Markov Chains, etc., which are beyond the scope of this text, are applied.

The structure of queuing approach can be seen as machines with breakdowns arriving at the service facility in the repair shop. The customers are served in the sense that the equipments are fitted with the repaired spare part, depending upon the number of rotable spares available in the service facility. Just as the spare parts consumption follows the Poisson distribution, in many cases the arrival rate and service rate follow the Poisson distribution. Here the probability of arrival of a customer/breakdown at any time is assumed to be independent of previous arrivals. The service time is assumed to be an exponential distribution in the queuing approach.

If the number of customers arriving per unit of time is greater than the number of customers served, then there will be an ever lengthening queue of customers and it is not possible to solve the problem. But if the arrival rate is less than the service rate, the system will settle down to a steady state. The solution of a queuing problem consists in selecting those factors which can be controlled. The queuing approach is not only helpful in identifying the optimum rotable spares to be maintained at the service station, but also the number of servicing staff in the area of service after sales.

Chapter 29

Overhauling and PERT (Programme Evaluation and Review Technique)

OVERHAULING SPARES CONCEPT

The term *overhauling* signifies giving a new lease of life to an equipment or part of it, when organisations undertake periodic overhauling of equipments on planned basis after predetermined hours of working. The timing and extent of overhauling are normally predictable with reasonable accuracy. The estimate of the overhauling spares requirement is confined only to those spares which are to be replaced. This uncertainty in estimating the requirement can be minimised by proper maintenance of history cards and log books to watch earlier replacements. The suppliers' complete catalogue on overhauling practices will also serve as a starting point for estimating the requirements. The question whether a complete module or assembly has to be replaced has to be answered in advance, based on technical knowledge. It is evident that the time and extent of overhaul are normally predictable with reasonable accuracy. The main difference takes place when system fails. Overhauling is a preventive action to avoid failure of components resulting in loss.

REQUIREMENT OF OVERHAULING SPARES

The corporate policy with regard to repairing/discarding the defective part has an important bearing on the procurement of spares. It is advantageous to change the mating part even if some parts are serviceable, Rubber items such as fan belts, gaskets, and hoses are replaced fully. It is preferable to replace slightly worn-out, cheap internal components during the overhaul as their replacement later will be costly. The repair kits, wherever provided, should be replaced *in toto*. The knowledge, skill, experience and expertise of maintenance staff should always be utilised in determining the requirement planning of the overhauling spares. This procedure of consulting maintenance staff in advance is helpful for resolving issues such as standardisation of hardware, problems of replacement versus repair, undersize bearings, etc.

Admittedly, the requirement planning of the first overhaul of an equipment is rather difficult, and the maintenance staff lend to play safe by giving conservative estimates. Perhaps visits to similar plants may reduce the error of estimation, it will be noted that every item is critical in the overhauling spares. Hence ABC analysis based on estimated consumption rupee value is done in order to enable the maintenance engineer to have a second look on the requirements. If the lead times of the different items are estimated in advance, then the procurement department can schedule the deliveries only when they are needed. Such a system will ensure minimum locking up of the working capital.

The overhauling schedules and turnaround time are usually determined from the preventive maintenance reports and planned at least six months in advance to coordinate with production scheduling and procurement planning. The timing of overhauling is informed well in advance to all concerned. It is not uncommon to subcontract the overhauling of motors, engines, compressors and pumps to reputed repair contractors. In order to effectively plan, coordinate and carry out overhauling, PERT/CPM/Network analysis techniques are used as analytical tools. These techniques enable the purchase department to carry out studies on make/buy/lease/repair, and the marketing department to slick to delivery schedules for supply of capital equipment and for service after sales. These are useful follow-up measures and show how critical the completion of the overhauling is. These techniques highlight the inter-dependence between activities, review the dynamic nature of overhauling problems, and indicate the criticality of the jobs with uncertainties inherent in the time estimates.

The overhaul spares are those which are made use of when the machine is completely stopped periodically and a new lease of life is given to the equipment after assembly with replacement. During overhaul some parts, such as track links of track can be built up, some parts such as track bushes can be rotated and adjusted and some require outright replacement.

The requirement will be only for those parts which are to be replaced completely but this requirement cannot be assessed correctly. This uncertainty can be minimized by proper maintenance of history sheets and log books to watch earlier replacements and arrive at a pattern of replacement. Manufacturers' complete catalogue on overhauling practices will also serve as a starting point. The question as to whether a complete assembly/module or a part will only be replaced during the overhaul, should be answered technically in advance. Admittedly this will depend upon the degree of stripping the assembly as well as the skills available in the company. Whenever the repair kits are provided, it will be replaced in toto.

Sometimes it is advantageous to change the making part even if some parts are serviceable. For instance, textile and rubber parts, such as fan belts, gaskets, hoses, washers, etc. are usually replaced 100%. It is definitely advantageous to replace slightly worn out cheap internal components during the overhaul as their replacement later will be time-consuming and will be very costly. The knowledge and expertise of the maintenance staff as well as past data should be used to supplement the above steps for the requirements planning of the overhaul spares. This procedure is helpful for resolving issues such as standardisation of hardware, prevention of oversize bearing to be used, problems of replacement for repair etc.

The expected requirement of overhauling for all the items is obtained to the above fashion. It will be noted that every item is vital in overhauling, hence normally VED analysis is not readily applicable. However, ABC analysis based on estimated consumption rupee value is done so that the maintenance engineer can take steps to reduce the consumption of A category spares in the long run by improving the repair technology. If the lead times of the different items are known then the purchasing department could plan that 'A' items reach the overhauling spot only when they are needed. This will ensure minimum locking up of working capital.

The overhauling schedules are usually determined from the preventive maintenance reports and planned at least six months in advance to coordinate with production scheduling and market planning. The timing of overhauling is informed well in advance to concerned. In large organization and multiplant corporation, long-range planning for capital expansion is done by centralized planning wing and then meshed into the overhauling schedules of individual units. It is not uncommon to subcontract the overhauling of motors, engines, compressors, and pumps to reputed repair contractors.

PERT'S CONCEPT

In order to effectively plan, coordinate, and carry out overhauling, PERT is a very useful analytical tool. PERT—Programme Evaluation Review Technique—also helps the purchasing department to carry out/make/buy studies and the sales department to tender for delivery periods for supply of capital equipment and for after sales service. PERT is very useful to follow up as it pinpoints the critical areas threatening the completion of any project.

PERT highlights the interdependence between activities, reviews the dynamic nature of project problem and indicates the criticality of the job with the uncertainties inherent in the estimates.

Any project/overhauling programme is first split into identifiable activities showing all major jobs of the project. This is called work-break structure. The activities—consuming time and resources are detailed so that the work content is clearly known. The culmination of the activities is known as event, which is a point in time. PERT network consists of events and activities. It is also observed that the activities can be further splitted—as purchasing, receipt of enquiries, comparative study, placement of order, chasing order, inspection, transport to site, etc.

In PERT, the activities are interlinked logically, i.e., order precodes follow up whole receipt succeeds ordering. The responsibility centres are fixed for each activity and usually 3 time estimates are collected from the centres. These are optimistic, most likely and pessimistic. The expected time is arrived at as a + 4m + b/6 for each activity.

Based on theoretical as well as practical consideration, the network is then cast and in the forward pass earliest possible times are calculated. From this, the start and the finish dates are specified. Then from the backward pass, the latest possible dates are obtained. The difference between the latest and earliest completion times gives the slack. The path linking activities with zero slack is called the critical path. The network is then circulated for acceptance. Modification, if any, is incorporated in the network to fix the overhauling schedule. Feedback and updating is very important for proper utilisation of PERT. There should be a regular review of the progress on activities. Delays may cause more activities to go critical necessitating reallocation of resources.

An actual example of using PERT for overhauling the aircraft is given in the following two exhibits. The events are numbered and the major activities are shown in Table 29.1 while Table 29.2 shows the network with critical path indicated by two lines. The dotted line indicates the dummy activities which do not consume resources, but completion of which depends on the beginning of the next activity.

Table 29.1 Aircraft Overhaul

Activity Number*	Activity	Expected Time† (Hours)
1–2	Prepare for inspection	1
2–15	Unpanel engines	3
15–16	Clean port engine	1
16–17	Inspect port engine	24
17–30	Rectify port engine	36
30–31	General check	1
15–20	Clean starboard engine	1
20–23	Inspect starboard engine	20

(Contd.)

Table 29.1 Aircraft Overhaul (*Contd.*)

Activity Number*	Activity	Expected Time† (Hours)
23–30	Rectify starboard engine	30
2–12	Unpanel for lubrication	1
12–14	Check instruments	18
14–30	Rectify instruments	24
2–3	Inspect airframe	10
3–10	Rectify airframe	28
2–4	Clean and remove wheels	4
4–5	Inspect main wheels	7
5–7	Rectify main wheel	9
	Test retraction-mechanism	1
7–10	Rectify retraction mechanism	3
4–6	Inspect nose wheel	6
6–7	Rectify nose wheel	8
10–11	Recharge emergency system	1
21–22	Inspect fire extinguisher	2
22–30	Rectify fire extinguisher	11
12–13	Lubricate	3
13–30	Check lubrication	1

* These numbers are used in the PERT network in Figure 29.1.
† The expected time in hours is arrived at after getting the three time estimates and applying the formula (a + 4m + b)/6.

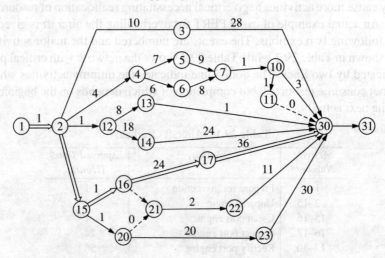

Figure 29.1 Critical path (1) (2) (15) (16) (17) (30).

PERT Programme evaluation review technique demands 3 times estimates research activity called optimistic, most likely and pessimistic time and expected time is calculated as o + m + p which is used. Hopler drawing the net worth, the longest path is called critical path (CP) is identified. You have to buy time as cheap as possible by crashing, in the critical path.

DETAILS OF PERT

PERT (Programme Evaluation and Review Technique) is a management tool which can be used very profitably for spares planning. It pinpoints the critical activities, resource allocation and the cost of the project. This technique (i) highlights the interdependencies between activities (ii) reflects the dynamic nature of project progress (iii) indicates criticality of jobs and (iv) indicates uncertainties inherent in these estimates.

PROCEDURE FOR PERT

Any programme or project is split into identifiable activities showing all major jobs of the project, i.e. develop a work breakdown structure. The activities are detailed so that the work content is known. For example, if purchasing is a major job it can be split as receipts of enquiries, comparative study, placement of order, chasing of order, inspection, transport to site, etc.

The activities are interlinked, i.e. ordering precedes chasing while receipt succeeds ordering, etc.

The responsibility centres are fixed for each activity and the time estimates for the activities are collected from the respective responsibility centres.

The network is cast and in the forward pass earliest possible times are calculated. Then the start and finish dates are specified, and from the backward pass the latest possible dates are obtained.

The difference between the latest and earliest completion times gives the slack. All the activities which have zero slack are in the critical path. (It may so happen sometimes that in the first iteration some of the slacks are negative. This implies that the project cannot be completed in the scheduled time. Hence either the project completion date is extended or the activity times are reduced— (i.e. activities are crashed by paying an extra price).

The network is then circulated for acceptance. Modifications, if any, are made.

Feedback is very important for proper utilisation of PERT. There should be a regular review of the progress on activities. Delays may cause more activities to go critical. Here a reallocation of resources will be needed.

Usually 3-time estimates are taken (a = optimistic estimate, m = most likely estimate, and b = pessimistic estimate).

Then the mean is given by

$$\frac{a + 4m + b}{6}$$

and often this is resorted to as the expected time. On theoretical as well as practical considerations, the above formula holds good. (Readers can refer to standard textbooks on PERT for details.)

EXERCISE ON PERT

Telengana Petroleum Limited is an oil refinery with an annual capacity for processing 2.5 million tones of crude. In the annual schedule 3 weeks are set apart for a shut down during which period major overhauling of all the process plants is undertaken.

There are 13 plants in the refinery. The crude which is pumped to the refinery is first allowed to settle in the tank. The crude is split into light, medium and heavy distillates in the atmospheric and vacuum distillation plant. The processing of light distillate yields Naphtha, liquid petroleum gas and motor spirit. The middle distillate yields high speed and low speed diesels, kerosene and aviation turbine fuel. The heavy distillate yields lube stock and asphalt.

During the continuous process there may be leakages, mild bursts etc. which will render a particular section unworkable till the faults are rectified.

The annual shutdown has to be carefully planned since the whole plant is shutdown and any extension of the 3 week duration will increase the losses. In fact the attempt is to keep this period as short as possible. PERT is used very effectively to help the follow-up. The planning starts 10 to 12 months ahead. The attempt is to stock all possible requirements at least 2 months ahead of the scheduled shutdown. During this period the materials manager may have to procure anywhere between 6000 to 8000 items, out of which 40 to 50% may be imported. Items procured for the shutdown are not issued for regular use.

The whole shutdown operation is planned and coordinated by the planning cell (responsibility code—Plg). The maintenance wing gives the necessary requirements along with specifications (responsibility code—MTCE). The Purchase Section undertakes procurement action (responsibility code—PVR) and the stores holds the material in stock (responsibility—STS).

The major jobs are:

		Responsibility
(i)	Initial Plan	Planning
(ii)	Specifying the needs	Maintenance
(iii)	Verifying the needs	Planning
(iv)	Procurement	Purchase
(v)	Storage	Stores
(vi)	Overhauling	Maintenance

The skeleton network will be as follows:

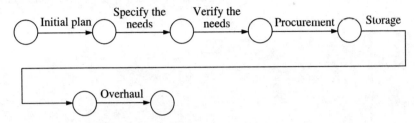

Figure 29.2 Skeleton network.

Now to draw the PERT network each job has to be split into activities which can be identified and whose progress can be monitored on a time scale. The following table gives an idea of the activities and the expected time.

Table 29.2 Data for Network

Major Job	Activities	Duration Expected (in weeks)	Resp.
Planning	Prepare past data	1	PLG
	Meeting	1	PLG
	Draw up plans	1	PLG
	Meeting	1	PLG
	Get approval of final plans and information	1	PLG
Specifying the needs	Study the plants/history	4	MTCE
	Specify lead items	4	MTCE
	Specify other items	4	MTCE
	Study the requisition and stock positions	2	PLG
Verify the needs	Release purchase requisitions	2	PLG
Procurement	Call for enquiries/Tender	4	PUR
	Discussions	4	PUR
	Order	2	PUR
	Procurement	Variable depends on items	PUR
	Shipment	4 (Local)	PUR
		14 (Foreign)	PUR
	Import License	4	PUR
	FE clearance	4	PUR
	LC Raising	4	PUR
Storage	Receipt and inspection	1	STS
	Storage	2	STS
Overhauling	Overhauling	3	MTCE

Now, draw a PERT network for monitoring the shutdown schedule.

Drawing the PERT (Programme Evaluation and Review Technique)

The skeleton network will be as follows:

Figure 29.2 Skeleton network.

Now to draw the PERT network, each job has to be split into activities which can be identified and whose progress can be monitored on a time scale. The following table gives an idea of the activities and the expected time.

Table 29.2 Data for Network

Main Job	Activities	Duration Expected (in weeks)	Resp
Planning	Prepare part chart	1	PLG
	Meeting	1	PLG
	Draw up plans	1	PLG
	Meeting	1	PLG
	Get approval of final plans and information	1	PLG
Specifying the needs	Study the plants/activity	2	MTCE
	Specify lead items	1	MTCE
	Specify other items	1	MTCE
	Study the requisition and stock positions	2	PLG
Verify the needs	Release purchase requisitions	2	PLG
Procurement	Call for enquiries/Tender	2	PUR
	Discussions	1	PUR
	Order	2	PUR
	Production	Variable, depends on item	PUR
	Shipment	4 (Local) / 14 (Foreign)	PUR
	Import license	4	PUR
	LC issuance	4	PUR
	LC Advising	4	PUR
Storage	Receipt and inspection	1	STS
	Storage		STS
Overhauling	Overhauling	2	MCE

Now, draw a PERT network for monitoring the shutdown schedule.

Section IV
Related Issues on Spare Parts

30. Reliability and Quality
31. Procurement of Spares
32. Logistics and Warehousing
33. Pricing and Marketing of Spares
34. ASS (After Sales Service)
35. Multiechelon Distribution
36. Management of Obsolescence
37. Reconditioning
38. Information System for Spares
39. Evaluation of Spares

Chapter 30

Reliability and Quality

NEED FOR RELIABILITY

Today, India has at its disposal sophisticated equipments, complex industrial systems, and advanced, precision weapon systems, which require complex spare parts. This process of increasing sophistication sometimes leads to production of defective parts while manufacturing, due to omissions and commissions at various stages. In this context of increased malfunctioning of the manufactured spare parts, particularly in electronic, aeronautic, and astronomic equipments, reliability engineering enables a logical and indepth study of part failure.

The first attempt to formulate reliability studies was made in the U.S. department of Defence. An advisory group on reliability of electronic equipment (AGREE) was formed to study reliability requirement of the critical hardware, and the report gave an impetus to reliability movement in all spheres.

CONCEPT OF RELIABILITY

Reliability is the technology concerned with production, control, maintenance, measurement and continuous reduction of equipment failure. It is the probability of a device performing its function adequately for the period of time intended under the stipulated operating conditions. The elements summarising the major provisions of a reliability specifications are: definition of the device, information system at production stage, modification status, criteria of adequate performance, basis for computing time description of operating conditions, maintenance conditions, definition of failure, description of sampling procedure etc. Reliability is the capability of an equipment that it does not breakdown during operation.

Introduction of reliability into a part is basically a design feature which has to be implemented at the manufacturing stage by strict quality control measures. In order to increase the reliability of a system, usually the principle of redundancy or using a parallel system which will take over in the event of a failure is introduced. This implies the existence of more than an item for a given task and hence a cost-benefit analysis has to be carried out before increasing reliability.

FAILURE ANALYSIS

Most equipments fail as some parts in it fail. Failure rate is also known as hazard rate. The failure analysis identifies weak components as also the patterns of occurrence of failures. Most of the weak components are the result of defective manufacturing for the following reasons:

(a) Underestimating (by the designer) the operational and environmental stress the components are subjected to
(b) Overrating the ability of the component to withstand stress
(c) Inferior quality of raw material used in manufacturing
(d) Overlooking or not locating the defect in manufacturing
(e) Poor maintenance
(f) Normal wear and tear
(g) Catastrophic sudden failures
(h) Random failure
(i) Human failure.

The following factors may help understand the defects leading to the failure:

(a) What element or material is the component made of?
(b) What is the component's function? What does it do? How does it move? How does it transmit?
(c) Maintainability and stress conditions
(d) What is the function of the equipment in which this part is a component?
(e) What are the stresses this weak component is subjected to?
(f) Environmental conditions, power supply, skill mix of operations, atmospheric conditions, pollution control etc.
(g) Time at which the stress takes place—starting, during job changes, continuously, frequently, infrequently.

The objective of failure information system must be to do the following:

(a) prevent future defects, early recognition, efficient repair,
(b) modify operation, and
(c) introduce design changes and modify the maintenance.

For this purpose the failure information system must answer the queries given below:

(i) When did the failure occur?
(ii) When was it attended to?
(iii) When did it become operational?
(iv) What was the defect?
(v) Which component caused the failure?
(vi) What caused the defect?
(vii) What was the material used?
(viii) How many man hours were spent?
(ix) What was the interval between two similar failures?
(x) What were the warantee claims?
(xi) What was the fault diagnosis?

FAILURE PARAMETERS

Failure rate, $r(t)$ is the fundamental failure parameter. It is stated as failures per hour and usually has a low value like .001 or one failure per 1000 hours. Mean average time between failures (MTBF) is the reciprocal of failure rate and the larger the value of MTBF, the greater is the reliability. The reliability of a system will depend upon the length of time it has been in service and hence the distribution of the time to failure of a spare part becomes important. Good failure analysis is the most important step in formulating a sound, economical maintenance programme and the failure analysis depends upon determining the failure rate of an equipment at every point of time during its useful life. The total reliability of a process equipment consists of the product of process, design, equipment and construction reliability. The user has very little control in the first two components for operating systems.

Reliability refers to a single equipment such as a motor, pump, vessel, boiler or furnace or a part like shaft, impeller, valve or tube. The reliability of an equipment system is known by the term *equipment availability*. Equipment availability is defined as the ratio of MTBS to (MTBS + MTD), where MTBS is mean time between shutdowns and MTD is mean downtime. The more the downtime, the lower will be the equipment availability.

When an equipment is made more complex, by assembling several parts, each with its own survival rate $P_s(t)$, then the probability of survival of the total equipment is $[P_s(t)]^n$. Since $P_s(t)$ is expressed as a decimal fraction and is less than one, the probability of survival of the equipment decreases rapidly with an increase in the number of components and the probability of failure increases at a corresponding rate. The objective of reliability engineering is the need to make very complex systems operate with satisfactory reliability. In most cases Weibull distribution is applicable and we have $P_s(t) = \exp(-t/v)^k$, where t is time variable, v is the location parameter or characteristic age at failure and k is the shape parameter. It is noted that increasing v implies improved reliability; k measures the fluctuations and is used to calculate the variance when $k = 1$, $P_s(t) = -(t/v)$ or $\exp(-rt)$, where r is $1/v$ and v is MTBF.

If a number of items of a system are connected together in such a way that the failure of any one item causes system failure, then the items are functionally in series. Each item must survive for the system to survive and the system is no better than the item with the lowest survival probability. On the other hand, if a number of items are connected together such that the system does not fail until all items have failed, then the equipment items are in parallel. As long as one item operates, the system does not fail, and the reliability of a parallel system is the probability that one or more of the items will survive a specified operating time.

It is clear that grouping equipment items in series combination reduces system reliability, while grouping them in parallel combination increases system reliability. System reliability is always less than the least single component reliability when in series, while the parallel reliability is always larger than the largest single component reliability.

MEAN TIME BETWEEN FAILURE (MTBF)

MTBF mean time between failures and MTTR mean time to repair are used by manufacturers and customers without the same concept. Customer misapplication or wrong use must not be considered a failure only when inbuilt factors are overlooked by the designers—which is very tough to point out—is a real failure, risking the product usage. If an LED (Light emitting diode) of a computer fails, it is considered a failure, even though it has not impacted the operation of computers. Is expected wear out of a consumable item such as battery—considered a failure? Are shipping damages considered a failure? Question like the above arise before taking reliability issue. So MTBF has been a daunting subject for decisions over hundred years. An area, where this is evident, is the design of rocket/space/equipments using computers for critical facilities, as its impact is felt all over. The above discussions lead to the following definition of MTBF. The termination of the ability of any individual component to perform its required function without terminating the equipment.

MTTR is obtained as total down time number of breakdowns. For example, if a redundant in a rail array fails, the failure does not prevent the rail array from performing its required function of supplying critical data at any time. How can the disk failures not prevent a component of the disc array from performing its required function of supplying storage. According to definition (1) it is not a failure, but it is according to definition (2). The MTBF is typically part of a model that assumes the failed system is immediately repaired—zero elapsed time. (Reference may be made to Chapter 4)

MEAN TIME TO REPAIR (MTTR) AND LOG NORMAL DISTRIBUTION

We have another useful parameter known as MTTR (mean time to repair). This gives the estimated average disposal time, required to perform corrective maintenance, after isolating the fault. Fault correction consists of disassembly, interchange reassembly, alignment, and check out for repairs. The long down time would occur when diagnosis is difficult or removing a complex part is complicated may be due to rusting of nuts/bolts. Collection of such data is useful. The system MTTR may also dictate changes in the design of the system, to meet the turn around time. Criteria for life support systems of space station, in calculating extra vehicular capacity activity time for astronauts to repair a system. MTTR is defined as the average time necessary to trouble shoot, remove, repair and replace a failed system, component. An interval estimate for MTTR can be developed from the mean of the sample data within a lower and upper limit with a confidence band. For example, from a sample data set, one can find 90 per cent confidence that the range 3.2 to 4.2 will contain the population mean.

The most commonly pattern used to predict MTTR is the log normal pattern given in Figure 30.1 (derivations found in statistics). Let us examine some of the factors associated.

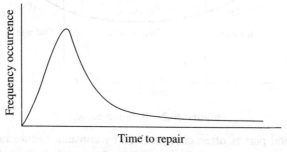

Figure 30.1 Log normal curve.

The reassembly shows time associated with the equipment, after interchange is over. Alignment in time associated with aligning the system of replaceable item after correcting the fault. Check out indicates time for verification that the fault has been corrected and the system is repaired. Constant failure rates occur, in the useful period of a unit. MTTR does not include the overhead. All equipments usually have constant failure. The prediction depends upon data of reliability and maintainability and the designer plays a crucial role in the process. It is a basic measure of the maintainability.

BATH-TUB CURVE

The useful life of an equipment is divided into three separate periods known as (a) wear in, (b) normal operation, and (c) wear out. Figure 30.2 shows the typical bath-tub-shaped failure rate curve, which shows the failure rate experienced by an equipment during its life time. This is similar to the fact that mortality of human beings is high during infancy and old age, with a steady state in between.

The wear in period, also known as the infant mortality period or the initial teething trouble phase, is characterised by decreasing failure rate, i.e. if the equipment gets through today without breakdown, it will be in a better shape tomorrow. In the teething trouble or initial debugging period of high breakdown rate, reliability is low and nothing performs like it should. The poorly manufactured items are weeded out in this phase. Most equipments are usually debugged by the manufacturer before delivery. These failures are caused by material defects and human errors during assembly, and supplier quality control programmes are used to detect and eliminate these bugs. Failures during this period are described by the Weibull distribution pattern.

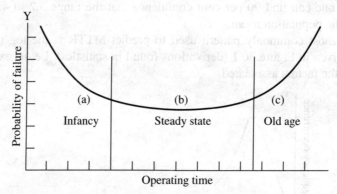

Figure 30.2 Bath-tub curve.

The second part is often characterised by constant failure rate and here only chance failure occurs. The useful life of a part is indicated by the flat and uniform failure rate in the bath-tub curve and the normal operation period covers the major part of an equipment's life. The probability of failure at one point of time is the same as at any other point of time. The passage of time does not decrease or increase the failure probability; failure during this period is indicated by the Poisson and Weibull distributions.

The third part is characterised by an increasing failure rate and it is the period during which the components fail because they are worn out. It is noted that with time all equipments and all materials degrade. After a long duration of normal operation period, the probability of failure begins to increase in the wear-out period. The passage of time increases the probability of failure when an equipment enters the wear-out period. With $r(t)$ beginning to increase, the equipment must be overhauled. Failures during the wear-out period are described by the Normal and Weibull distributions.

INSPECTION OF SPARES

The inspection of spares is carried out to see that the part conforms to the stipulated reliability and specifications of dimensional and metal characteristics. The spare parts executive is concerned with the incoming quality of the parts as he is responsible for the supply of right quality of spares to the maintenance function. For prudent spares planning, better buyer–seller relations, vendor rating and removal of rejections from the stores, he should be aware of various inspection plans, even though the parts may be examined by the users or independent quality control department. Since the objective of inspection is to identify the causes of defective items and prevent substandard items reaching the user, the inspection may be carried out at different stages of manufacturing, outgoing stage at the supplier's plant, and at the incoming stage of the user's factory. As pointed out earlier, the aim should be to detect various factors causing defective production and take appropriate remedial action to improve the quality by preventing recurrence of the defects.

INSPECTION METHODS

Before a batch or shipment of spares produced under homogeneous conditions reaches the buyer, three alternatives are available to the parts executive:

(a) Accept the lot without any inspection. This depends on the confidence placed in the supplier or on the assumption that the items have been inspected at supplier's end—or items bought from sister organisations, wherein pre-shipment inspection has been carried out.

(b) Examine each and every part in the batch. Such cent per cent inspection is not always possible due to considerations of cost, time, fatigue and nature of destructibility of the item. Even when cent per cent inspection is insisted on, in case of destructive tests, the efficiency of cent per cent inspection is critically examined by a sample check.

(c) Choose appropriate representative samples of the batch on the basis of well known sampling techniques—formulated by the Bureau of Indian Standards—and undertake planned inspection to segregate defective parts in order to decide/accept/reject/rectify/repair the items in the batch. This is known as acceptance sampling schemes, and quality control books deal with the number to be inspected for specific sampling schemes. Inspection is usually a separate department under quality manager who constitutes quality circles in which parts executive is a member.

It is well known that all kinds of inspection cost money in the form of staff salaries, depreciation of inspection equipments, rent for place of inspection, etc. Admittedly, this depends upon the number of items examined, but beyond a particular stage only the inspection cost increases without commensurate benefit in the quality of decisions. The use of quality control sampling schemes allows the management to decide about tolerable levels of risk of undesirably rejecting good parts or accepting defective parts when information is based on sampling schemes for a given inspection budget. In this context, the DGS & D—Directorate General of Supplies and Disposals—inspecting for Defence, Railways, Post and Telegraph, charges the clients 1% of the value of parts as service charges.

INSPECTION INFRASTRUCTURE

It needs hardly emphasising that the inspection of spares must be entrusted to only technically qualified personnel to avoid use of substandard/spurious items in the organisation. Though manufacturers do carry out stagewise inspection during the manufacture of spares, and supply is effected under warrantee claims with certificates, the parties must be critically evaluated for quality at the users' end. It is also possible to engage the services of specialised inspection agencies, like Lloyds, particularly at pre-shipment stage of costly critical parts.

The inspection department should have access to all types of gauges and the following measuring instruments in order to carry out the tests adequately: (a) vibration meters to check vibration of bearings/slides/spindles; (b) technometer to determine movements of various units and their speeds; (c) meggan for insulation resistance; (d) oscilloscopes to check the frequency, amplitude etc. in electrical circuits; (e) dial gauge for end play backlash in spindle slides; (f) ultrasonic leak detector for leakages in hydraulic, lubrication and water lines; (g) instruments to check condition monitoring and to carry out signature analysis; (h) multimeter for electrical voltage and current continuity; (i) dial temperature gauge and thermalik for liquid presence in lubrication and hydraulic pneumatic circuits; (j) bearing testers for all types of bearings; (k) radiography—detection of flaws in castings; (l) hardness testers for metal hardness; (m) instruments to check fatigue on rotating/vibrating equipments; (n) ultrasonic testers to detect flaws; and (o) dye preservation to detect flaws in forgings or cast spares. This list is only illustrative and not exhaustive as suitable additions are needed for individual industries. Reference to technical literature and inspection manuals may be helpful to augment the list.

Chapter **31**

Procurement of Spares

BUYING RELEVANCE

Organisations using spare parts in their own plant or selling spares for outside customers have to procure spares either from their own stores or from outside vendors. Provisioning or supplying spare parts to the customer at the right time plays an important role in the gamut of spare parts activities because everyone in the organisation comes to know about the consequences of a stockout. Since customers consider the buyer as the first point of contact, he contributes to the corporate image of the company by his cordial relations with outside vendors. The goal of the parts executive is to ensure that the right spare part is available to the consumer at the right time in appropriate quantities. This involves considerable planning, and several parameters, like right time, right quality, right price, right quantity, right material, right place of delivery, right contracts, right transportation and handling and right source, affect the decisions in procurement process. We now discuss these parameters, though it is difficult to answer precisely the meaning of the term 'right'.

RIGHT QUANTITY

A budgetary provision of 10% of the cost of the plant and machinery is made towards the initial procurement of spare parts. For this, the manufacturer is asked to give his recommended list of spare parts and the buying organisation makes a selection from the list and only in very rare cases all the items in the list are procured. When the manufacturer changes the model, which may result in non-availability of spares for older models, the lifetime requirement of spares must be met. Similarly, the selling organisations, introducing technological upgradation, must advise the clients about non-availability of the spare parts for the older versions, before introducing changes.

Obviously the parameters—right quantity and right time—are interrelated. Concepts such as economic order quantity, economic batch size at manufacturing (see Chapter 31) may serve as the initial broad guidelines. The required quantity will usually be less than the economic order quantity given by the formula discussed in Chapter 24.

$$EOQ = \sqrt{\frac{2(\text{Annual consumption})}{(\text{Unit price})} \frac{(\text{Ordering cost per order})}{(\text{Carrying charges})}}$$

This is applicable only for fast moving maintenance spare parts. The quantity requirement will be different for each category of spares such as maintenance, overhaul, insurance and rotable. There are several mathematical formulae on minimum, maximum, reorder point, reorder quantity, etc. But these formulae are based on several assumptions which are not satisfied in practice. The simplest formula is [$KLTC$], where LTC stands for lead time consumption sales and K is a factor depending on cost-criticality-availability analysis and not zero For low cost highly critical items, K can be 2 and for non-critical high cost items it can be 0.5. If the sales or consumption is zero for insurance spares, then $KLTC$ is zero and to avoid zero stock—subject to minimum of 1—$KLTC$ has been introduced. If total lead time is six months and monthly consumption is 100, LTC is 600. For low-cost high critical spares, K is 2; we get 1200 as the stock inventory to be kept in the warehouse, and steps should be taken to bring it to this level as and when spares are issued from the stores. Since the inventory has a bearing on working capital commitment, care should be taken to avoid stocks becoming obsolete items.

RIGHT TIME

The most critical parameter in the spare parts field is the availability of the part at the right time, as the stockout cost, due to non-availability of spares at the right time, in the event of failures, is indeed very high. Our experience indicates that many orders/indents describe that they should be immediately available without considering the lead time aspect. For determining the right time, the buyer should have information on all the elements of the total lead time. The internal administration lead time of converting an indent to a purchase order may go well beyond one year for procedure-oriented government departments/ public sector organisations who may have to go through advertised tender or buy through centralised agencies. In selling organisations this is known as order processing time. The manufacturing and transporting time depends upon the nature of the item; even this component may exceed one year in a few cases. Follow-up procedures and uptodate status reports can reduce this component. The inspection lead time of approving the quality of spares by inspection staff may be high if adequate test facilities are not readily available. The total lead time, particularly for imported spares, may be as high as three years and needs advance planning. Network analysis technique like PERT—CPM may be used for planning long lead items. Steps must be taken to reduce the administration lead time by giving adequate powers of delegation and resorting to limited lenders as the stock level increases with the lead time. Contingency plans must be developed to meet unforeseen problems like strikes, floods, earthquakes, and transport bottlenecks, and rush purchases must be authorised only in real emergencies. The total lead time thus depends upon order processing time, manufacturing time, inspection time, powers of delegation, ordering systems, procedures, etc. and there is scope to reduce this by carrying out O&M studies.

RIGHT PRICE

The manufacturer knows that the stockout cost of spare part is always higher than the cost of a spare and hence the margin is always high in case of spare parts. The price depends upon a large number of factors such as demand, supply, availability, negotiation, imports, urgency of requirement, standard spares, quality of spare parts, competition, dealers commission, taxes, etc. Price is what you pay while cost is a reality unknown to buyer. Cost consists of manufacturing cost, material cost and transportation cost. The lowest price need not be the right price, as John Ruskin says: "There is hardly anything in the world that someone cannot make a little cheaper and the one considering price alone becomes the legal victim". Hence government departments are using lowest responsible price or lowest acceptable price consistent with the specified quality. In case of specially manufactured spares with long manufacturing cycle time part payments are made during the manufacturing stages. We shall discuss selling price of spares in Chapter 33 also. Here we confine to buying price.

Some items have price variation clauses, and price negotiation before placing the order is crucial. The technique of learning curve is used for negotiation for labour-intensive items. The feasibility of producing spares or buying from outside should also be examined. In this context, it is learnt that the Ministry of Industrial Development has fixed the dealers discount at 18% and the DGS&D negotiates for a discount of 27% for electrical spares. The prices payable in respect of imported spare parts are arrived at on the basis of the foreign manufacturers net prices in dollars as shown in their parts list and the net FOR destination ceiling prices arrived at by multiplying the dollar prices by the rupee conversion factor. Usually, catalogue prices, cost plus pricing, administered prices, price preferences for small scale industries etc.

are in vogue. The government departments have fixed 10% of procurement from small and medium industries.

The cost of order processing in selling organisations or cost of purchase in using companies is an important parameter to be considered. This includes salary component for the activity, space, communication, telex, telephone, FAX charges, stationery, advertisement, data processing and developmental cost. The DGS&D—the largest central purchase organisation in India—charges about 1% on its indentors. Hence the above cost should be around 1% and if the cost is prohibitive, then control is needed for the procedures. Ordering cost is another parameter and is about ₹600 per order for a typical Indian organisation. In Chapter 33, we will examine pricing from supplier/marketing point of view.

RIGHT QUALITY

Quality spares are what the consumer needs as the machine breakdown will be more if inferior quality or spurious items are used. The quality, expressed in terms of tolerances, should be meaningful, tangible, understandable and available. Our experience indicates that a substantial portion of indents or orders is incomplete with regard to specifications. The spare part identification details like part number or codification must be given and not described as "as per sample" in order to avoid rejections. In the case of quality specification of spare parts, cost reduction methods such as standardisation, simplification, questioning the specification, value analysis, pre-design value engineering, development of new materials etc. play a vital part to ensure that quality is neither overspecified nor underspecified. The standards for inspection, right sampling schemes, stage-wise inspection, and decisions on rejected parts should be clearly specified in an inspection manual. The inspecting authority must examine the acceptability of the spare parts on the basis of full details of specification/drawings furnished by the manufacturers. Deviations must be allowed only after carrying out full tests according to inspection criteria and price variations on account of these modifications are made known. Allowances may be given for instrumental errors, usually with a tolerance of 1% on maximum or minimum values of test performance.

RIGHT SOURCE

The basic principle of purchasing is that at least two sources must exist for each item but it becomes difficult to practice in case of a spare part as many users depend only on the original equipment manufacturer to supply the spare part. It is noticed that the spare part can be obtained from various sources such as (a) original equipment manufacturer, (b) authorised dealers of the OEM,

(c) after sales service contract, (d) making or reconditioning in the workshop instead of buying, (e) subcontracting to a local small scale industry, (f) by loan from similar industry, and (g) spare parts bank run by industry association. In this process, if the parts are to be procured from the suppliers, then the task of development, selection, registration, preservation, and rating of source is an important one, as suppliers are intangible assets of consumer organisations. The important factors that have to be considered while selecting a source for spare parts are: (a) prompt delivery, (b) high reliability, (c) low price, and (d) service facilities offered. These factors depend upon suppliers' technical capability, financial strength, labour motivation, and standing in the market. In this process, a dealer may be preferred on the strength of the services rendered, whereas the original equipment manufacturer may not even respond to small orders.

For determining the right source, the purchase department should have a well organised data base on spare parts suppliers. This information can be compiled from press advertisements, *Indian Trade Journal,* Bureau of Indian Standards Institution, visiting cards of salesmen, industry associations, industrial exhibitions, displays in office, DGTD publications, etc.

It is essential to visit the source in order to assess the production capability, labour motivation, financial strengths, inspection facilities, etc. If necessary, in the initial stages, the source may be developed by assisting him in providing working capital, drawing, technical help, raw material, etc. Switching off a source or cancelling orders with ancillary units must be avoided from the buyer-seller relations point of view. If there are more than one source for the same spare part, then the vendors must be rated on the basis of their performance quality, reliability, adherence to delivery schedule, price, service, efficiency, industrial relations, technical know-how and financial strength.

IMPORT SUBSTITUTION

In order to keep the imported machines in working condition the country imports spare parts worth about ₹10,000 crore. In view of the trade gap—presently about ₹20,000 crore—and to achieve self reliance over a long period, efforts for import substitution must be taken by all, including the parts executives. For this purpose, it is suggested that 1% of the turnover must be spent on R&D efforts, development of drawings, and modifications in the design. Besides the research efforts of individual organisations, national research establishments, scientific research institutions, industry associations and Directorate General of Technical Development (DGTD) can play a positive role in import substitution efforts. This process is a long one and admittedly the initial quality may not be up to the mark and the prices are likely to be higher during the developmental process.

The difficulties involved in indigenisation are as follows:

(a) Lack of infrastructure! facilities, including know-how and precious machines for manufacturing critical spare parts.
(b) Strategic high quality raw material and special steels not available.
(c) Quantity needed is too small for economic lot size manufacturing.
(d) Non-availability of drawings of spare parts and detailed technical specifications.
(e) Indigenous suppliers quote higher price for inferior quality.
(f) After contracting to supply at a specified price, the supplier resorts to *force majeure* and other methods such as power cut, strikes, lockout, non-supply of raw materials, etc. and demands increased prices.
(g) Many imported spares are patented and it is difficult to violate patents and copyrights.
(h) There is no incentive for indigenisation as the list of permissible spare imports in import policy includes most of the required spare parts; also, it is easy to get non-permissible spare parts through imports.
(i) The import policy is very liberal for emergency spares for actual users up to a value of ₹100,000, and organisations do not want to take the risk of relying on local suppliers.
(j) Whenever foreign collaboration is involved, the collaborators approval is required for indigenisation; the collaborator, in his own interest, does not encourage indigenisation efforts.

RIGHT CONTRACTS: A–Z ASPECTS

The purchase order is a legal document that binds the selling company with the buying company through contracts. The objective of procurement contract function in any organisation is to make available the required equipment/ spare parts/services of right quality in right quantity at the right price so as to reach the specified place at the right time. For a contract to be valid, the preconditions stipulate the following:

(a) Two or more parties
(b) Legal subject matter
(c) Consideration or a sort of *quid pro quo*
(d) Consent of the parties.

The technical and other aspects in the contract must include the following:

(a) Intent of specifications
(b) Scope and terminal points

(c) Technical conditions for specific contract
(d) Standards
(e) Design criteria
(f) Performance parameters
(g) Performance guarantee and warranty
(h) Shop and site tests
(i) Special inspection requirements
(j) Drawings
(k) Construction details
(l) Special tools and tackles
(m) Conditions of contract
(n) Liquidated damages for delay
(o) Buy back clause for non-usable spares
(p) Deviations
(q) Rejections
(r) Transportation details
(s) Transfer of title
(t) Taxes, octroi, customs, excise
(u) Transit insurance
(v) Settlement of disputes and jurisdiction of court
(w) Arbitration
(x) Insolvency
(y) Trade marks and patents
(z) Basis for escalation of price and validity period.

The order should contain purchase order number, description of materials, quantity required, delivery schedule, shipping instructions, location where the materials are to be delivered, and details of terms. While the parts executive may not be thorough with all the above a–z aspects of the legal contract features lack of knowledge can never be an excuse for an illegal act of commission/omission. It is advisable to consult a legal expert, particularly for large and imported spare parts in order to avoid bickerings and for ensuring smooth buyer–seller relations.

DELIVERY/TRANSPORTATION

The main thrust of the spare parts supplier is to ensure the delivery of the spare part to the shop/stic/project, which may be far away; transporting the items from the supplier to the user is a critical activity. It will be advantageous for the manufacturer to develop a manual of 'do's and don'ts' of packaging similar to that of Indian Railways in order to avoid breakages, damages and repeated handling during transit. If the spare parts are to be delivered to remote areas which are far away from national highways or railway lines, the cost of transportation may be as high as 25% of the spare parts cost. The movement may be local, regional, national or international, and may be through own resources or through contracts. For this purpose, it is necessary to know alternatives, rates, schedules, and merits of different transport systems such as rail/road/air/ship/courier/own messenger. When the movement of parts is sizeable and regular, sophisticated mathematical techniques like linear programming and transportation models may be useful in the distribution of spare parts from factories to retailers depots spread throughout the country. For movement of short distances up to 500 kilometres and small consignments, it is desirable to resort to road and prefer railways for movement of goods for longer distances. In multi plant operations, supplier lands the spares in another factory, instead of the requisitioning base.

RIGHT SYSTEMS

Formal systems, procedures and manuals are essential for smooth working of any organisation. In the case of selling organisations, procedures for acquiring, processing the customers orders, priority allocation, raising invoice, packing, despatching, bill raising and settlement with conditions of sales must be clearly spelt out. Similarly, formal systems have to be developed for pre-purchase, ordering, and post-purchase systems. Initiating the purchase through indents, requirement planning, selection of suppliers, obtaining quotations, and evaluating them constitute pre-purchase system. A thorough organisation and methods study must be done in order to minimise the lead time and paper work both for order processing and order placing.

The merits and demerits of different systems such as global tender, open tender, limited tender, single tender, reciprocal buying, systems contracts, stockless purchasing, just in time etc. should have to be carefully weighed as per the delegation of powers. Organisations must send to the supplier an acknowledgement copy with the purchase order and try to obtain a written acceptance from the suppliers. The post-purchase system includes follow-up with the suppliers, for which special systems have to be established for spare parts. It is also essential to scan the economic details and other relevant technical journals in order to improve the market research/purchase intelligence in the organisation for better knowledge of the demand and supply situation of the spare parts. Readers will appreciate one can discuss at great length each of the above "rights", but you are provided with salient points of purchasing relevant to spares.

Chapter 32

Logistics and Warehousing

STORAGE OBJECTIVES

In order to serve the customers/user with the spare part immediately on request, the parts have to be stored in the warehouse. Warehousing function is important as a service function in which the parts executive acts as a custodian of spares carried in the stores. The Warehouse Manager acts as a buffer between the user/consumer and the buyer/manufacturer. The entire spare parts management revolves round the quantity of spares available in the warehouse. The parts executive cannot order quantities for spares beyond the warehousing capacity nor can he afford to have nil stocks in the stores, particularly for critical spares. Good stores system can greatly assist the Warehousing Manager in accurate stock status reports, timely detection of discrepancies, prompt clearance of goods inward notes to expedite bill payment, reduction in demurrages, and better claims management. He should ensure that postings of receipts and issues are uptodate and accurate, because the Warehouse Manager is the custodian of crores of rupees worth of spares. His functions include: (a) receipt of customer orders or indents from different users; (b) order processing to decide on the priority of allocation; (c) receipt of spares; (d) getting the spares inspected; (e) issuing to user or selling to the customer; (f) maintaining accurate stock records; (g) valuation of stock; (h) inventory control and fixing levels; (i) stock verification; (j) storing the items, (k) preserving the parts; (l) disposing of obsolete/scrap items; (m) developing codification and other methods for easy identification; and (n) coordination with other departments. It is desirable to have a manual outlining the warehousing activities clearly. We must admit that the service function maintenance is least glamorous spare parts used by maintenance is service to service function and storage is service to service resulting in poor quality of staff.

PHYSICAL STORAGE: A–Z WAYS

In carrying out the objectives, the warehouse will have to gear up its activities of receipt, storage, handling, issues and disposal of spares. A large number of factors influence the process, and some of these are now listed:

(a) Warehouse/substores location
(b) Proper cooling and ventilations of warehouse space
(c) Shift timings of issue or sales to customers
(d) Type of spare parts
(e) Number of spares
(f) Quantity stored
(g) Minimum/maximum/reorder point/reorder quantity levels
(h) Cost of storage which is about 30%, including the cost of capital and storage
(i) Cost of order processing or ordering which is about ₹600 per order, inclusive of salary, advertisement, follow-up, and communication
(j) Stockout cost or damages due to non-availability of item
(k) Service level or meeting the consumer demand
(l) Number and quality of staff
(m) Use of codification/part number
(n) Layout of stores and marking aisles gangways, and working areas
(o) Application of preservatives
(p) Procedures of receipt, issue, indent, order processing
(q) System of binning, stocking, loading/unloading
(r) Safety fire extinguishers, pelletisation
(s) Classification of spare parts
(t) Control of main warehouse and departmental stores
(u) Communication and computerisation
(v) Documentation
(w) Toilet facilities and smoking areas
(x) Lighting arrangements
(y) Security and restriction of entry of unauthorised persons
(z) Cleanliness and easy accessibility and identification of the spare part for better customer service.

PRESERVATIVES: A–Z WAYS

Since spare part is slow moving in nature, the parts executive should be familiar with application of various preservatives to reduce the damage, rusting, obsolescence etc.

(a) If the spare parts are to be stored without depreciating their physical properties, they will have to be guarded against the effects of

dampness, dryness, heat, cold, temperature variation dust, dirt, and attack by insects by necessary healing or air conditioning.
(b) Protection against fire hazards must be provided.
(c) Special categories of items, such as metals, corrosive items, explosive devices, acids, chemicals, paints, gas cylinders, cement, timber, leather goods, rubber tyres, tubes, uniforms, and stationery, have to be properly stored and preserved.
(d) Obviously, the value of preservatives to be applied on a part should not exceed the cost of the part.
(e) All ferrous materials should be given a protective paint varnish and stored. Precision items like instruments, electronic items, electrical spares and ball and roller bearings must be covered in polythene bags enclosing moisture absorbent chemicals like silica gel.
(f) These materials should be maintained in dust-free air conditioned rooms without sunlight and moisture.
(g) Electrodes must be kept intact in original packing and in dry storage room.
(h) Sintered bush bearings must be soaked in warm oil for 24 hours once in a year.
(i) Pipes over 2" must be flushed/cleansed and filled with dry air. (In such cases exterior painting is done by protective paints.)
(j) Vertical stocking of grinding wheels with partitioning in between is necessary so that faces do not come into contact with each other.
(k) Castings with machined faces, particularly with threaded portions and holes must be protected with grease plugged in wherever necessary.
(l) Strip heaters in all high tension motors and LT motors must be cocooned.
(m) Copper parts must be protected against ingress of ammonium salts.
(n) Silver and lead parts must be cleansed with fresh water.
(o) The compressors and turbines of multistage pumps should be rotated on their journals every quarter to prevent sagging.
(p) Shafts, gears and impellers must be stored vertically after painting with dewatering protection films.
(q) Fasteners and screws, kept on the shelves, must be treated with hard preservative films.
(r) Perishable spares, with a low shelf life, must be identified, and first in first out method of issue must be practised.
(s) Brushes must be protected from moisture and insects by keeping in original-cartons and sprayed with insecticides.
(t) Battery cells have low shelf life in humid and damp conditions and must be stored properly.

(u) Expensive instruments should be delicately preserved in their boxes with small cotton bags of dehydrated silica gel.
(v) Glass sheets should be stacked upright and protected from rain and inundation as they will disintegrate if they absorb water.
(w) Cloth and paper should be protected from moisture, rats and moths by using insecticides.
(x) Rubber goods, with slow shelf life are affected by sunlight, moisture, air, heat, dust, dirt, etc. and must be covered with a tarpauline or heavy woven fabrics for protection.
(y) Tyres should not be piled near radiators or other sources of heat.
(z) Atmospheric temperature in the warehousing should be kept as low as possible, with air current circulating in the building replenishing the supply of oxygen.

RECEIPTS MANAGEMENT

The input mechanism in the warehouse consists of activities such as: (a) determining the requirement; (b) raising of purchase requisitions; (c) chasing the purchase orders; (d) receiving the material; (e) liasing with quality control inspectors; (f) stocking the accepted materials; (g) accounting the rejected items; (h) coordinating with purchase and finance for bill settlement; and (i) completing the documentation. These activities have to be systematised to avoid peaks of workload caused due to suppliers trucks arriving at their 'own' time. Parking places of trucks and wagons from outstations have to be clearly marked and cordoned off to facilitate unloading activities. The incoming spare parts should be counted and checked with the invoice, and one copy of challan returned with the receivers signature to indicate the acceptance, shortages and excesses should be duly accounted for. Miscellaneous receipts such as samples, cash purchases, components from feeder shops, scrap for disposal, safe custody items, etc. have to be regularised through appropriate documentation systems. The receipt section then liases with the quality control section or maintenance department for inspecting the spare parts. The goods receipt note is used to regularise the material inflow in the stores and it indicates the name of the supplier, quantity received, quantity accepted, rejects and remarks on shortages. The goods receipt note and other documents are forwarded to the accounting section by the Warehouse Manager for payment to the vendors.

DOCUMENTATION OF OUTPUTS

The issues to the different departments or sales to the customer constitute the output from the warehouse and this is the most critical function since the performance of warehouse is judged on the basis of its ability to meet the demands of the consumer. The stores have to adopt different procedures for (a) regular consumable spares, (b) tool kit requirements, (c) issues to subcontractors, (d) issue on imprest, (e) sample items, (f) issues on a returnable basis, (g) issues to servicing departments, and (h) sales to consumers. In the case of sales to customer, an invoice is prepared at the time of selling. The authorisation is in the form of a material requisition voucher for internal use and this indicates work centre, quantity, material code, date and signature of authorised person. If the item is not readily available in the stores, purchase is expected to expedite it. It is desirable to restrict issue timings for different departments for better coordination and the common place of delivery is the stores window, particularly for small items. Bulky spare parts are delivered directly to the shop floor. To enable proper stock rotation, first in first out principle is followed, for emergency issues, like in night shift, the executive controlling the operations has the necessary powers to open the stores and take the item in the presence of the security staff. Control on indent quantity and issues act as effective cost reduction devices and for better forecasting of future consumption.

VALUATION AND VERIFICATION

The Warehouse Manager should have an idea of the value of spare part stocks he is holding in order to formulate the future budget and also to price the issues. Methods of valuing the materials in stores range from conservative practices such as market force or cost of procurement (whichever is lower) to modem methods such as replacement cost. Each method has its own advantages and disadvantages. For example, the performance during dull periods can be offset by increasing the value of closing stock by suitably altering the method of valuation. Hence the government has introduced statutory controls on the frequent changes in methods of evaluation by stipulating that any declared system must be followed for a duration of three years and any change therein must be approved by the Board of Directors. The different methods of valuation of inventories are:

(a) First in first out—FIFO
(b) Last in first out—LIFO
(c) Highest in first out—HIFO
(d) Next in first out
(e) Base stock
(f) Simple average
(g) Periodic simple average
(h) Weighted average

(i) Standard cost
(j) Replacement price.

Under the existing tax laws only the first in first out and weighted average methods are accepted.

The annual stock verification on the movement of materials and checking up with physical stock is usually done at the end of the year. Some organisations do the perpetual inventory or checking a few items daily, particularly when there is an issue. Discrepancies between physical stock and book balances are checked to identify various causes like over-issue, improper documentation, fraud) theft, late posting, etc. and suitable remedial action, particularly updating the records, are taken to rectify the situation.

MULTIPLE WAREHOUSES

Organisations selling spare parts have to cater to a multiplicity of needs and cover large areas with established distribution channels with warehouses located at different levels. The supplier sends the spare parts to the central warehouse as well as the regional warehouses directly. Sometimes, the stocks are sent from the central warehouse to regional warehouses which cater to zonal warehouses which in turn cater to subregional warehouses and so on up to the ultimate depot that caters to the consumer. This phenomenon is present when forecasting at each individual level is not accurate and the company does the total forecast for the entire area of operation. Obviously, each warehouse will have a safety stock; there will be orders from the lower warehouse to the upper warehouse and there will be in-transit stocks. Instead of treating each individual warehouse as a separate entity for inventory optimisation decisions, it is better to have a centralised order processing system, centralised monitoring of stock movements and centralised stock of high value items. This obviously requires a well established information system, but the reduction in total inventory cost of carrying stock in different warehouses amply justifies the increase in information costs. In a multiple warehouse system that has n warehouses, the safety stock can be reduced by a factor equal to the square root of n, if it is held at the central warehouse, instead of at all the subsidiary warehouses. It can also be proved that if central procurement or bunching of orders is done, the total inventory cost is also reduced by a factor, viz. the square root of n. We will discuss multiechelon systems—Multiple Distribution in Chapter 35 in greater detail.

LOGISTICS MANAGEMENT

In the context of selling spares, the term "business logistics or physical distribution management" is often used. The spare parts management has to ensure that the parts reach the customer in time in order to ensure a reasonable return on his investment. This distribution can be undertaken either by the manufacturer himself by having his own channels or by establishing a distribution network consisting of stockists, wholesalers and retailers. The distribution of all finished spare parts always demands that the spares be packed in such a way that they can be easily handled, transported, and delivered to the customer in an acceptable manner. Storage will automatically creep into this process because of time lags—between the completion of manufacturing activity and packaging, between the packed state and loading into carrier, and between the unloading at the destination and final customer. Besides, the products may be detained while waiting for connecting transport arrangements *en route* to their destinations. Efficient management requires that these time elements as well as the inventories are reduced to as minimum a level as possible. Unnecessary investment in finished goods will necessitate increased working capital commitments. Thus the Stores Manager is often confronted with problems of keeping the distribution channels full and at the same time minimising the relevant costs. A judicious balance between the cost of holding the spares in the channels of distribution and the cost of lost sales must be struck, by studying factors such as types of spare part, geographical pattern of demand, location of warehouses, transportation lead time, distribution network, inventory holding charges, transportation charges, demand pattern, obsolescence, etc.

BUSINESS LOGISTICS

Since 2010, business logistics has been growing at a vertical level in India. The logistics process involves selection of warehouses, freight transportation, receiving the spare part, inventory control, handling, market forecasting, protective packaging, etc. The logistics management attempts to satisfy the consumer's demand with the least cost after considering all relevant costs and constraints in the system. Some of the internal constraints are location of plan to customer service standards, product pricing policies, length of order cycle, product profile, levels of technology, and organisation culture. The major external constraints affecting the business logistics system are: completion strategy, customer profile, government regulations, and capabilities of different transport modes. As against the traditional functional organisation structure, business logistics provides a total approach after considering the constraints and interfaces. For instance, the warehousing executive is often confronted with the problems of keeping the distribution channels full and at the same time minimising the transportation costs. A mere reduction of transportation costs will often result in the drying up of supplies at the consumption centres. In a few cases it may even be advisable for the organisation to have its own vehicles.

The location and size of raw material sources, and inventory levels at different points of manufacturing constitute the central elements forming the logistics system. The transportation sub-system consists of the selection of the mode of transport, viz., rail, road, air, shipping—and the rating for movements of goods from plants to consumers through warehouses and substores. The inventory subsystem pertains to all types of inventory of spares, in the logistic network. The material handling subsystem deals with loading and unloading, dock facilities, containerisation and handling facilities. The information subsystem consists of the flow of the information including processing of customer orders after obtaining number, location, and size of customers as well as information on part characteristics like value, weight, measurements and competing products.

Thus the system definition in terms of product, geographical territory, raw material source etc. as well as the interrelations between the subsystems is an important step in the study of business logistics. The basic belief that the integrated system performance will produce an end result which would be then the traditional non-coordinated approach is the focal point in the development of the integrated business logistics concept. Business logistics of the supplier reaching the consumer in this competitive world is developing very fast and the challenges must be converted into opportunities.

Chapter 33

Pricing and Marketing of Spares

PRICING

Spare parts sales constitute an important activity in a very large number of private and public sector organisations like TELCO, Ashok Leyland, TVS Group, Mico, HMT, Blue Star, Escorts, Eicher, CMC, Larsen & Toubro, BHEL, MAMC, BEL, BEML, Greaves Cotton, Kirloskar, Voltas, Josh Engineering, Jyoti, General Marketing and Manufacturing, HEC, Allwyn, Bajaj, etc. A large number of industries manufacturing components such as bearings, compressors, motors, rotors, fasteners and valves have been set up on the basis of replacement market. Some of the above companies have their spare parts sales more than ₹1000 crores per annum. The total spare parts sales in India are estimated around ₹50,000 crores, the imported value of spare parts per year is over ₹10,000 crores. Usually, the marketing executive is directly involved in determining the pricing strategies and in executing the orders in most of the organisations. The spare part sales executive works in close liaison with those involved in equipment sales and after sales service staff. You must appreciate that marketing is the most glamorous while spare part is the least glamorous jobs.

Of all industrial goods, pricing spare parts is probably the most complex problem. Price is a quantitative expression of what the seller thinks the buyer will buy. It assumes a certain level of sales volume and market. A market is a system of information on cost, demand, and competition, with uncertainly being the feature of the market. Here it is useful to consider the nature and characteristics of customer, product and market as well as appraise the customer of the supplying company's objective and standing.

The major objective of organisations selling spare parts is to ensure the availability of right quality of spares at the right time to the user with minimum investment at the seller's end. From the seller's point of view, the maximum satisfaction to the user through steady supply of spares to retain his patronage in future, is only one of the many methods. The other methods include: (a) pragmatic pricing to meet the profitability requirements of the organisation; (b) reliable forecasting of the demand for spares and efficient handling of customer orders through a good communication system; (c) reliable and timely delivery of genuine spares; and (d) a sound after sales service system. It should be noted that customer satisfaction is subjective while pricing is objective. India is spinning at the same rate as the rest of the world. The youth of the nation are of course leading and often dictating the way a brand speaks to them. Music, films, art, sports, internet are the biggest drives in India, thanks to technology and social networks, like facebook, Google, Turtle, user generated sites such as Youtube, Wikipedia and smart applications on smarter phones, the realisation is every thing is a medium interested in all new experiments, change and new models. Markets are on the hunt for newer, faster, and more targeted way of reaching the consumer.

PRICING: A–Z FACTORS

A large number of socio-economic, political, natural, regulatory, and financial factors affect the demand and supply of spare parts, which in turn determine the marketing strategies and pricing of spare parts. We now give an illustrative list of factors which influence the price of spare parts in varying degrees.

(a) Importance of customer
(b) Urgency of demand when demand exceeds supply in shortages/scarcity; monopoly supplier or imports
(c) Standard/non-standard in specification
(d) Single-to-multiple usage or flexibility in usage
(e) Relation between part sales and equipment value
(f) Economic life/age of equipment, model, technology upgradation
(g) Maintenance, replacement, rebuilding, reconditioning nature of the part
(h) Competition, including sale of spurious parts
(i) Indigenous versus imported/landed cost or price of parts
(j) Nature of sophistication of industry, franchising, technology upgradation and product life cycle
(k) Criticality of customer/equipment/part
(l) Number of equipments—similarity and variety in use—small orders
(m) Credit worthiness and future potential equipment sales of the consumer
(n) Buyer behaviour, purchase practices, accuracy of demand forecasts

(o) Customer service location/expectation, transport difficulties, communication bottlenecks
(p) Nature of industry—public private/defence/export/small/foreign/government department/DGS&D, etc.
(q) Demand profile—regular/irregular, custom-built, one-off, universal, predictable, unpredictable, bogus, genuine, seasonal, urgent, slow moving, fast moving, and so on
(r) Overall business potential for equipment and part sales to the consumer
(s) Price elasticity, volume discounts, handling freight charges, advertising and promotional expenses
(t) Cost of manufacturing material cost, labour charges, inventory carrying charges, selling cost, advertising cost, order processing, computer cost, obsolescence stockout cost, etc.
(u) Governmental regulations, import policies, customs duty, taxes, levies, penalties, octroi, excise, sales tax, entry tax, commission, other levies
(v) Inflation rate in India and at global level
(w) Production capacity, exclusively allocated to spares manufacturing and market mechanisms, including channels of distribution
(x) Lead time requirements from subcontractors and order processing time
(y) Raw material availability and technical competence of vendors
(z) Objectives of customer servicing and priority allocation as stated in the corporate objectives.

Huge advertising even in educational field, in ads, media, increases the capitation fees of students and customers in industrial scene.

PRICING STRATEGIES

While the spare parts seller exists in business only to make reasonable profit, the customers invariably think that organisations make huge profits even up to 200% in spare part sales, though they face uncertain lead time, high stockout costs, lack of drawings, poor customer service, etc. This can be easily shown if you buy individual parts of an engine and assemble it and compare the cost with the cost of total engine. The cost of buying individual parts and assembling will be twice or even thrice that of buying the total engine. However, it must be admitted that during the warranty period, the manufacturers make substantial profits in view of very few initial breakdowns. In subsequent periods, the stockout cost is much more than the price of the spares. The buyer feels that spare parts are over priced and unwanted spares are being dumped on him, particularly at the time of procurement of capital equipment. The average margin of profit, according to response to our survey is 60%, depending upon value of part, importance of customer, imported indigenous part, volume of purchases, criticality of spare, competition.

One way of determining the right price is the mechanism of cost plus pricing. The sales price of a spare part contributes to the direct costs of manufacturing the product, bears a certain amount of overhead expenditure, and yields a reasonable profit as return on investment made in the form of capital employed by the organisation. The direct cost of manufacture includes the cost of material, manpower, fuel, power and utilities used in the manufacturing of a product. To estimate the direct cost, we must have an idea of the material content and manhours needed to manufacture the parts. Allowances for price negotiations must also be included in the catalogue price. Normally the vendors try to increase the price by 10% of previous year's and attribute this to general inflationary spiral witnessed in developing economies. Policies such as administered prices, dual price, levy price, transfer price, retention price, etc. must also be considered. The pricing of a spare part can also be done on the basis of mark-up pricing or target return on investment. Sometimes, in order to beat the competition, a flexible approach in pricing is also adopted. The mark-up, on an average may be even up to 50%, depending upon supply and demand. The price discrimination can also be based on customer, criticality of part, location of customer, urgency of requirement, and quantity needed. To save themselves from being deceived, the buyers must carefully check the seller's strategies and analyse their own buying procedures.

MARKET SKIMMING STRATEGIES

Generally prices of spare parts will be exorbitant if the parts are critical, patented and short supply items. The manufacturer of new equipment, while supplying spares, may adopt the high pricing strategy to recover design, engineering and developmental costs. Such a method of pricing is similar to skimming price strategy and will work out until competition develops. Price skimming is used by the firm to introduce a high price for skimming the cream of the demand. On the contrary, for equipments which are general purpose in nature and for which the user has several options, the pricing of spares is more analogous to the market penetration pricing. Basically, in arriving at the pricing policy, the seller considers factors such as service level, customer satisfaction, stock levels, and costs of manufacturing, transportation, warehousing, distribution, preservation, handling, selling, and administration.

In the marginal pricing method, fixed costs are ignored and prices are determined on the basis of marginal cost. This method deals with the anticipated revenues, expenses, and costs. With the marginal pricing, the firm seeks to fix its prices so as to maximise the total contribution to costs and profit. In the *going rate pricing method,* the firm changes its policy to the general pricing structure in the industry where price leadership is established. Charging according to what the competitors are charging may be the only safe policy for some spares. In the *rate of return pricing method,* the targeted return on sales is expressed as a percentage of rupee sales. The company prices its inventories with sufficient high mark-ups so that the sales revenue will cover total cost and yield the desired profit on the operations during an year, with different spare parts carrying different mark-ups. For general purpose spares, the seller would try to offer lower price to attract customer and penetrate the market.

It is obvious that the pricing policy will be such that it covers all the relevant costs mentioned. The seller may mix different profit margins for different categories. The margins will be high in case of slow moving, critical, monopoly, or patented spares. In industry the profits are found to vary from 25% to 100%. For spares with higher sales, the margins will be low but the inventory turnover will be high and the seller prefers to make up by increasing the sales volume of the part. However, minimum stocks mean close following, accurate forecasts, and excellent communication, without any transport bottlenecks. Many organisations aim at a turnover of 4–5, implying that the spares inventory should be less than three months average sales. A higher inventory turnover means that the distributor is trying to operate with minimum stocks of parts and hence he is losing because of frequent stockouts, thereby encouraging spurious manufacturers to enter the market. But lower inventory turnover implies locking up huge working capital and increased obsolescence. Various control measures are necessary to identify movement of various parts and to institute suitable ordering pattern for various categories.

The steps involved in pricing are as follows:

(a) Select pricing objective.
(b) Determine the demand by market survey.
(c) Estimate fixed and variable costs.
(d) Analyse competitors' price.
(e) Choose a price method.
(f) Bring about price modifications and changes.

It is learnt that bulk buyers like ministry of industrial development, DGS&D—are negotiating up to 27% particularly on electrical spare.

FORECASTING AND PLANNING

Usually the equipment manufacturers have several customers each with their own policies and practices. While assessing the total requirements of spare parts for different customers, several factors have to be considered. The part sales record indicating the average offtake for each region/customer and fluctuations in usage is a good starting point to throw light on future demand. Factors, such as operating conditions, terrains, maintenance policies, overhauling requirements, equipment population, age of equipment, repair shop facilities of the user, communication system, planned equipment sales, recommended spares list, competition in spares market, technological changes at customer's end, customer's instruction for non-standard/undersize items reports from the field and after sales service staff at the site, play a crucial role in better forecasts. Changes in supply profile such as entry of new spares manufacturers, or expansion of existing suppliers with the prices of spares must be considered to know as to what extent the share of the market will be affected.

An important prerequisite in this context is the existence of a sound system to collect the feedback data from the end users. Normally the sellers contact the maintenance staff in user organisations through the after sales service representatives, or dealers for ascertaining the future requirements. The manufacturer also knows the total number of equipments in operation, the age of the equipment, maintenance policy, and spares requirement at the individual plant level. They also make a subjective estimate of their operating conditions to find out the deviations from the normal practices. An all time requirement is indicated by the users for the equipments which are expected to be phased out or when the manufacturer opts for a particular model. The requirement figures thus submitted by various end users are then analysed and consolidated. Since in India the tendency is to give conservative estimates for avoiding stockout and service, there is no guarantee that the estimates will be turned to selling quantities; hence the adjustments are made by the seller. He relates the projections to previous years sales history and, after making suitable adjustments, the total demand is ascertained under various policy constraints such as inventory turnover, service level, and tight money conditions.

Forecasting methods like moving average, trend analysis, exponential smoothing and adaptive correlation methods are found to be useful in this regard. The demand figures obtained in the above manner form the basis of manufacturing programme or procurement from vendors. This again falls into two categories—indigenous spares and imported spares. The problems of manufacturing or procurement of indigenous spares are not many. However, in case of imported spares, factors such as availability of foreign exchange, technological obsolescence, and lead time come into the picture. The demand figures are decided on monthly or quarterly basis to ensure planning out the manufacturing stage and to solve working capital problems.

ORDER PROCESSING

Once the spare parts seller has determined the demand and formulated procurement policies consistent with service level and inventory turnover criteria, the other important aspect is the manner the customer orders are processed. An efficient handling of customer's orders can not only cut down internal lead time and administrative delays but it will also enhance the seller's image in customer's mind and win his continued patronage. Usually, spare part sellers are big organisations who manufacture original equipment itself. To them spare part sales must constitute an important after sales market. In the rest of the cases the spare parts are sold by dealers, traders, stockists and specialist spares selling organisations who sell spares of several equipments.

The volume of sales and variety of spares are so much that processing of customer order becomes complex. As soon as the customer places the order, it must be entered into a register or log book without delay, irrespective of whether or not such an order can be met. This will provide crucial historical data in future. Several customers place orders for the same parts and very often whenever dealers place an order, it will be in bulk, covering several parts. It is essential that the manufacturer allocate 10% of manufacturing capacity in order to meet the replacement market orders.

The demand for different parts has to be consolidated individually as well as through customer groups. This necessitates the establishment of scientific codification in the form of part numbers and customers. When the demand exceeds availability of spare parts, a system of priority allocation has to be worked out apart from creating a back-order system and adequate follow-up; where there are adequate stocks of spare parts, the customer orders can be immediately met, the necessary invoices paid, and the consignment detached to the customer. The speed with which the above activities can be performed will entirely be dependent upon codification of parts, warehousing, handling methods, and the system of documentation used. Special situations may arise when the seller may have to import spares against customer's licence.

When there is a shortage of supply in relation to the demand, which is a common feature in critical spare parts, requirements are made prioritywise. The priorities for allocation will be determined by the selling organisation on the basis of factors such as importance of customer, potential equipment sales, current volume of business, long-term relationship, natural disasters, credit worthiness, solvency, previous purchasing pattern, and past relationship. Once the available stocks have been allocated on the basis of priority, the balance orders are backlogged. The backlogged orders can be taken up periodically say, once in two months, for stock allocation.

DISTRIBUTION OF SPARES

In marketing the spare parts, distribution constitutes an important function. In the distribution policy of spares, the seller is faced with the task of ensuring a particular inventory turnover ratio, but at the same time he is called upon to meet reasonable demands of many customers for all spare parts. In spares distribution, the management has to decide the number of levels such as central warehouse, stockists, regional warehouse, field warehouse, retail outlets, and dealers. He has to determine the inventory levels to be stocked at these locations and ensure reliable delivery of spare parts from the manufacturing place to the consumers' site. Selection and control of transport carriers is an important step in ensuring reliable delivery. Basically, these aspects relate to identification of shipment, building freight charges, insuring the packages, inspection for quality, replacing rejected items, and evaluating and settling the claims. The carriers have to be evaluated on the basis of their delivery and transit time. Reports are useful in this area and on the basis of the conditions of goods delivered, rate contracts are entered into carriers for transporting the spare parts to different regions.

A continuous analysis of expenses of total distribution cost is necessary to control such expenses. It is desirable to establish norms for transit times between two points. This will be a significant factor in winning the patronage of the buyer as standardised transit times to distribution imply a high reliability with regard to delivery of spare parts.

SPARE PARTS SALES: A–Z PROBLEMS

A large number of problems are encountered by the sellers in the field of spare parts marketing, as well as in spare parts pricing. We shall now discuss some of these problems and their solutions in detail.

(a) The image of the sellers' organisations as perceived by the buyers/maintenance engineers, and other customers, is not very good and leaves much to be desired, particularly in the availability of the spare parts at the right time with the right quality at the right price.

(b) The seller must improve this image by adherence to the delivery schedule and to the specified quality and treating it as his corporate objective.

(c) The seller must study the difficult environmental, social, political, financial, and governmental changes in the economy and adapt his marketing strategy to suit the uncertainties in technological changes, inflationary pressures, erratic economic growth, import policies, taxation policies, power cut, nonavailability of quality raw material and unstable aggregate demand, through a dynamic market research approach.

(d) The original equipment manufacturer admits that it is his responsibility for a satisfactory supply of spares to keep the equipments in good operating conditions for a reasonable period. He is not sure how long he is responsible for the supply because of rapid technological advances; new models are introduced and old ones are phased out.

(e) In view of the scarcity of capital, the user invariably prolongs the life of the equipment and complains when parts are not readily available. In this context a 10-year time horizon is considered as a reasonable period for the supply of spares by many manufacturers.

(f) The problem of power cut in summer months in many parts of India forces manufacturer to use generators, resulting in increased production cost of spares.

(g) In view of uncertainty in demand, sometimes exaggerated demand, the suppliers find requirement forecasting a difficult exercise.

(h) The manufacturer does not plan or allocate separate capacity to produce spares for replacement market and usually concentrates more on sales of the original equipments.

(i) The spare parts are demanded only in small quantities and it is difficult to manufacture economic lot sizes, thereby rendering the process as uneconomical in most cases.

(j) The recommended spare parts list at the time of selling the original equipment and the suggested periodic replacements as outlined in the manual are not usually given due importance by the customers, making the requirement planning a difficult exercise.

(k) Problem arises because of non-availability of special high quality raw material indigenously and considerable lead time involved in importing.

(l) The different cost elements, like inventory carrying charges, discounts, selling charges, advertising costs, order processing costs, credit, selling costs, communication costs, freight charges, stockout costs, loss of profit, loss of customer goodwill, loss of image etc. are not readily available.

(m) The customer's requirement of a critical spare part is always urgent, with very little planning at his end, without giving sufficient lead time for order processing, manufacturing, transporting, etc.

(n) The sales turnover and stock rotation and requirement determination of different spares vary considerably, so that the margins cannot be fixed on a scientific basis.

(o) The customers are dispersed throughout the country, their demands vary, requiring special features and it becomes very difficult to open warehouses, stocking points and service centres within optimum investment policies in far flung areas.

(p) It is almost impossible to compete with the spurious spare parts manufacturers who quote even below the materials cost at times.

(q) Sometimes the customers insist on non-standard spare parts even in common items such as hardware items, pipe fittings, beltings, etc. and it is difficult to supply on the basis of cost-benefit analysis. Spurious manufacturers jump into fill the vacuum in supplying such items.

(r) Some government organisations insist on lowest quotations and it is difficult to compete with sub-standard suppliers offering lower price.

(s) The original equipment manufacturers have to meet proper quality specifications and maintain adequate facilities such as precision instruments, machine tools, testing facilities, inspection set-up, etc, resulting in higher cost of production as compared to the sundry manufacturers.

(t) The original equipment manufacturers may not be in a position to extend discounts, concessions, credit facilities, etc. which may be offered by the small scale manufacturers.

(u) The original manufacturer has to observe all the formalities in taxation such as sales tax, octroi, excise, other statutory taxes, forming up to 40% of the spare parts cost which can be evaded/reduced by non-registered small scale suppliers, who also get benefit up to 15% from sales to some government organisations.

(v) Authorised and registered suppliers have to be covered by labour legislations, all of which may not be applicable to the small scale sector.

(w) In view of the liberalised import policies and lower price of landed cost for spare parts, consumers prefer imports to indigenous spares.

(x) In spite of all efforts, the original equipment manufacturer is not able to get complete information at the customer's end on the number of equipments in use, the expected usage, maintenance policies, re-conditioning facilities, repair policies, replacement schemes, equipment discard policies, customer buying behaviour, operating terrain, technology policy of the consumer, customer service expectations, etc. for all the spare parts.

(y) The original equipment manufacturer is not able to get reliable estimates on overall business potential, extent of competition, presence of spurious manufacturers, reaction to fluctuating prices, distributors' influence on price, pressure on upgrading technology, governmental regulations and so he classifies demand as unpredictable, genuine, and bogus.

(z) Often the original manufacturer, because of the above discussed uncertainty in demand, has to face problems such as low turnover, large inventory, high carrying cost, low shelf life, increased obsolescences, small quantity orders, highest customer service expectation, unpredictable demand, lack of comparative data by other manufacturers, and lack of standardisation of drawing structure for effecting cost reduction techniques.

However, in spite of the foregoing problems, it is necessary that the marketing executive take steps to improve the corporate image and build a better buyer–seller relation, by providing the required spare parts at the right time to the customers.

The marketing executive must beware of service tax, value added tax (VAT), goods and services tax (GST), sales tax, octroi, as applicable. The marketing executive should be familiar with various rules, regulations, amendments, corporate laws, corporate social responsibility, company image as he alone represents the company in meeting a very large number of customers, advertising agents, etc. He should be aware of the day-to-day problems relating to central excise customs procedures, invoice based assessment with procedures and documents, modvat on capital goods and others maintenance of records central excise valuation etc. as these will affect the prices. The slogan must be "sell one and make one", "sell today and make today".

Chapter 34

ASS (After Sales Service)

IMAGE OF AFTER SALES SERVICE

In today's Indian context, many consumers feel that the after sales service aspect has received only scant attention. The importance of after sales service becomes obvious if one accepts that the consumers buy an equipment to derive certain continued value and benefit from it. However, in those sectors where the sellers market operates or spare parts are supplied in smaller quantities to the customers, the after sales service needs considerable improvement. The service department is the best point of contact between the ultimate user of the company's products and the company itself. Hence the image of the selling company and goodwill of the user greatly depend upon the calibre of the service department.

Service to others is indeed a noble thought but the consumer, whether in domestic life or industrial sphere, invariably complains about the poor quality of the service and the inordinate delays in the servicing process, e.g. domestic servant, telephones, plumbing, restaurants, transport services, repair work, communication services, contract services, railways, airlines, etc. The complaints of the consumers are more in case of monopoly suppliers and imported machinery. If sellers dominate the market, selling companies take shelter under one of the numerous clauses of items and conditions of sales. The after sales service problem becomes acute if the sellers refuse to take cognisance of the fact that their duty and commitment to the consumer extend far beyond the date of the sale; in fact, till the possible expiry of the anticipated life of the product. When supply shortages and demand gap exist due to poor infrastructural facilities like power cuts, the user is treated like a step-child, which, needless to say, is the contradictory philosophy of marketing, viz., "consumer is the king". This philosophy can be guaranteed only by a mix of equipment reliability and timely after sales service, which is absent in many cases. This is perhaps due to the poor consumer awareness of any movement to protect consumers.

No wonder, the maintenance engineer, who is hit hard by the poor after sales service, describes it as ASS. The poor, cynical and frustrated image formed by the consumer must be changed. For this, the image of the service function should be improved in our country, particularly because of the arrival of increasingly complex and sophisticated high technology equipments, by considering the consumer as the focal point in the total marketing approach. One can well imagine one's frustration when the costly air conditioner or refrigerator does not perform due to poor servicing. However, the service dimension loses its role and gets a low priority when the consumer is not in a position to dictate his requirements in the context of the sellers' market. Service, the backbone of marketing in a competitive environment, should also be extended adequately even in monopolistic environment for gaining a proper image. This is particularly essential in continuous process and spares intensive industries, where substantial loss occurs due to poor servicing. The problem arises as the supplier upgrades his technology to meet the competition once in 5 years and forgets the original buyers. AMC (annual maintenance contract) is to ensure prompt service.

SERVICING GOALS

While the customer service should be provided to all, it needs flexibility in its operationalisation so that it can be rendered more generously to some than to others. Customer services should be administered so that they are offered to those present and potential customers at a suitable time and in a way that is likely to meet the Firm's objectives. A package of customer service should be built around a mix of the following aspects: (a) performance enhancing services; (b) life prolonging services; (c) risk reducing services; (d) effort reducing services; (e) capital reducing services; and (f) efficiency impoving services. Although services represent intended benefits to the users, needless to say, the organisation offering the services should also stand to gain on subsidized basis and self-liquidating or profitable basis. At the end of the guarantee period, a detailed technical report, indicating expected failure time, estimates of quantity of spares needed, estimated time of overhauling, and other aspects for the duration of life of equipment, must be prepared and given to the customer for better customer relations. On completion of warranty, the seller may offer a service contract—with spares and labour—against prefixed premium.

The role of after sales service in the overall marketing strategy can be discussed only with reference to the specific market. The concept is not relevant in the category of consumables like soaps, while in the service sector, such as hotels, airlines, banks, and transport, the objective is to provide service to customers throughout the duration of the sale. After sales service must be given a prominent role in case of consumer durables, industrial equipments, etc., and the reliability of the product performance is enhanced by providing prompt facilities such as supply of spares and doing repairs. Marketing executives admit servicing is always a problem, because of non-availability of spares for old models sold by them. They do not provide drawings of even critical spares.

PROFESSIONALISING CUSTOMER SERVICE

An effective and genuine after sales service, both during and after the normal guarantee period, is the most important link between the original equipment manufacturer and the user. The manufacturer who ignores the after sales service aspect will lose his image and reputation in the long run. On the contrary, good after sales service can enhance the image and also the sale of original equipments. Through prompt service, the seller wins the confidence of the user and is placed in an advantageous position in relation to the competitors. Service department is the best point of contact between the manufacturing company and the equipment users by quick response, prompt repair and minimum repeated failures.

To a large extent, the image of the selling company and the goodwill of the user depend upon the calibre of the service department, the reliability and quality of service, as well as the top management's support to the servicing aspects. The service engineer should realise that he is a representative of the company, and this feeling should be prevalent even when the customer is not happy with the equipment. This demands a high degree of self discipline and understanding on the part of servicing technician. The objective of every serviceman, by organising proper service and satisfying the customer, should lay the ground work for the sale of further equipment. The marketing, technical, sales, R&D come into contact with buyers while servicing. It is difficult to bring legalities in view of image problems. The warranty clauses are usually in small letters, that it is difficult to read.

ORGANISATION AND MANUALS

For effective functioning, the servicing department maintains a close liaison with marketing, production, quality control, procurement, maintenance, warehousing, distribution, design and spare parts division. While every department may claim to serve the customer, a few organisations like BHEL and Voltas have a spares and service division to cater to the customers' requirements. The spares requirement for diesel locomotives, for all the zonal railways is met by the spares stores division of the diesel loco works at Varanasi. The organisation of service department must be based on the estimated demand, span of control and territory factors. The main service activities are repair and maintenance, both in the field and at the company's centres for the trade and end users. The number of technicians/officers in service department depends upon the above factors.

The service department should be so organised as to include the main areas of associated responsibilities like pricing for repairs, service activity, market research, commercial liaison with customers, advertising promotion, credit control, cash inflow, branch efficiency, customer satisfaction, computerisation, salary, training, procuring components from different sources, access to spares, handling customers correspondence, inspection facilities, budget for servicing, preparation of service manuals, etc. Servicing manuals and operating instruction manuals are invaluable documents in reducing the time spent by service engineers at the site. Well documented manuals can help overcome the lack of trained technicians and sometimes these will enable the users of the equipment to employ their staff for minor repairs.

The choice of organising service function on a centralised or decentralised basis is difficult and depends on a large number of factors. In a geographically vast country like India, it is necessary to organise and coordinate regional, territorial, district-wise service centres, necessitated by the equipment populations, with adequate investment infrastructural facilities like machine shop, workshop, welding equipment, lubricant equipments, drilling machines, generators, testing instruments, and inspection facilities. These facilities should be designed to attend to on-the-spot repair and servicing for reducing production stoppages. A country-wide network of distributors and service dealers staff with trained technicians or franchisees, is a basic necessity to achieve quality servicing. Very few large organisations have introduced mobile service units to satisfy consumer needs located in different areas.

Since after sales service is linked with commissioning, installing, and maintenance of the plant, a few organisations include it in maintenance function. In this process, the local contractor—who is a specialised agency—takes charge of the maintenance of all equipments including the service function. It is not always easy for the original equipment manufacturer to carry out all the servicing required by the users. He must, to a large extent, depend on the retailer's support, and the servicing may be faster and economical. The dealer may draw upon the OEM for infrastructure facilities like capital, test facilities, and literature. In addition to the usual sales promotion literature, service support to dealers is given in the form of exhibition displays, demonstration and infrastructure manuals for the user.

COST OF SERVICING

While it is recognised that service is an important function, it must be stressed that administration of service is a costly activity if the quality and reliability of the service has to be ensured. The service team should be well qualified and trained to ensure the quality of service. They should visit the field and this involves a great deal of travelling from the headquarters. The original equipment manufacturers must determine the optimum part of operating by striking a judicious balance between customer satisfaction and profitability. It is obvious that servicing involves high investment in technical manpower, skilled labour, buildings, testing equipments, inspection facilities, workshop, and spare parts. Hence the manufacturer has a right to expect a reasonable return on the investment. The cost of service beyond the warranty period should not be normally attributed to the original purchase price but to the nature of repair job and the replacement needed. For this purpose, standard rates, based on labour charges, spares used, administrative cost, travel, communication and other charges may be worked out and given to customers in advance. As a policy, many firms do not service the equipments, if they have discontinued manufacturing relevant models or if the age of the equipment is beyond 15 years. The dealers have independent mechanics who usually service such old equipment.

MARKET INTELLIGENCE

The after sales service department, in order to be efficient, should have at their disposal experts who have a knowledge of the equipments sold by the company, age of equipment, operating conditions, workers and their operating conditions, spares manufacturing capacity, design charges in future equipments, pricing policy for shares, cost of servicing, etc. The service engineers, on their part, observe the performance of the equipments under different operating conditions. This process enables them to bring in useful data pertaining to the equipment performance in the field, failure rate, etc., and this feedback constantly improves the design features. The reasons for breakdown correlated to diverse working conditions provide a reference base for product improvement. In the Indian environment, a positive attitude to after sales can enhance the equipment reliability and provide a basis for future competitive edge.

The seller normally provides the customer with complete details about the equipment and its various users. The information contains range of operations, limitations, catalogue of spares, part numbers, servicing methods, modifications to be carried out to suit changing needs of the consumers, inventories of parts to be carried, etc. The servicing department also goes to the extent of giving a servicing/overhauling schedule and periodical replacements to be carried out throughout the operating life of the equipment. For this purpose, the service department sometimes goes to the extent of thoroughly studying the requirement of the buyer to ascertain the operational aspects of the equipments and the backup spares they need. The after sales service is reasonably streamlined in the case of automobiles, tractors, commercial vehicles, typewriters, air conditioners, motors, TV sets, and electrical motors. In the transformer industry, the service department is concerned with installing and commissioning. In the chemical industry, the service department assists the buyers in bringing about suitable process modifications. In the computer field, the seller trains the buyer's employees in maintaining the hardware and developing the software. However, the consumer desires that the after sales service in all sectors must improve in India if it is to be compared with that prevailing in the developed nations and for this purpose faster, improved communication is a must.

LEGAL ASPECTS AND PLANNING ASS

Promptness and courtesy are the foundations on which after sales service function should be built. Planning the service function includes planning of right quantity of spare parts, ordering for spares after considering the lead time, planning of maintenance staff, wage checking, and preparation of budgets on the basis of standard time for each activity. For effective planning, information must be generated on pending service jobs. A visit from the service engineer is planned to check the service job (which needs materials) priorities on service jobs, jobs awaiting customer's approval, work in progress, defects in inspection after completion, customer's complaints, etc. Service jobs can be coordinated at the headquarters by means of a computerised network in order to allocate resources on an optimum basis. Information on the service report, like customer's complete addresses, telephone, location of equipment, equipment model, reasons for breakdown, reasons for servicing, etc. may be recorded in order to improve the future product design.

If the delivery time is stipulated in the conditions of service contract, it should be adhered to in order to avoid breach of contract. The customer is entitled to replacing or damages for faulty equipment only if the contract conditions are not broken. The sale of goods act stipulates that statements made in advertisement as well as warranty must correspond to the description. Items must be of a reasonable quality and must be fit for the purpose for which they are bought. If the buyer asks for a product by its trade mark, he gets no assurance from the seller that it is fit for the particular purpose. The legal responsibility, if any, for after sales service lies with the manufacturer only, for the equipments sold by him, and not for the assemblies bought directly from the traders. Where there is no contract, there is no legally enforceable agreement between the manufacturer to service the ultimate customer.

A warranty should leave no doubt in the customer's mind as to what is a mistake and what is an accident. Any restriction on labour, transportation and the equipment itself must be clearly stated. If necessary, the system must cater to the usage of items which may not be available with the warranty team but may be required for the rectification of defects attracting warranty clause. The original equipment manufacturers cannot avoid the responsibility altogether even when the equipment is passed through several customers. Under the common law of negligence, the manufacturer owes the customer a duty to take reasonable care in manufacture. If he fails, and faults due to negligence in manufacture cause injury to the customer or any other person, he is liable to pay damages to any one injured, although a guarantee can and often does not exclude the responsibility to the customer for this. Further, a retailer who incurs loss in respect of defective products can of course attempt to recover the loss from the manufacturers.

SPURIOUS SPARES AND A–Z PROBLEMS

The discussion on servicing will not be complete without touching the issues related to the spurious spares and other problem areas related to ASS.

(a) The spurious spares manufacturers thrive in markets where there is a good demand for servicing and replacement spares with huge profit margin. Such suppliers are mostly local manufacturers with the ability to meet emergency requirements immediately, with less transport costs.

(b) These manufacturers copy the original product, incurring negligible cost in developmental expenses and this enables the spurious sellers to price the item at very low levels compared with genuine suppliers. Made in USA—Ulhasnagar Sindhi Association—is a typical case in point.

(c) The problem has become acute recently due to the liberal licencing in scooters, TV, automobile sectors and airconditioners where the consumer runs from place to place to get discounts, rebates and for price haggling.

(d) The Indian buyer, who may not be a discriminating buyer, becomes a victim of spurious manufacturer by paying exhorbitant prices eventually.

(e) Thus the consumer-conscious manufacturers have to face with unethical competition from unscrupulous manufacturers who offer gullible buyers an apparently similar product with inferior quality at a very low price.

(f) The original equipment manufacturers, may tackle the problem by adopting the following measures.

(g) First, they can enter into annual contracts with major users for supply of spares, and second, they can plan separate capacity only for spares manufacturing with an assurance of continued supply.

(h) The spare part pricing may be done in a pragmatic way so that the spurious suppliers do not have any incentive to sell them.

(i) The original suppliers can contemplate on legal remedies for patented suppliers.

(j) A ready availability of genuine spares at a reasonable price will eradicate spurious spare parts.

(k) The firms marketing consumer durables are usually faced with another problem: whether to go in for their own costly modern service centres or ensure that the dealers invest in a service set-up after considering aspects such as overhead expenses, pilferage, working capital for slow moving spare parts inventory, wage bill, labour indiscipline, technical staff, quality of service, etc.

(l) Service managers agree that instances of the technical staff doing private work covertly are not uncommon.

(m) If the after sales service job is assigned to dealers, it may mean compromise of quality at some stage.

(n) In order to tackle the aforesaid problems, great emphasis must be laid on customer education.

(o) The customer must become quality conscious and demand proper after sales service by paying the correct price.

Some other problems experienced by the servicing department of original equipment manufacturers include the following:

(p) Investment of working capital in slow moving spares.

(q) Bottlenecks in transportation—customer is located in far-flung areas.

(r) Communication gap with the user.

(s) Wide range, large variety and sizes of spares, leading to problems of identification and availability of correct spares at the right time.

(t) Difficulties in carrying out cost-benefit analysis.

(u) Centralised versus decentralised service points.

(v) Inability to assess the correct demand for services as the consumer does not adhere to generating instructions.

(w) Long lead time associated with imported spares.

(x) Inability to know how long the OEM should continue to serve only one customer when models change due to technological upgradations.
(y) Low morale of the service staff.
(z) Customer switching over to spurious manufacturers due to non-availability of critical spares from the OEM on an emergency basis.

The above list of problem areas is only illustrative and not exhaustive. But, keeping in mind the competitive environment of today's market, the strategy of a good seller would be to adopt a balanced approach in pricing, backed up by appropriate after sales service in order to achieve customer satisfaction, by providing right spares, with right way of working, by rightly trained personnel and with right dealings.

Chapter 35

Multiechelon Distribution

MULTIECHELON CONCEPTS

In the field of marketing and after sales service in very large organisations, it is common to stock the same item in a large number of depots/stores/places/echelons. This phenomenon occurs as the user expects that the manufacturer of a machine to be responsible for the satisfactory supply of spares of the machine when the equipment is in operation. Similarly, firms selling spares or involved in after sales service, often operate the spare parts depots nearest to the consumer organisations. This holds also for large organisations consuming a lot of spares, such as road transport undertakings and the different branches of the defence services, viz. Army, Navy and Air Force. In the above cases one stocking point at a higher level serves as a warehouse to another stocking point at a lower level, and such systems are known as multiechelon spares inventory system. There is bound to be interaction between the different stocking points as demands from one or more substores for the same item will arise on the main warehouse. Here the stocking points would be the factory, main warehouse, wholesale depots, stockists, regional stores, retail outlet, etc. The number of locations of the substores and quantity to be stocked in each store can be optimally arrived at after considering different parameters such as demand pattern, order processing cost, carrying charges, stockout cost, and lead time.

The conventional inventory model is a major simplification of the multiechelon model as we deal with only one facility at a time. In this process we are concerned with the optimisation of the performance of a single store only, without considering the effects of those parameters on other stations within the system. With individual factories supplying spare parts to customers, the basic assumption made is that the supply is unlimited and from a continuous source. But, if we work out the inventory policy for each station, assuming that the supply is from an infinite source, then the policy would no longer be optimal in practice since the supply station at the higher echelon is not an infinite source and there is bound to be shortages during some occasions. The inventory policy in such cases should aim at minimising the system costs as a whole, the multiechelon theory deals with problems of this nature where the aim is to determine optimal policy for the system as a whole and not for any individual substores in a location.

In inventory systems of practical nature, a relationship exists between two or more locations. These relationships may take many forms like sharing of common resources, cost interactions, supply and demand interactions, and manpower skill-mix interaction. In this context, items under common control, of which there exist internal supply and demand, are said to belong to an echelon. Multiechelon problems can be solved by mathematical models to minimise the shortage, inventory carrying, order processing, and other parameters, since external demand greatly affects the highest echelon closest to external supply. The optimal solution is based on the single echelon model with the introduction of a penalty for insufficient stock in highest echelon and discounted cash flow.

MULTIECHELON CHARACTERISTICS

As discussed earlier, a multiechelon system can be defined as a series of production on supply facilities where any changes in the policy parameters in one facility will affect the other facilities either directly or indirectly. A typical multiechelon spares inventory system is one involving factory, central warehouse, regional warehouse, regional depots and retail outlet, in the marketing distribution channel. Here the retail outlets are defined as the first echelon, the regional depots the second echelon, the central warehouse the third echelon, and so on. It is possible to have more than one facility at each echelon. While many inventory systems in industrial products are multiechelon in nature, due to operational difficulties and coordination problems, the system is controlled by a single organisation. In such situations the inventory policy should be so designed as to consider the system as a whole. A multiechelon system can be looked upon as a network with the nodes used to represent the various facilities and flows represented by linkages. The networks usually considered in practice are those which have at most one incoming link for each node and the flows are a cycline, i.e. the flows do not return to the same node. Such a structure is termed as *arbove essence* or an inverted tree structure.

CLASSIFICATION OF MULTIECHELON PROBLEMS

The multiechelon models have been extensively researched by Clerk (*Management Science,* Vol. 6, 1960) in his article on optimum policies for multiechelon problems and classified them as follows:

(a) Deterministic/stochastic. In a deterministic model, external demands of each facility are known in advance with certainty. For a stochastic model, the demands are assumed to be known with a given probability distribution or conditional distribution, as used in Bayesian approach.

(b) Single product/multiproduct. A single product model deals with only one product at a time, assuming no interaction between products, i.e. all products are dealt with independently. In the multiproduct model, there is at least one interacting variable between the products like formation of budgets.

(c) Consumable product-repairable product. In a repairable product model, some or all items issued to meet external demands are regenerated in the system after repair or failure. In the consumable product case, the item is permanently lost to the system.

(d) Stationary-non-stationary. In the stationary model, parameters used to define external demands are assumed to remain the same over a period of time. In the non-stationary model, this may vary from time to time.

(e) Continuous review-periodic review. In a continuous review model, stocks are reviewed continuously and orders placed whenever a certain level is reached. In a periodic review model, stocks are reviewed after a fixed period of time as practised in a large number of cases.

(f) Backlog or no backlog. In a backlog model, backlogging of demand is assumed in case of short supply. In the no backlog model, unsatisfied demand is assumed to be lost.

INTER-ECHELON INTERACTIONS

If there are three echelons, then stocking point 1 at echelon X receives the spare part from stocking point 2 at echelon Y, which in turn is replenished from stocking point 3 at echelon Z. This process can be continued if there are more echelons. Sometimes it may be possible that X not only receives stocks from Y but also directly from Z. For an organisation selling spare parts, Z, Y and X may represent the factory manufacturing spares, the warehouse, and retail outlets. This implies that many inventory systems in actual practice are multiechelon in nature. Since each echelon has to get the stocks from the higher echelon, there will be interactions between item. These interactions are needed to minimise stockouts at a given stocking point. For this purpose, the time taken at stocking points at higher echelons to meet the demand placed on them by the stocking point in the lower echelon is important. This time is known as the response time and is a function of stocks held at the higher echelon.

It is quite conceivable that in the same echelon different stocking points may be owned by different organisations. In such a situation each organisation can determine its area of operation and its own policy of controlling inventories it possesses. It is logical that such organisations are concerned about their own inventories and stockouts rather than the entire system. Thus, before studying a multiechelon system with a view to formulating policies pertaining to stock levels, the implications of different policies must be carefully evaluated. A special situation may well arise, where the entire system with all the echelons may come under the control of one organisation. But even when a single organisation controls stocking points at different echelons, the amount of freedom enjoyed by the executives at different levels is usually high. The multiechelon is similar to suppliers factory—suppliers godowns, suppliers wholesalers, suppliers, regional/retail outlet-buyer, channel of distribution.

Similarly, the repair scheduling at depots interacts with stock requirements, at different echelons. Another interaction arises from the determination of base level order quantity for spare parts independently of the higher echelons. The frequent ordering directly affects the operating costs in the higher echelon and the size of the order affects the apparent variance of depot issues and increases the required stocks. The interactions between the echelons become more complex if the spare parts are interrelated and if we have to deal with multiple items. From the above discussions, it can be observed that the problem becomes one of allocating successive units of each spare part to one of the bases so as to achieve the maximum improvement in performance. In the multiechelon system, the performance is measured at the lowest echelon by expected number of outstanding back orders.

PARAMETERS IN MULTIECHELONS

The first step in analysing the system of multiechelon inventories is to define the parameters: these are lead time between the echelons, demand distribution function, inventory charges, ordering costs, order processing charges, shortage or backlog costs, etc. Exhaustive computations are necessary to arrive at optimal provisioning policies at different echelons. An important aspect that we will be considering in detail is the effect of changes in the demand of one echelon and its implications on other echelons. Multiechelon problems are tackled by mathematical techniques and also by computer simulation models.

MATHEMATICAL APPROACH

Let us now consider the interaction between two intermediate echelons, where the lower echelon is the field warehouse and the higher echelon is the regional warehouse with repair facilities. We are here considering a situation where there is a consumption of spare parts in the field warehouse which is being replenished by the regional warehouse. The spare parts to be repaired are sent to the regional warehouse where they are repaired to the extent possible. It is assumed that for a specific duration of time period, T_1 (i.e. a month), consumption takes place during which period there is no supply from regional warehouse. The replenishment is done through either repair of the parts which have failed during the T_1 period or by purchase. In this approach we are interested in arriving at optimal purchase quantity for various rates of failure or repair.

For different failure rates, repair rates and service levels, the optimum purchase quantities have been shown in the next section. A sample print out using a computer for optimum stocking policies is given in Table 35.1 for 99% service level. It can be seen from the table that for a given service level and repair rate, the optimal purchase quantity increases as the failure rate increases.

Table 35.1 Optimum Stocking Policies

\multicolumn{14}{c}{Months}
1	2	3	4	5	6	7	8	9	10	11	12	13	14
\multicolumn{14}{c}{Demand in retail level}
100	200	200	200	200	200	200	200	200	200	200	200	200	200
\multicolumn{14}{c}{Projected demand through 4 months moving average at dealer level}
100	125	150	175	200	200	200	200	200	200	200	200	200	200
\multicolumn{14}{c}{Required operating safety stock at dealer's level–2 months}
200	250	300	350	400	400	400	400	400	400	400	400	400	400
\multicolumn{14}{c}{Spare stocks at dealer's level after meeting the demand}
100	0	50	100	150	200	200	200	200	200	200	200	200	200
\multicolumn{14}{c}{Demand at regional depots}
100	250	250	250	250	200	200	200	200	200	200	200	200	200
\multicolumn{14}{c}{Required operating safety stock level—Four months}
400	480	560	640	700	760	800	840	900	880	840	820	800	800
\multicolumn{14}{c}{Projected demand through 12 months moving average method at plant level}
100	120	140	140	160	175	190	200	200	225	220	210	205	200

(Contd.)

368 Maintenance and Spare Parts Management

Table 35.1   Optimum Stocking Policies

\multicolumn{14}{c}{Months}
1	2	3	4	5	6	7	8	9	10	11	12	13	14

Months													
1	2	3	4	5	6	7	8	9	10	11	12	13	14
\multicolumn{14}{c}{Stock at plant after meeting demand}													
300	150	240	310	390	500	560	600	640	700	680	640	620	600
\multicolumn{14}{c}{Order to manufacturing at purchasing}													
100	330	330	310	260	240	260	180	160	180	160	180	180	200
\multicolumn{14}{c}{12 months moving average of projected demand for plants ordering}													
100	120	140	160	175	190	200	210	225	230	230	240	250	240

For a given failure rate and service level, it will be seen that the amount to be stocked steadily decreases and ultimately reaches zero with increasing repair rates. This is so because with the expanded repair facilities, the need to stock through external purchase will be minimum. From the analysis of the computer print out, it can be concluded that the optimal quantity is more sensitive to the rate of failures than on repair rate, indicating that consideration of failure rate is crucial in arriving at a decision as to whether repair facilities are required at the concerned echelon.

MATHEMATICAL DERIVATION

Let x be the number of parts failing in at the field warehouse and y be the number of spare parts repaired in regional warehouse during the lead time of supply from the factory. Obviously, x and y depend upon the failure rate m and repair rate n. Let K_o and K_u represent the overstocking and stockout cost of the particular spare part with $P(x)$ denoting the probability of demand of x spares.

$Q(y/x)$ is the probability of repairing y spares given that there are x spares available for repair $= (x, y)$.

$$R(t) = \sum_{x=t}^{\infty} Q[(x-t)/x][(P/x)]$$

Assume that $(x - t)$ spare parts failure rate is a Poisson distribution with m as the mean failure rate and the repair time varying with $1/n$ as the expected time required for repairs, n being the mean repair. Then $R(t)$ can be derived as

$$P(x) = \frac{e^m m^x}{\lfloor x}, \quad Q[(x-t)/x] = \sum_{n=x-t}^{\infty} \frac{e^n n^{x-t}}{x-t}$$

$$R(t) = \sum_{n=x-t}^{\infty} e^{-m} \left(\frac{m^x}{x}\right)\left(\frac{e^n n^{x-t}}{\lfloor x-t}\right)$$

If x represents the purchased stock of spares, then the expected surplus stock is

$$\sum_{t=0}^{x}(x-t)R(t)$$

The expected shortage is

$$\sum_{\dot{x}}^{\infty}(t-\dot{x})R(t)$$

The term $\sum_{t=0}^{x} R(t)$ represents the cumulative probability of surplus stock up to a stock x^0 for various values and can be represented as $F(x^0)$ The term $\sum_{t=x+1}^{\infty} R(t)$ is therefore equal to $[1 - F(x)]$. The system can now be analysed for various stocking policies after introducing the understocking and overstocking costs.

If the stocking policy is x^0 purchased parts, then the expected cost is given by the following equation: $E(\dot{x}+1)$ can be arrived at by putting $(\dot{x}+1)$ instead of $\dot{x}$. We can get $E(\dot{x}+1) - E(\dot{x})$ by simplification. Thus,

$$E(x^0) = K_o \sum_{t=0}^{x^0}(x-t)R(t) + K_u \sum_{t=x+1}^{\infty}(t-x^0)R(t)$$

$$E(x^0+1) - E(x^0) = K_o \sum_{t=0}^{x^0} R(t) - K_u \sum_{t=x^0+1}^{\infty} R(t)$$

$$E(x^0+1) - E(x^0) = K_o F(x^0) - K_u[(1 - F(x^0)] = K_o + K_u F(x^0) - K_u$$

Hence,

$E(x^0+1) < E(x^0)$ if $F(x^0) < K_u/(K_o + K_u)$

$E(x^0+1) > E(x^0)$ if $F(x^0) > K_u/(K_o + K_u)$

Thus, it is clear that the cost is minimum when x^0, i.e., the stock of purchased spares is such that $F(x^0) > K_u/(K_u + K_o)$ or service level. The optimum stocking policies are given in Table 35.2 for different failure rates and repair rates.

Table 35.2 Optimum Stocking Policies

Failure Rate m	Repair Rate n							
	0	0.1	0.10	0.50	1.00	2.00	3.00	4.00
0.01	0	0	0	0	0	0	0	0
0.10	1	1	1	0	0	0	0	0
0.50	3	2	2	2	2	1	1	1
1.00	4	4	4	3	3	3	3	3

(Contd.)

Table 35.2 Optimum Stocking Policies (*Contd.*)

Failure Rate m	Repair Rate n							
	0	0.1	0.10	0.50	1.00	2.00	3.00	4.00
2.00	6	5	5	5	5	4	4	3
3.00	7	7	7	7	6	6	5	4
4.00	9	9	9	8	8	7	7	6
5.00	11	11	11	10	9	9	8	7

OPTIMUM STOCK LEVELS

Let us consider a multiechelon system constituted by plant warehouse, regional depots, local dealers and retail outlets. The direction of supply will proceed in this order while the direction of demand and information feedback will be in the reverse order. In this system, the plant warehouse is the topmost echelon which supplies to the regional depots which in turn meet the demands made by the local dealers who are responsible for supply to retail outlets. Let us consider a situation where the demand at the retail outlet is steady at 100 units. However, when a change occurs in the demand, the system behaviour is far more complex than one thinks, A small increase in demand in the lowest echelon is progressively magnified as it passes through successive higher echelons and ultimately when the highest echelon is reached, the demand may be several times the anticipated increase in the lowest echelon. Thus magnification is due to the fact that change in demand affects a part from operating stocks, safety stocks and order quantities. In addition, the change in demand also affects the pipeline quantity and overtime. This is due to the forecasting techniques used to assess the demand at different echelons. If the demand at the retail level changes from 100 units a month to 200 units a month, then it introduces several changes. Let us assume that at the regional depot level, 8 monthly moving average is used to forecast the dealer's demand, and as a policy, four months of the forecast requirements are carried as demand, and as a policy, four months of the forecast requirements are carried as operating stocks. Let us assume that the plant uses 12 monthly moving average to forecast the demand and plans its manufacturing and purchasing activities.

Table 35.1 details the effects through a computer simulation. We see that the dealers perceive the demand hike of the retailers in a steadily increasing manner from the initial value of 100 to the ultimate value of 500 in the fifth month. However, the regional depots find that the demand placed on them by the dealers increases from 100 to a maximum of 225 in the ninth month before it stabilises to the correct demand as it increases from 100 in the first month to 250 in the 13th month before it stabilises in the 25th month to the correct demand of 200 units. It should be noted that the increase of 100 units in the retail level has blown the operating safety stocks to 400 units at dealer level and to 800 at the regional level. It will also be noticed that the operating safety stocks themselves stabilise over a period of time.

When the number of levels increases and forecasting methods vary, the multiechelon systems will become complex and computers are needed to simulate the behaviour of multiechelon system over a period of five years or more into the future. In practice, a dealer caters to the need of many retailing units and a demand change in each retailer can give rise to tremendous application at dealer level and above. The computer simulation will enable the decision maker to adopt the right kind of optimum stocking policies at different echelons.

Chapter 36

Management of Obsolescence

OBSOLETE SPARES PROBLEMS

Obsolescence, non-moving, slow moving and SOS management *scrap,* are the words not liked by any prudent manager, even though they occur commonly in the spare parts field of every organisation in all the sectors of economy. Obsolescence is observed in several areas such as management, technology, spare parts, stores items, and even human life. The challenge before the spare parts managers in the context of obsolete and non-moving spares is how to reduce if not eliminate the incidence of obsolescence. This is same as human being who grow from childhood to youth, old and disappear.

According to the former Union Finance Minister, Shri C. Subramaniam, about ₹2500 crore worth of spare parts are obsolete. The present estimate of slow moving spares will definitely be more, i.e. over ₹10,000 crore, due to inflation, economic growth and industrial development. The problem of obsolete items can be well appreciated considering the fact that the inventory carrying capital is 30% per year. The Warehousing Manager has the unpleasant task of getting the supply of the items every day and he has to provide space for stocking these items. Our research studies indicate that about one-third of spares inventory is non-moving in nature and has a tendency to become obsolete. Since the obsolete materials are not issued from stores, they cannot be hypothecated, to get the working capital from the banks. It is well known that spare parts have a tendency to become obsolete at a faster rate than the other stores materials. Our experience and research studies show that every organisation, irrespective of the size, sector or structure, contributes in varying degrees to this national loss.

SOS ITEMS CONCEPT

The commonality of surplus, obsolete, scrap (SOS) items is that they have to be disposed of from the stores and hence let us identify them by defining the terms. A distinction can be made between surplus and obsolete items in that the former can be consumed at some future time, while the obsolete item is unlikely to be consumed in future. Surplus materials arise because they are in excess of a reasonable rate of consumption due to wrong judgement at the procurement stage. Scrap can be defined as the residue from a manufacturing process which cannot be economically used within the organisation. Obsolete spares are those spare parts which are not damaged and which have common economic worth but which are no longer useful for the company's operation due to several reasons. Sometimes, the maintenance engineer, who is responsible for indenting, tends to escape his responsibility by declaring an item as obsolete before writing off, by categorising it as "slow moving" and avoids risks by asking the Stores Manager to keep them in stock for future. Our research studies indicate that the insurance spare parts constitute a major portion of obsolete spares.

OBSOLESCENCE A–Z ISSUES

In order to minimise the accumulation of obsolete items, let us now analyse the causes for obsolescence.

(a) One of the most common reasons for obsolete spares is the transfer of the project surplus items—increased quantity purchased due to the overenthusiasm before commencement of project and which may not fit regular consumption in future—to the main stores.

(b) The most important cause for obsolete stores is that all organisations do not pay sufficient attention to initial provisioning of spares at the time of procurement of capital machinery. This is observed as the most error prone area in all cases.

(c) Introduction of non-planned sudden technological changes or design modifications, without adequate preparation, renders spares of old machinery obsolete.

(d) Maintenance staff tends to categorise many non-moving items as insurance spares, which are not likely to be used for a very long time. It is desirable to keep only one such standby assembly to avoid incidence of obsolescence.

(e) Adoption of standardisation and elimination of non-standard varieties have led to obsolescence of the non-standard spares.

(f) When a machine breakdown occurs, it is rectified by using parts of an identical machine. This process of cannibalisation is quite common

in many agricultural/irrigation/power generation projects which have poor transport and communication facilities. When continued unchecked, the remnants of the cannibalised machinery become obsolete.
(g) The recent concept of modular approach of replacement, or removal of a total assembly, often renders obsolete a few parts in the replaced assembly.
(h) Sometimes, the marketing department may have projected a sales forecast which might be on the higher side. Any materials planning has to be based on sales forecasts, and this could result in surplus items which will become obsolete in job shop industries.
(i) Suboptimising decisions like bulk buying to take care of discounts or freight advantages to take care of economic batch quantities of manufacture, without adequately considering factors such as shelf life, storage space requirements, and technological changes, leads to obsolete spare pans.
(j) In some organisations spare parts continue to be ordered on the basis of previous year's operation even though the equipment has been phased out of commission, due to communication gaps.
(k) The stores continue to keep spares even though the equipments using the spares have been sold.
(l) The stockout costs or consequences of non-availability of spare parts are invariably higher than the spare part cost and organisations take the line of least resistance by stocking more, resulting in surpluses and obsolescence.
(m) The Purchase Manager feels that the spare parts may not be readily available from the OEM in future due to technological upgradation at his end, and hence there is a tendency to overbuy the requirements, leading to surpluses and obsolescence.
(n) The buyer feels that the future price of the spares may be greater than the cost of equipment itself and hence there is a tendency to hoarding spares, leading to obsolescence.
(o) Sometimes, unwanted spares are procured as there is fear that funds may lapse and future import quota may be reduced, or to take advantage of the existing liberal licensing.
(p) Importing spares with original equipment involves less customs duty, whereas the duty will be more if spares are imported separately. Hence lifetime requirements is usually imported—within permissible limits—and the surplus spares become obsolete later.
(q) Improper material handling and incorrect codification, by using only suppliers' part number, lead to obsolescence.

(r) Faulty store keeping methods, improper binning, incorrect records, and inadequate application of preservatives in a few cases cause obsolescence.

(s) Adopting an overcautious approach by treating every spare part as insurance item results in accumulation of obsolete inventory. This phenomenon is particularly witnessed in spares intensive industries like transport, process, and mining.

(t) Fear of nonavailability of good quality imported spares and price rise often leads to unnecessary accumulation of non-moving items.

(u) Excess forecasting by marketing departments without checking the veracity of orders, and overindenting by users results in non-moving stocks.

(v) Purchasing spares in anticipation of expansion, which may not materialise, may lead to obsolescence.

(w) Provisioning of spares on the basis of increased capacity utilisation leads to obsolescence in a few cases.

(x) If the usage is very low, then some spare parts have to be carried as in the stores, which invariably become obsolete later on.

(y) Obsolete stock calculations are done by companies to determine how much of their stock on hand is unlikely to be used in future.

(z) The financial value of stock obsolence can be entered into general ledger systems to create a 'stock obsolescence provision which can reduce the tax liability of a firm.

(z1) Changes in product mix, fashion, style, modernisation, rationalisation, etc. to meet market demands may render the existing stock of spares obsolete.

(z2) Inability to carry out the failure analysis and forecast the future requirement or spares sales, leading to non-moving and obsolete stock.

(z3) Typically a stock obsolescence report uses the value of "stock on hand as a starting point and then reduces this value based on that potential that stock will be used in future".

SOLUTIONS TO OBSOLESCENCE

"A problem well identified is half solved" goes the famous saying. In each of the above a–z reasons for obsolescence, one can infer the solution as better planning, forecasting, use of scientific methods, information systems, records and communication. For instance, the design changes, technological upgradations, and equipment discard policies, product mix changes, file, must be communicated in advance to the parts executive. This will enable him to freeze the pending orders and modify his plans accordingly. The buyer must try to introduce the 'buy-back' clause at the time of initial provisioning, clearly emphasising his right to exchange the non-moving spares with fast moving ones during the next 10 years.

In a few organisations, "effective point advice" or EPA has been introduced to communicate to all departments the proposed design changes, details of new material requirements, the time of changing, consequences on inventory, phasing out equipments, etc. EPA helps in reducing the stock of non-moving items, cancellation of orders for such items, placing orders for buying or manufacturing the required new items. It is claimed by these organisations that the EPA system helps in better coordination for profitable introduction of changes with minimum side effects such as accumulation of obsolete items. Life cycle costing and cost-benefit analysis for phasing out old equipments may be introduced. The suppliers should give adequate notice to the users of the equipments about the technological upgradation, resulting in non-availability of spares of old equipments. Standardisation of the equipments must be done at the industry level by industry associations, Directorate of Technical Development and the Bureau of Indian Standards, so that the interchangeability of spares from one plant to another in the same industry may be thought of. This will not only reduce the working capital locked up in slow moving spares but also reduce obsolescence at the unit level.

MOVEMENT ANALYSIS

The parts executive must periodically carry out the movement analysis in order to know the pattern of demand and items that are likely to become obsolete. The items in the warehouse are classified into fast moving, slow moving, and non-moving, depending upon whether the part has been issued or sold at least once within the last one year, once in the last five years, and not even once in the preceding five years. This is called FSN analysis and the year limitations as well as the number of categories may be changed to suit the convenience of the organisation, and the use of computer is availed of to carry out this analysis. The real insurance items, which would not have been issued in the last five years, are excluded from the analysis. It is also necessary to carry out periodic stock rotation of all the spares, and movement analysis enables to do the same. Movement analysis can be carried out separately for indigenous spares and imported items. A typical movement analysis carried out in a process plant is reproduced in Table 36.1.

Table 36.1 Movement Analysis

Category	No. of Months Elapsed Since Last Issue	No. of Items	Per cent	Spares Inventory Value ₹ Million	Per cent
F	0–12	24,800	25	248	25
S	12–60	47,000	48	353	35
N	above 60	26,200	27	395	40
Overall		98,000	100	996	100

XYZ ANALYSIS

The XYZ analysis enables the Parts Manager to know how the inventory values are distributed in the stores. The procedure is similar to the traditional ABC analysis, with inventory value replacing the annual consumption value. After the annual stocking and valuation of the materials, they are arranged in a descending order according to the inventory values. The cumulative total value is entered against each item; the descending number of the item is found as a percentage of the total number of items, and the cumulative total value is found as a percentage of the grand total value in the stores. In most cases it is observed that about 10% spare parts account for about 70% in terms of inventory value, with the middle 20% number containing about 20% in terms of value, and the remaining 70% number accounting for about 10% of inventory value. These categories are called XYZ analysis. The analysis is initially done for items not moved beyond five years, numbering 26,200 and valued at ₹395 million as illustrated in Table 36.2.

While further control is needed for Y and Z categories, immediate action is called for by parts executives/materials/maintenance personnel for the 2700 non-moving X category items as the working capital commitment is quite high for this category of spares. The Materials Manager should cancel all pending orders for these categories.

Table 36.2 XYZ Analysis

Category	No. of Spare Parts		Inventory Value (₹ Million)	
	No.	Per cent	₹ Million	Per cent
X	2700	13	280	71
Y	5300	21	78	20
Z	18,200	66	37	9
Overall	26,200	100	395	100

The users will be asked to indicate for all category spares as to whether they are likely to use the spare parts in the near future, and the probable date of requirement. Items classified as real insurance spares must be segregated from the group and only one set per machine must be retained in the stores. They will also be requested, wherever possible, to modify the spares in the workshops and free issue of these items may be made. Obviously, the user will take a very long time to decide in declaring an item as obsolete. Government departments and public sector undertakings usually compile huge list of such non-moving/obsolete items and circulate to other organisations. But the response invariably is not helpful as each organisation has got its own problems of obsolete spares. If the organisation has introduced a 'buy back' clause, then the supplier can be contracted for replacing them with required spare parts.

DISPOSAL OF OBSOLETE ITEMS

In most cases, the companies do not expect to realise the original value by disposing of the obsolete items. Needless to emphasise, non-ferrous spares and imported items may appreciate in value and care should be taken to segregate them if the utility is assured. For other categories of spare parts, financial sanction must be obtained to write off the value of obsolete items. The Finance Manager will oblige only during those years when the company makes adequate profits. If the management does not want to take decisions for disposing obsolete spares—as happens in government and public sector undertakings due to fear of audit queries and vigilance problems—then such non-moving items must be segregated to minimise clerical efforts on record keeping. Reader is advised to refer to Chapter 23 Music-3D for further details.

Disposal of surplus, obsolete and scrap items is not an easy job. On the one hand the parts executive has to deal with a set of hard bargainers, who invariably form themselves into cartels and on the other defend before audit with regard to realisation. Hence this subject is more of an art and no amount of theory can replace the experience of a manager who has the knack of convening obsolete spares and scrap into money.

In practice, it is profitable to break or mutilate the obsolete spare parts, so that they are not returned to the same organisation in new packing. It is essential that spares and scrap be segregated by metal and size, particularly when costly scrap such as copper, aluminium or tungsten is involved. Oil scrapped parts must be kept separately. A mixed lot will procure only the price of the cheapest item. If these items cannot be used within the organisation, then it is desirable to sell the scrap to companies who use scrap as inputs. A proper account of the scrap must be maintained as scrap also fetches money. Before disposing, the highly fluctuating prices in scrap market must be considered.

The normal procedure adopted in many organisations is selling by the common tender system. In the extreme case as in USA, organisations may have to pay money to remove the waste occupying huge storage space: Parties are normally required to inspect the scrap and deposit the requisite earnest money. At times, the finance department insists on a minimum value. It is interesting to note that the Director General of Supplies and Disposal (DGS&D), sells over ₹200 crore worth of scrap and obsolete materials every year.

The usual procedure here is to invite quotations from scrap vendors at periodic intervals to avoid accumulation. A vendor is chosen for each category of scrap based on best offer, financial capability and other factors. Finished goods and spare parts must be destroyed to comply with all excise formalities.

After getting written contracts, a security guard must be posted while loading the material. A gate pass/delivery challan giving details is prepared by the Warehousing Manager and the bin card entries corrected if necessary. Instead of the tender system, sometimes auction is conducted for selling the scrap. In all cases, central excise clearance and sales tax formalities must be adhered to.

Chapter 37
Reconditioning

NEED FOR RECONDITIONING

Reconditioning of equipments and spare parts has assumed significance in the field of spares planning and maintenance of equipments in a capital scarce country like India. Even in affluent countries, the ship repairing industry is well organised. The energy crisis has forced the recycling of used fuel in many oil-consuming industries, e.g. reuse of processed water in paper, sugar, petroleum, refining, chemical and pharmaceutical industries. Industries are even showing interest in converting millions of litres of sewage into reusable water.

Reconditioning of old equipment has a significant role to play in developing economies like India due to prohibitive costs of new equipment, lack of adequate foreign exchange resource for machinery imports, and scarcity of capital. Once the life span of a machine is over, its reliability decreases rapidly and planning of spare parts becomes a difficult task. The components have a low residual life and could fail without a warning. Reconditioning or partial repairs tends to strengthen certain localised areas, but the risk of breakdown in areas not reconditioned continues to remain. The power in the machine is limited to the weakest component and the reliability of the machine is defined by the most unreliable mechanism in the equipment and reconditioning or partial repairs tends to postpone the problems for a future period.

RECONDITIONING CONCEPT

Reconditioning of an equipment may be defined as a planned systematic engineering activity designed to restore the equipment to its original sound performance condition. It is a practical exercise intended to bring back the equipment—workout after long use—to its original reliability and performance state. This process involves repairs to certain problematic areas in the machine, like grinding and scrapping of guideways to remove slackness due to wear and replacement of certain parts depending upon the intended performance of the machine. A finer distinction is made in the context of rebuilding which is defined as similar to the manufacture of a new machine by stripping down of the complete machine componentwise, in order to perform to the original standard envisaged.

The process of reconditioning is very similar to the manufacture of a new machine and often network analysis techniques like PERT/CPM/CPA are useful in this process. This involves the following activities:

(a) Stripping the machine componentwise
(b) Degreasing and cleansing the equipment and parts
(c) Total inspection of components and determining their residual life
(d) Replacement of all components which have a residual life of less than 7 years and matching components
(e) Replacement of all mandatory components
(f) Inspection of all castings and repairs to castings
(g) Grinding of main grindways and hardening, if possible
(h) Scrapping of all making guideways
(i) Replacement of all jobs
(j) Assembly and testing of all sub-assemblies
(k) Reassembling the complete machine, and achieving OEM standards of alignments
(l) Load testing of the machine to OEM standards
(m) Complete repainting and change of name plate wherever necessary
(n) Inspecting the machine (in case of rebuilding) to perform to OEM standards after completion.

REPLACEMENT vs. REPAIR

The maintenance executive and the Spare Parts Manager often face the dilemma as to whether a part should be replaced or reconditioned. Such parts which require considerable degree of specialised skills and costly processes are best replaced on failure or provisioned adequately at the time of initial procurement. In some cases the advance in technology has been so rapid that replacement of a part can be more advantageous by way of increased output, greater reliability, and less variety of spares. For this, the technical personnel need to scan the environment thoroughly.

If the repair facility available is limited and optimal use can be achieved by replacement of certain machines with technologically superior equipments so that the other older machines can be repaired expeditiously, then such a selective replacement-cum-repair policy can pay rich dividends in the long run, such a step will also permit better use of repair shop according to the priorities and lesser variety of jobs to be tackled. If a cheap and fast replacement part like bearing is available, reconditioning is not the correct alternative. Normally, parts of capital equipment can be classified as repairable parts or replacement parts. So when a part fails and if it happens to be a repairable part, then decisions to repair manuals, utilities, shop facilities, and improve the quality of skilled manpower have to be taken.

There are also a number of advantages arising out of rebuilding. To start with, training of operators is not necessary as they are already used to the machine. The jigs, fixtures, tools, and tackles available with mechanical department can be continued to be used and the cost of capital of the equipment goes down. The economics of the decisions with regard to replacement vis-a-vis repair have to be worked out at the following stages of the capital equipment: At the design stage, the equipment can be designed for high maintainability, i.e. for easy repair by providing throw away replacement parts and easy accessibility to the critical part. At the ordering stage, initial provisioning can be done after considering repair facilities and maintainability aspects.

RECONDITIONING TIME

In general, reconditioning is resorted to in the industry when the following situations arise:

(a) Whenever the machine is imported, reconditioning is made compulsory for not only saving the foreign exchange, but also for ensuring economy in consumption of spare parts.
(b) Critical mechanical components like shafts, gears, pumps and valves must retain dimensional and structural integrity so that the performance is obtained for maximum periods of time before replacement becomes necessary due to breakage. Even if they wear out, they should be reconditioned and reused as long as possible.
(c) Supply of machines/spare parts is unreliable and patented.
(d) The OEM has discontinued the manufacture of the present model and hence is not in a position to supply spares.
(e) Replacement of existing machine is costly and the choice of the new machine introduces compatibility problems with the existing machines.
(f) The budget does not provide enough funds for covering replacement programme while reconditioning may cost less than half the price of similar equipment.

(g) Original model is a special purpose one and is tailor made to suit the one-time requirement of the specific organisation.

(h) Some organisations recondition the equipment if rejection/rework/scrap is beyond a predetermined level, although investigation of machine alignment is made before reconditioning.

(i) Planning for reconditioning should be given at the end of about seven years of trouble-free operating life for a machine tool. If it is far beyond this period, then the OEM may not provide the proprietary units, components of technical defaults due to technological developments at his end. In this context, very few organisations consider reconditioning cost-inclusive of material, labour, depreciation and outside machining cost—of up to 30% of the price of a new machine, as acceptable. This level of performance can be assessed from factors such as trouble-free service after reconditioning, rate of output, and expected life after reconditioning. The reconditioning will be compatible with economics only if due considerations are given to basic requirements like provenness of design, use of appropriate materials to ensure quality, reliability of performance, maintainability, safety aspects, lubrication facility, interchangeability of components, etc.

A critical cost-benefit analysis has to be done in deciding whether or not a part has to be replaced or reconditioned. The components of replacement costs are replacement spares cost, installation, other service costs, costs of manuals/drawings/catalogues, cost of ordering, transporting, handling, unpacking, and inspection. Reconditioning costs include depreciation on equipments, costs of test facilities, manpower, utilities, repair procedures, updating repair catalogues, and transportation.

There are a large number of advantages due to rebuilding. To start with, training of operators is necessary as they are already used to the machine. The jigs, fixtures, toolings and spares available with the machine can be continued to be used and the cost of capital equipment decreases.

RECONDITIONING A–Z FACTORS

A large number of factors influence decisions on reconditioning. The original equipment manufacturers usually specify the condemning limits for the wearable parts and individual industry's requirements since the capacity of infrastructure facilities is a crucial factor. The shop incharges are consulted in the planning of repair reconditioning and overhauling. In a multiechelon system like railways and roadways, location of facilities in a central place, region-wise in each depot, etc. is an important factor. The decision of centralisation will be complicated when the organisation makes several products with different equipments using a variety of technologies.

The other factors influencing the reconditioning process are as follows:

(a) Anticipated failure rate
(b) Susceptibility of part to damage in handling
(c) Operating conditions
(d) Manufacturer's experience
(e) Fabrication characteristics
(f) Nature of repairable item
(g) Salvage value
(h) Availability of standard repair kit
(i) Equipment model/family/age
(j) Frequency of breakdowns
(k) Rejection/rework due to machine defects
(l) Pollution
(m) Noise
(n) Innovations
(o) Technological developments
(p) Accuracy and precision
(q) Reliability
(r) Operators' safety
(s) Productivity
(t) Economics of operation
(u) Competition
(v) Profitability
(w) Availability of past data and history cards
(x) Logistics support
(y) Life cycle of equipment together with relevant costs
(z) Availability of infrastructure facilities.

ECONOMICS OF RECONDITIONING

The process of rebuilding requires facilities which are even more modern than the equipment required for building new machines. This is due to the fact that facilities are required for a large variety of components in one off batches to high accuracies. It is interesting to note that the replacement cost of an indigenously produced machine tool is about five times its original cost and that of imported machine tool about 15 times its original cost due to inflation, customs duty and difference in exchange rate.

Very few organisations consider reconditioning cost—inclusive of material, labour, depreciation and outside machining cost—of up to 30% of the price of a new machine, as acceptable, provided it is guaranteed that there would be no adverse effect on its performance. This level of performance can be assessed through the hours of trouble-free service after reconditioning, rate of output, expected life after reconditioning, etc. The reconditioning will be compatible with economics only if due considerations are given to basic requirements like provenness of design, use of appropriate materials to ensure quality, reliability of performance, maintainability, safety aspects, lubrication, interchangeability facility of components, etc.

A critical cost-benefit analysis has to be done in deciding whether or not a part has to be replaced or reconditioned. The components of replacement costs are replacement spares cost, installation, other service costs, costs of manuals/drawings/catalogues, cost of ordering, transporting on equipments, test facilities, manpower charges, cost of utilities, repair procedures, updating repair catalogues, transportation charges etc. There are a large number of advantages of rebuilding. To start with, training of operators is not necessary as they are already used to the machine. The jigs, fixtures, tools and spares available with the machine can be continued to be used and the cost of capital equipment decreases.

INFRASTRUCTURE FOR RECONDITIONING

For proper reconditioning, adequate infrastructure facilities in the form of reconditioning shops, uptodate machines, and skilled manpower must be made available. Appropriate coordination of design, planning, purchase, inspection, sales and engineering staff is essential for a successful reconditioning programme. It is noted that the reconditioning operation is a miniature version of the original equipment manufacturing process. Hence it calls for certain machining, heat treatment, tools, assembly, testing, and inspection facilities.

The equipments used for reconditioning process must cover a range of requirements such as shop equipments, auxiliary equipments, scrapping equipments, test mandrels, dial gauges, magnetic stands, high precision gauges, and other measuring instruments to ensure reconditioning work of varied nature. The concerned original equipment manufacturer must train and educate the individual users by holding demonstrations at major centres. Sufficient provision for capital must be made by each industry for research and development in repair technologies to optimise reconditioning time and cost.

At the national level, it is necessary to establish reconditioning companies for each industry in different regional industrial centres which can not only recondition the equipments, but also offer solutions to problems of reconditioning in respective industries. These corporations should also disseminate information on the results achieved in recycling techniques achieved by individual organisations.

TECHNICAL DETAILS OF RECONDITIONING

A complete command of all technical aspects is basic to successful reconditioning. For instance, the engineers involved in reconditioning should be thoroughly conversant with the design of the equipment, its performance, assemblies, components, control elements, lubrication system, and engineering material specifications. The period of use prior to taking up for reconditioning depends upon the operating conditions, maintenance system, number of working shifts, order of accuracies expected, frequency of breakdowns, and rejections.

Before reconditioning, one would like to explore the possibility of extending the life of mechanical sleeves, impellers of pumps, gear teeth, crank shafts, cylinder lines and shafts by using bronzo chrome and rotate eutecordes made by Larsen & Toubro. Advanced reclamation techniques like metallising, hardfacing, thermit welding, eutectic welding, metal lock process etc. can be applied to reclaim worn-out spares. Shrink fitting by liquid nitrogen is another technique used extensively in maintenance works where fittings with large interfaces are needed and heating of the mating parts to expand them is neither desirable nor possible. The refineries have evolved procedures for recycling and reusing valves, flanges, bearings etc. by welding, metallising, and spraying; other types of reconditioning techniques used are as follows: reclamation welding, submerged arc welding, gas welding, electro-deposition, metal spraying, removing the metal, use of adhesives, metal locking, use of plastic steel, using non-metallic guideways, machining oversize, and fixing linen. A critical problem encountered during reconditioning is distortion of castings, which has to be rectified by scrapping, and this calls for highly qualified scrappers.

Chapter 38

Information System for Spares

ENVIRONMENTAL SCANNING

Information is power, and the basic job of a Spare Parts Manager is to control spare parts through accurate and uptodate information. Reliable and relevant information is used to establish the objectives and to direct the attainment of these objectives by measuring the results. Needless to emphasise, business decisions on manufacturing capacity utilisation must be based on factual information as competition, rising costs, and rapid technological advances leave little room for error.

The increasing complexity of society, growth in the size of the organisation, and expanding information needs have necessitated for a systematic approach to gathering, storage processing, selection, and dissemination of information in the spare parts field. Spares management information systems are planned and organised approaches to supplying parts executives with intelligent aids to facilitate improved spares practices. The successful spares executive is the one who is most skilled in gathering and synthesizing quantity and quality of right kind of accurate information which must flow to him if he is to perform his function successfully. Any organisation can be viewed as a system consisting of various subsystems. The input of the material subsystem is converted into output by the manufacturing subsystem, which is sold by the marketing subsystem. Thus, the throughput of the entire system depends on the subsystem, their interactions, and the interaction with the environment.

The need for improved MIS arises particularly when the following conditions are noticed in an organisation:

(a) Managers are supplied with a lot of irrelevant information.
(b) Some changes are introduced within the organisation.

(c) The performance of the company is much below expectations, although no weaknesses can be spotted on the surface.
(d) Uniform goals and subobjectives are not clearly set up for managers and departments.
(e) There is little knowledge about competition, new supply sources and new customers.
(f) Information on decision variables reach managers very late and do not contain uptodate requirement in the right form.
(g) There is conflicting information from different sources about the same point.
(h) Information is neither properly summarised for top management nor detailed for middle level executives.
(i) It is not possible for managers to identify variances by cause and responsibility.
(j) Overheads costs are rising very fast.
(k) There is excess spare parts inventory, resulting in increased obsolete items.

Management Information System (MIS) introduces optimum controls by a regular supply of information on all aspects of the spare parts after considering the interactions of other subsystems and environment. The function of Management Information System is to collect data relating to the external world and the internal operations within the company and convert the data to useful information. It must be noted that this information necessitates appropriate selection of relevant data on real life situations after conducting a cost-benefit analysis.

MIS STEPS

A Management Information System is an organised method by which managers at all levels in the organisation are presented with needed and accurate information in right form and at the right time so that they are in a position to perform their tasks better. In developing the management information system, the following steps are recommended:

(a) Study the organisation carefully.
(b) Establish decision responsibility.
(c) Identify information flow to decision centres.
(d) Evaluate information flows.
(e) Develop a conceptual design of the subsystems and the system.
(f) Produce detailed system for creation of data bank and output from the system.
(g) Implement the system after careful consultations and communications with the concerned executives.

Needless to emphasise, the spare parts management information system will form a subsystem of total system of the entire organisation. The management information system can be on a manual basis, even though it is advisable to resort to computerisation if the volume of transactions is large.

DATABASE

Lack of relevant and accurate information is the critical deficiency which confronts the spare parts executives today in India. A good information system should lead to communication between managers to improve the overall performance in the spare parts field. The parts executive need not get into the intricacies of Management Information System or computerisation, but he must know how to intelligently use the information. The amount of information available on spare parts field is staggering as the number of catalogues, magazines, brochures, journals, books, and dailies containing information is unlimited. The need for constantly separating the wheat from the chaff must be always kept in mind. For dynamic review of information, a data base is needed. The data base is a carefully planned structure of basic data elements of the organisation. The data can be conveniently stored in a device that can be interfaced with the computer and used from several points. The data base should be flexible to provide for an orderly growth and the capacity to absorb data elements overlooked in the initial design. The objective of spares information system must be to help provide promptly the required spares to the needy consumers/users divisions, and to ensure that stocking policies reflect optimum utilisation of working capital. It may be observed that the information system of an organisation marketing spare parts will be different from that of a consumer organisation. Experience shows that the information system in a sales organisation is better organised with the help of computers as compared to a user industry.

MIS FOR SELLING: A–Z ASPECTS

The information system for a spare parts selling organisation should provide the following:

(a) Quick order processing
(b) Priority allocation of spares that are in short supply
(c) Optimum stock levels at central warehouse
(d) Turnover of each spare part
(e) Minimum obsolescence
(f) Feedback from customers
(g) Action on customer complaints

(h) Service staff strength
(i) Skill mix and quality of staff in each region
(j) Forecasting of demand for each customer
(k) Field/retail stock levels for each part in each region
(l) Rapid turnover of parts
(m) Performance analysis of parts from different vendors
(n) Despatching instructions
(o) Satisfactory credit control
(p) Outstanding amounts from consumers
(q) Invoicing
(r) Backlog or orders for each part
(s) Liaison with manufacturers on status of production of spares
(t) Follow-up with ancillaries for supply of spares. Besides, the following conditions must be fulfilled.
(u) The communication system must ensure that the time lag between despatches and billing must be minimum.
(v) The billing must be backed up by a close follow-up by way of collecting the money so that the accounts receivables are managed efficiently.
(w) The management information system should provide the ageing schedules of accounts receivables classified for each part and consumer in each region.
(x) The salesmen/servicing staff should send their orders on a daily/weekly basis to enable quick order processing.
(y) Warehouse should make arrangements to send the parts at the earliest.
(z) Spares should be available to the user according to his requirement at a reasonable cost.

SELLING: A–Z QUESTIONS

A good management information system must be capable of answering the following questions:

(a) How is the market intelligence organised?
(b) How does the customer order the part?
(c) What happens after the receipt of the order?
(d) How long does the customer take to pay?
(e) What procedures are adopted in the warehouse?
(f) Who are the competitors?
(g) What are the pricing and lead time strategies?
(h) Who are the spurious manufacturers?
(i) Where are the spurious manufacturers located?

(j) What is the difference in cost/price of spare part between the spurious manufacturer and the genuine manufacturer?
(k) How much manufacturing capacity is reserved for replacement market by the original equipment manufacturer?
(l) How do you tackle spurious manufacturers' competition?
(m) How many machines have been sold in the last years?
(n) Do you have the knowledge of age and operating condition of equipments?
(o) How do your customers' estimate their requirements of the spare parts?
(p) What is the average number of hours of operation per year?
(q) What is the channel of distribution from the supplier to the consumer?
(r) What is the distance between the client and the warehouse?
(s) What is the storage condition of spares in central regional/retail/site levels?
(t) What is the priority for discounts, and customer order processing, particularly for shortage item?
(u) How are the parts arranged and identified in stores?
(v) What are the different components of the lead time?
(w) What steps have been taken to reduce the lead time?
(x) What is the customer service level for different spare parts?
(y) What are the documents used in connection with spare part transactions?
(z) Has the spare part function been computerised, and how is it evaluated?

USERS: MIS A–Z

(a) The objective of the information system in case of spare parts used in an organisation must be to provide proper forecasting of requirements of different categories of spare parts, inventory levels and consumption control.
(b) At the users' end, the spare parts inspection system must cover the entire gamut of the spares management function.
(c) Many of the points discussed in the data base for the sales are also applicable to the MIS at the user's end.
(d) The failure rate resulting in the replacement of a spare part is known as the *usage rate* and forms the basis for application of scientific techniques.
(e) In obtaining the usage rate, one has to consider the three different stages: (i) initial provisioning, (ii) normal middle period, and (iii) discarding or phasing out the equipment.

(f) For effective spare parts management, the following aspects are relevant in developing the MIS for inventory planning and control.

(g) Records of monthly receipts and issues from stores and consumption data on spare part, which will be useful in the determination of inventory policies for different spare parts.

(h) For items which are infrequently used or only during overhauls, the rate of replacement for each overhaul of the information system.

(i) Service level or stockout rates for different categories.

(j) Categorisation into maintenance, overhauling, rotable, capital and insurance items.

(k) Machine down time due to non-availability of spares.

(l) Ordering, carrying and stockout costs.

(m) Budgets and forecasts.

(n) Lead time and its components such as internal administration lead time, manufacturing time, transporting time, and inspection lead time.

(o) Reliability data and failure analysis.

(p) Criticality of the part.

(q) Availability.

(r) Identification details.

(s) Purchase order status.

(t) Parts master file containing part number, code, description of the part, location code, etc.

(u) Invoicing/purchase costs.

(v) Number of sources for each item.

(w) Minimum.

(x) Maximum.

(y) Reorder point.

(z) Reorder quantity details.

MIS OPERATION

As a control and feedback system, the programme commences with maintaining the stock ledger incorporating the receipts, issues and balance stock on hand at any time. Purchase orders must be entered into the system so that follow-up can be initiated and further ordering done as necessary. This may follow the reorder level or periodic review system. When manually processed, the reorder level can indicate items at the time of issue. The same is also true of random access computer systems when the issues are being recorded. Additional information regarding the department and equipment needing the same can also be fed into the system so that usage history data can be built to enable projections of demand, taking into account replacements made. This would enable cost accounting as well as guide the management to plan replacement of equipments when the spares requirements become prohibitive.

When similar equipment is used at different places by one organisation, the same spare part is likely to be required at many places and it will be necessary to stock the same item at different locations. But the safety stock required will be less if the total demand is considered to occur at the same place with a centralised store. In such cases, the information system should indicate the position of spare part availability at different locations at any time. An efficient information of this kind can bring about a reduction in stock holding by arranging for the transfer of spares from one location to another involving minimum lead time.

Different manufacturers have different catalogue numbers of the same item. In order to readily find a source of supply, when a particular spare part is needed, it is necessary to maintain a cross reference listing, giving different catalogue numbers of various manufacturers for the same product. Such a cross number besides speedy procurement, ensures that products of different manufacturers are not treated as different items in the same organisation. A similar procedure will be used to indicate the various parts needed for each equipment and the various equipments for which the spare part is used. Such an arrangement will help compute the demand, based on the number of different types of equipment being used, taking into account recent acquisitions and the scrapping of old ones. This is particularly essential when certain models are being phased out of use.

ROLE OF THE COMPUTER

In Chapter 14, we have discussed Computers and Maintenance. The electronic computer, which burst upon the world scene in 1950, symbolises, more than any single device known to man, the spirit of change, and the enormous power it possesses can be used in numerous fields. Computers are useful for computer-based management information systems. Computing devices fall into two categories—analog and digital. The analog computer is essentially a mathematical device that makes elegant use of the circuit art. The digital computer can be instructed to carry out speedily fundamental mathematical processes such as addition, subtraction, multiplication, division, and can solve mathematical equations.

The first generation computers used vacuum tubes; the second generation computers using transistors established higher levels of speed and reliability. The third generation computers use integrated circuits. With the advent of large scale integration (LSI) in electronics, thousands of electronic gates constituting entire subunits can be built into a silicon chip. The fourth generation computers, the computers of today, use large scale integration and very large scale integration (VLSI) of circuits. These have enabled the advent of microcomputers which are low-cost desk top computer models appropriate for small scale organisations, for managerial control and for personal use by managers and executives. The fifth generation computers use the technologies of microelectronics and artificial intelligence.

The information processing required for spare parts management in any large organisation is too large to be left to manual operations. The computer can quickly analyse the data to answer questions relating to customer service and backlog of orders through the help of input documents known as *master file* comprising the part number code, type of part, description, unit of quantity, manufacturing cost, price, discounts, warehouse, location code, order quantity, etc. The customer order processing updates, and the cumulative department takes care of updating cumulating receipts, issues and stock status data.

MIS REPORTS: A–Z ASPECTS

The scope of computer applications in the spare parts field is indeed very wide. The computer can also be used to generate a large number of reports. The computer application areas include the following as well:

(a) ABC analysis
(b) SDE or availability analysis
(c) Failure data analysis
(d) Issue from stores
(e) Consumption trend
(f) Minimum-maximum levels
(g) Reorder point, reorder quantity and other inventory levels
(h) Movement analysis
(i) Price forecasting
(j) Quantity forecasting
(k) Purchase budget
(l) Stock status in transit
(m) Pending orders
(n) Variance or differences with the purchase budget
(o) Advance warning on shortages and stockouts
(p) Data on overstocking of spares
(q) Vendor rating
(r) Delivery reports
(s) Overdue inspection reports

(t) Values of receipt and issues
(u) Updated lead time information
(v) Requirement list and value of spares for overhauling
(w) Spare parts used by temporary codes or part numbers
(x) Obsolete spare parts value
(y) Number of stockout and overall inventory value
(z) MUSIC-3D or cost-criticality-availability analysis.

The frequency of the above reports can be on a daily, weekly, monthly, quarterly or on an annual basis, depending upon the levels in the hierarchy. The reports can give comparative information of the preceding period and highlight the causes for the differences. These reports can form the basis for evaluating the spare parts activity in any organisation. This chapter must be read in continuation of Chapter 14 on computers.

Chapter 39

Evaluation of Spares

ORGANISATION PROBLEMS

A well-organised spare parts division is essential for any organisation using/selling spare parts in order to achieve maximum capacity utilisation with optimal investment in spare parts inventory. However, in spite of the challenging nature of spare parts problems, the function is not considered to be a significant one and is usually delegated to the lower levels in the hierarchy. Supply of quality spare parts at the right time in required quantity to the user/consumer is a service function to the maintenance department, which in turn is a service function to the production department. Hence service to service function is not considered as the problem of elite or high fliers who prefer finance or marketing areas.

The typical maintenance engineer's attitude in user organisation is "it is your job to get me what brand of spares I want at the time of my requirement". The fear of stockout looms large, resulting in obsolescence of spares. For firms selling spares, the performance of after sales service department is judged by the users' opinions who always complain about poor quality and long shutdown period. The attitude at all levels has to change and due recognition given to spares function by having qualified engineers.

SPARES PLANNING CELL

Organisationally, the requirement planning for different categories has to be done by the maintenance department as identification of critical spares and supply of quality spares play a crucial role in reducing the maintenance cost. In the process of planning, the purchase department and commercial intelligence should play an active supporting role in explaining the lead times and manufacturers' difficulties. The warehousing section also comes into the spares planning picture by suggesting ways for easy identification, codification, preservatives, and development of drawing. The plant engineer has a wider knowledge on spare parts than any one else in the organisation.

There should be suitable powers of delegation for purchase/store/issue/servicing/inventory/sales/staff recruitment/development of drawings/application of preservatives/use of cost reduction methods. A manual must be prepared to indicate all relevant aspects and control in spares field. The manual must also indicate the interlinks, interfaces and interrelationships between the different departments. It is ideal to have a spare part planning cell reporting to the Chief Plant Engineer of the organisation as prevalent in some steel plants. The procurement of high value spares and spares bought with the capital equipment must be cleared by a steering committee. The spares planning cell must look after the requirement plant, both in the short term and in the long term.

The corporate objective of spares planning committee should be to improve the availability of quality spares at the right time at optimum cost for obtaining best results, the spares should be classified by a group of executives from design, spares planning, production, purchase, stores, and maintenance. The sales people familiar with service after sales, finance managers acquainted with costing and depreciation policies, and industrial engineers knowing cost reduction techniques must share their expertise at the appropriate time in the periodical coordination committee meetings.

SALES AND SERVICING

In the case of organisations selling spares, the marketing department plays a crucial role and the Chief Sales Executive of spares usually reports to the Marketing Director. The after sales service department is also attached to the marketing department which handles the original equipment sales. The marketing department exercises adequate controls on the geographical distribution, such as regions, areas, districts, depots and channels of distribution like wholesalers, stockists and dealers. The depots are conveniently located to cater to customer needs of spares, after sales service, and has the necessary spares and staff. The spares planning cell attached to the marketing department in the head office usually has access to a computer for analysing the field requirements and reports. A cost-benefit analysis has to be done before opening additional service/spares depots near the customer organisation.

Contract maintenance, as required by the customers, is also handled by the after sales department. It should be noted that in service-oriented business, it is difficult to estimate what a unit of given service is, much less the cost of service. The price of service is based on value rather than cost and usually depends upon whatever the market can bear. The number of staff engaged in after sales service depends upon the volume of sales, location of customers or products, variety of equipments, and availability of spare parts.

ROLE OF TRAINING

Proper selection of staff with adequate skill mix in the challenges of spare part is a must before training them for the requirements of the organisation. It is desirable to identify the role of responsibilities clearly by adequate job description. In this process, the peculiar problems faced in the spares field must be pointed out to them. Hence it is necessary that the executives involved in spares— marketing, design, after sales service, maintenance, purchase, stores, finance, costing, inventory, planning, and computer—must be acquainted with the problems, spare part categories, and scientific management techniques. Besides lectures, case studies, games, computer simulation, and problem solving situations must be dealt with. The primary focus of the training programme should be to minimise investments in spares and at the same time reduce the working capital investment. The training programme must help formulate stocking policies at different levels for the categories of spares such as maintenance, rotables, capital, project commissioning, insurance, and overhauling. Operations research techniques like inventory models, simulation, queuing, linear programming, transportation model, exponential smoothing, and forecasting techniques must be adequately covered in the training programme, besides giving technical knowledge and managerial skills.

NEED AND PROCESS OF EVALUATION

Sound corporate planning involves forecasting, planning, coordination, control, evaluation, review monitoring, and integrating various functional objectives into an overall corporate strategy in the field of spare parts. The basic strategy of evaluation must be to simultaneously minimise stockout and the working capital in spares. In this context of spotting and rewarding competence, the Chief Executive, always in search of excellence, is in need of guidelines on evaluation. The steps involved in evaluation are:

(a) Set acceptable standards for parameter, reflecting the objectives.
(b) Check observed performance against the standards.
(c) Spot deviations.
(d) Identify causes of deviations.
(e) Evaluate the costs of such deviations.
(f) Review and take corrective steps for improvement.

The word "control" in this context has several connotations such as (a) to check or verify, (b) regulate, (c) compare with the standard, (d) exercise authority, and (e) restrain. The control mechanism in the spares field helps the organisation to secure accuracy and reliability of records to provide operational efficiency, thereby adhering to policy guidelines, particularly in the areas of maintenance and spares management.

Evaluation can be done by internal agencies like spare parts cell, stores, maintenance, purchase, audit, finance, design, after sales service, costing, and top management. External agencies like suppliers, customers, subcontractors, vigilance, bankers and government departments also evaluate the spares department in their own ways. Evaluation can be done by an organisation by comparing the relevant performance characteristics with those of similar organisations, known as inter-firm comparison, or with its own performance over a number of years.

The evaluation is usually done, based on criteria to be developed by reports prepared by individual spare parts cells. The frequency and contents of the report depend upon the level to which it is intended, but the reports should give an uptodate and accurate picture. Large number of criteria can be developed in the form of ratios linking consumption, inventory, purchases, production, output, machinery value, number of sources, costs, etc. for each category of spare parts.

EVALUATION OF PURCHASING: A–Z ASPECTS

The purchase activity of spare parts must be properly evaluated on the basis of the following:

- (a) Value limits for requisitioning materials
- (b) Determination of specifications, quality and delivery time
- (c) Periodical review of actual lead time with buying by fixation and accuracy of maximum, minimum and reorder levels
- (d) Monthly stock control and review, including obsolete, surplus, and slow moving spares
- (e) Review of very low stocks
- (f) Ensuring that the purchase requisitions are correctly authorised
- (g) Obtaining quotations, tenders and determining the price
- (h) Procedure for internal assessment of prices quoted
- (i) Purchase orders to clearly state quantity, quality, specifications, price, delivery, taxes, payment and contractual obligations
- (j) Recording reasons where lowest quotation is rejected
- (k) Authority limits for buyers
- (l) Procedure for authorisation of subsequent price increases
- (m) Mode of transport and cost of freight
- (n) Control of work done by third parties

(o) Commercial soundness of long term arrangements
(p) Effect of raw material price changes on product margins
(q) Control over open market purchases and monopoly suppliers
(r) Rejections and rebates for substandard material
(s) Price comparison of spare parts where the company has more than one unit or similar units
(t) Out-turn control and claims for shortages
(u) Material intelligence system, particularly to avoid dumping of spares in the initial stages
(v) Application of cost reduction techniques like standardisation, simplification, variety reduction, value engineering, reliability analysis, etc.
(w) Policy of developing ancillary industries for spares
(x) Policy of reconditioning, recycling, and reusing the material
(y) Policy of import substitution and indigenous development on spares
(z) Obtaining the right quality spare parts in right quantities at the right time at the most favourable prices.

EVALUATION OF SPARES SALES: A–Z ASPECTS

Spare parts sales and purchasing are two sides of the same coin and the evaluation criteria for spare parts buying, with modifications, are relevant to the spare parts marketing activities as well.

The evaluation of spare parts marketing involves the following:

(a) Determination of the quantity, price and total value of recommended spares initially at the time of equipment sales
(b) Pricing strategies discounts from catalogue price, and non-pricing factors, and turnover margin for replacement market for different spares
(c) Preparing list of acturian population or equipments sold and the conditions of working
(d) Forecasting the demand for replacement market
(e) Provision of in-built special manufacturing capacity spares for replacement market
(f) Providing data base on failures, reconditioning, recycling, replacement of spares for equipments supplied to customers
(g) Having a feedback system on performance, reliability, complaints, satisfaction, etc. of the customer
(h) Formulating a policy for order processing, mechanism and priority allocation among customers
(i) Gathering market intelligence on environmental changes, technological upgradation, competition, spurious spares manufacturers, credit

worthiness of customers, technology policy of customers, and criticality of part of the user
(j) Rotating the stock for optimum turnover ratio with minimum obsolescence and maximum service
(k) Having a pricing mechanism for after sales service and the infrastructure needed in different places
(l) Building a basis for giving part numbers for spare parts
(m) Formulating a policy for distribution, warehousing, stocking, etc.
(n) Application of scientific cost reduction principles, like linear programming for transportation, inventory control, value engineering, packaging, multiechelon storage, centralisation, decentralisation and dealer network
(o) Sharing of benefits of cost reduction programmes with consumers
(p) Producing report formats on spare parts sales and after sales service from customer sites, fields, depot zone, region, etc. and the follow-up action
(q) Estimating relevant costs like costs of order processing, inventory carrying, stockout, rejection, taxes, sales, and transportation
(r) Policy of obtaining spares from subcontractors and ensuring their quality by suitable vendor rating schemes and channels
(s) Making known policy aspects on stock rotation, obsolescence and movement analysis of spares
(t) Assessment of production capacity, lead time and inventory levels of different categories of spares
(u) Formulating risk aversion policies which enable to change prices due to competition
(v) Identification of customer services, customer requirements, customer satisfaction, and efficiency improving services to customer
(w) Building a communication mechanism which informs about price increases, model changes, life cycle costing, phasing out the items, last supply, and use of buy back clause for various spare parts
(x) Having a policy relating to long-range contracts response to tender enquiries, bogus orders, and performance guarantee
(y) Having a mechanism to sort out disputes on quality, delivery, price, service, guarantee, warranty, rejections, damages, frauds, and handling losses
(z) Giving importance to spare parts marketing and after sales service departments in the total marketing strategy of the organisation in terms of quality of personnel, number of staff, procedures, manuals, turnover, etc.

STORES EVALUATION A–Z ISSUES

Warehousing or storage of spare parts is a critical but neglected activity in the field of spares management, as the entire spares activity revolves around the accuracy of the two basic stores documents, namely, receipts and issues. We now give the illustrative list of some parameters for the evaluation of the stores function:

(a) Positioning of receiving function vis-a-vis its relation to stocking, issue order processing, ordering and follow-up activities
(b) Security and gate check of vehicles moving in and moving out
(c) Effect of restricting deliveries to office hours and arrangements for receipt outside office hours
(d) Earmarking of separate area for goods pending inspection or quarantining the parts
(e) Adequate control on receipt and issue of material sent to subcontractors
(f) Policy of direct receipt by user department
(g) Documentation of such items
(h) Segregation and labelling of each lot received
(i) Periodic verification of accuracy of weigh scale and other measures
(j) Receiving description of authority levels for certification of goods receipt notes
(k) Comparison with despatched number, weight, volume against invoice figures
(l) Shouldering responsibility for shortages, rejections, damages and demurrages
(m) Issue of material only against authorised requisitions and accounted accurately for costly items
(n) Having an approved list of persons authorised to draw spare parts from the stores and exceptions thereof
(o) Adoption of suitable preservative techniques to increase the shelf life
(p) Control of issue of spares outside office hours as well as adherence to norms of consumption for individual spare part
(q) Providing security of materials in godowns located inside and outside factory premises
(r) Control on theft, pilferage, and insurance costs
(s) Segregation of slow moving, non-moving, insurance and obsolete spare parts
(t) Safe custody of expensive material as well as materials on loan to user department and ancillary industries
(u) Ensuring accuracy and uptodate nature of all records for goods receipt, indent, issue voucher, invoice, kardex, bincards, transfer advise, and entries on write off

(v) Differentiating between physical balance with book balance and timely investigations to minimise the faults in the system
(w) Drawing up periodic stock verification procedures and corrective actions thereof
(x) Use of computer terminal and/or personal table top computers for checking data accuracy
(y) Review of stock levels of critical items and biaising with purchase for follow-up, and removal of rejected items
(z) Cancellation of outstanding orders overdue for delivery and reporting of receipt of excess quantity.

A–Z INVENTORY EVALUATION

The statement "uncontrolled inventory is industry's cancer", is particularly true in the case of spares as the spares inventory accumulates at a faster rate than other items, leading to accumulation of obsolete spares in all firms. Given below is an illustrative list of parameters for evaluation of spares inventory:

(a) Identification of fast moving, slow moving, non-moving, insurance, obsolete spares
(b) Formulating inventory policies for rotables, maintenance, over-hauling, insurance, project, and capital spares
(c) Review of minimum, maximum, reorder point, reorder quantity, safety, reserve, buffer quantities for different categories of spares
(d) Identification and control of different elements of inventory carrying charges, order processing costs, acquisition costs, and stockout costs for spare parts
(e) Identification and reduction of internal administration lead time, manufacturing lead time, transportation lead time, and testing and inspection lead time for various spare parts
(f) Application of scientific techniques like VED, ABC, XYZ analyses, MUSIC-3D analysis, EOQ/JIT/ZIN systems
(g) Relating inventory to consumption, sales, lead time, criticality, and availability
(h) Having a basis for inventory evaluation like FIFO, LIFO, average price, and standard price
(i) Application of critical path method and network analysis for overhauling spares
(j) Application of Poisson distribution to maintenance spare parts
(k) Application of queuing principles to rotable spares
(l) Determination of cost-benefit analysis for insurance spares

(m) Evaluating the capital needed for costing, insurance and capital spares
(n) Application of quantitative techniques for forecasting the quantities and price of major spare parts
(o) Use of reliability principles to determine life of spare parts
(p) Review of stockout items and preventing their recurrence
(q) Identifying the role of maintenance department in inventory control and development of consumption norms for each spare part
(r) Clarifying the role of after sales service in inventory control of spares and sales target for replacement market
(s) Exercising monetary norms on the total inventory value as well as obsolete spare parts
(t) Development of norms for spares inventory value to consumption value, sales value, machinery value, output value, overall inventory value, maintenance budget, and other relevant aspects
(u) Formulation of inventory budgets and identification of reasons for budget variances
(v) Adoption of policy of avoiding stockout at any cost and providing 100% assurance level as stockout cost is always greater than the cost of spares
(w) Showing relationship between inventory and failure analysis or reliability data of a spare part
(x) Ensuring accuracy and uptodate nature of documentation relevant to inventory
(y) Delegation of authority to review inventory levels, budgets, obsolete item declaration, writing off, charging off, etc. of spare parts
(z) Having a policy manual on inventory control, procurement, sales, warehousing, etc.

MANAGEMENT AUDIT

In the context of evaluation, the management audit of spares function can be done in the following fashion:

(a) Evaluating the long-range plan of spare parts management to ensure that these plans support effective achievement of organisation goals
(b) Evaluating the total spare parts strategy on aspects such as procurement, order processing, sales, inventory, indigenisation etc. to ensure that these strategies promote achievements of the spare parts management goals
(c) Evaluating the effectiveness of organisation structure for attaining spare parts goals

(d) Reviewing the systems and procedures of spare parts management to ensure that the systems are capable of achieving the spare parts management goals
(e) Ensuring that the spare parts department's staffing policy provides adequate skills in line with organisation needs
(f) Ensuring that sufficient skills are obtained or developed in the organisation to man various positions
(g) Ensuring that the managerial styles of individual managers acceptable of effectively achieving the targets which result in overall improvement in all spheres of spare parts activities.

This chapter must be read with Chapter 12 and Chapter 19.

Section V
Caselets/Short Case Studies

40. A Hazare Fertilisers Ltd. (AHFL)
41. Vasanth—ASS Limited
42. Bhushan Refineries Limited (BRL)
43. Middle East Air Transport
44. Aruna Oil and Gas Limited (AOGL)
45. Mahatma Gandhi Road Transport Corporation (MGRTC)

Chapter 40

A Hazare Fertilisers Ltd. (AHFL)

CORPORATE SCENARIO AND STRUCTURE

A Hazare Fertilisers Ltd. (AHFL) is a public limited firm producing urea and 4 grades of complex fertiliser located in Pandoor, the delta region of south east coast. AHFL is an American international company and was committed at a cost of ₹500 crore in 2008. When it started it was the largest, most modern, single stream plant with an output of 3000 MT of urea, and 1800 MT of complex fertiliser per day. The plant has been running at 98% load with an annual turnover of ₹500 crore. The company also has plans to install a coal fixed power plant of 20 MW shortly. AHFL is also expanding its facilities by new projects.

The manufacturing facilities may be classified into Ammonia, urea, complex fertilisers and utilities for all the three plants. The marketing is done by its own marketing department through wholesalers, stockists and retailers who also provide freely soil testing, fertiliser application advice, water management and soft agricultural loan. The managing director has got general managers in charge of finance, works, human relations and marketing. The board is brought into picture only in case of raw materials, when the value exceeds 25 crore at one time. All other items are handled by the works general manager (WGM) with well-defined powers of delegation. The maintenance manager (Hazare) is incharge of spares, planning, provisioning, stocking and ordering. He is well acquainted with all materials function, having attended materials management and other courses conducted by Prof. P. Gopalakrishnan in ASCI. Mr. Crane Bedi was the supply chain manager (SCM) with 3 purchase offices and 2 stores offices—one officer incharge of planning, implementing, purchasing materials and spares excluding capital equipments, import of spares, direct materials, and indigenous space. Another was incharge of insurance of materials in transit, customs clearance, excise duty, and clearing and forwarding contractors. The third was incharge of laboratory chemicals, 'direct charge items' appliances and packing bags. It is well known that in the fertiliser industry packing bags are a major 'A' items, accounting over ₹200 crore every year. The central stores has two offices rotating in shifts assisted by stores assistants. All officers have laptop systems and transactions by paper has been reduced only to minimum. Each officer has been given by cell phones. Monday scheduled meeting takes place in every department while monthly review meetings are conducted by the MD on Ist of every month.

SPARES PLANNING

The case writer had a series of discussions with all officers, MD, staff and workers, but intensively interacted with Hazare and Crane Bedi. Mr. Hazare remarked "ours is a single stream plant, where equipment downtime has to be prevented at any cost". Most of the parts were new and imported. The initial provisioning of mandatory spares, after detailed technical visits by WGM and SCM, and appropriate screening by technical services under WGM. The board fixed a maximum of 10% of the plant on initial spares value after detailed discussion with inspection, tqm, maintenance and operating personnel. WGM informed the case writer "since our plant is high capacity, new generation, single stream Ammonia and urea plant, global tenders were called for design, fabrication, erection and screening were done in two stages. In the first stage, we blanked out the prices and the comparison was purely on technical and maintainability basis. This ensured selection of technically suitable bids. Subsequently a limited rebidding was done on technical screening results and prices were considered for the final decision. Except for the rotors for the centrifugal compressors, which needed spin testing, in the casing, other spares were handled by ourselves.

Crane Bedi added "after initial provisioning of spares, we always used budgetary control in ordering and stocking spare parts. The production plan is finalised, based on sales forecasts and plans. As a policy the board has fixed that the maintenance budget will be 5% of the production budget. This maintenance budget is subdivided into 80% for spares and 20% of labour. The spares budget is then divided into imported and local, with the emphasis of minimising imports if technically suitable quality parts are available. Some suppliers complain that inspection and quality department reject passable items when they do not need and even accept low quality spares when badly needed. This statement has been rejected by Hazare.

Hazare added that there is a constraint that requirement of imports cannot exceed $2^1/_2\%$ of CI and value of the installed imported machinery plus $1/2\%$ of the value of indigenous machinery and it has worked very well till this day for AHFL.

SPARES CONTROL

Mr. Crane Bedi informed to the case writer "we divide spare parts into two categories namely insurance spares and bulkable spares, as per the classification done by operating, maintenance and technical departments. Spare parts vital for plant and machineries with long lead times and insurance for us. We invariably carry a stand by spare for each case. However we do exercise a control in the sense, that once a consumption of insurance takes place, we do not automatically replace it, but bring it to the notice of the management and only after approval procurement is made.

'As far as bulkable spares are concerned, we have fixed certain inventory norms based on past experience as well as judgement and any consumption is automatically replenished. Data relating to inventory of materials and spares are furnished in Table 40.1 and Table 40.2. As stockouts leading to a shut down are very costly for a single stream plant, we tend to go slowly by the requirement figures furnished by the operative, technical and maintenance departments.

PROVISIONING OF SPARES

Table 40.1 Consumption and Inventory (₹crores)

Description	Annual Consumption	Inventory
Naphtha	6000	300
Imported spares and equipment	160	800
Indigenous spares and equipment	150	315
Common consumables	70	105
Coal fuel oil	285	60
Chemicals and consumables	105	45
Lubricants/Petrol/Diesel	210	20
Catalyst	180	225
Stores and spares	280	995

Table 40.2 Category-wise Analysis of Spares (₹crores)

Product Group	No. of Items	No. of Nonexisting Parts	No. of FND/ Moving Items	%	No. of Non-Moving Items	%	Annual Consumption	Year end Inventory	No. of Monthly Inventory
Ammonia spares	1750	489	128	28	1133	65	24	248	102
Urea spares	1170	344	777	15	640	56	26	63	28
NPK spares	579	77	52	9	450	78	5	23	52
Utility spares	1448	310	100	7	1020	71	3	169	149
Mix spares	3338	429	363	11	2546	76	4	39	128
Electrical spares	1693	314	265	16	1114	66	10	47	58
Instrument spares	2529	295	275	11	1950	78	10	105	153

INCREASED SPARES INVENTORY

Budgetary control system forces the resource available to spare parts and the requirement and issues are entirely controllable by the user departments. The only area where he could exercise control is purchase. He pointed out that over 100 officers are entitled to draw materials and felt that issue control by logistics and supply chain management is out of question. We also have direct charge materials. By this we mean such of those materials for which we do not anticipate repetitive consumption. These are directly charged to the respective accounts heads and are kept in a separate enclosure in our stores. Only for the stock items, we maintain detailed stores records. This simplified our system a great deal. Any request to comment a direct charge item to a stock item is viewed carefully.

The reasons for increasing spares inventory are as follows: There were a services of heat exchanger tube bundles and these being lead time items, it was decided to keep spares bundles for vital critical units. Since the NPK plant started earlier than others, spares were ordered from that year. A mini plant was installed at a cost of ₹10 crore for pollution and carbon emission control. This has equipments like pumps, control valves, heat exchangers, spares around 5–10% of the total cost were ordered. To improve the production a debottling project was initiated. The overall cost is around ₹15 crore and additional 5–10% worth spares were ordered. Because of the unexpected failures of some turbines and compressor rotors, it was decided to keep one spare order of each type fitted. This included rotor for the mainsteam turbine which drives the generator. It was decided to keep full set of aerodynamic spares like diaphragms.

TYRE FOR 30 TONNE TRUCKS

Crane Bedi continued "we have four stainless steel road tankers of 30 tonnes capacity each used for transporting phosphoric acid from pot to plant. Since phosphoric acid is highly corrosive it has to be transported in stainless steel containers. These trucks are used very heavily and each of them makes several trips to the port every day. The types for these trucks need frequent replacements. We, in the materials management, as a policy, decided to buy tyres directly from the manufacturers for the following reasons: We have to buy not infrequently special tyres for our forklifts from manufacturers which no dealer stocked. So buying a bulk volumes of tyres for our trucks will enable us not only get 15% discount, but also will get certain preferential treatment for odd requirements of forklift tyres. We felt directly from manufacturers will free us from short term fluctuations in the market. During the lean period, the dealers formed a beeline and offered heavy discounts, but we did not budget. The purchase manager periodically suggests that buying from the dealer in lean period, when prices were attractive. But such short term price advantages and short sighted advances will not help for sustaining good buyer-seller relations which must be maintained in thin and thick periods. Who is going to guarantee that the tyre is brand new and of good quality and will have a long life. No one knows when the dealer has acquired stock and these are fast moving one. Each 30 tonne truck chassis has 16 tyres and our consumption is over 200 tyres per year. These tyres are high unit value 'A' items.

ANCILLARIES FOR IMPORT SPARES

Import substitution and ancillary development, Crane Bedi—added—are the catch words, but not easily accepted our maintenance engineers due to inordinate delays and inferior quality. The requirements are so low and low economies of scales drives the locals crazy. Even in indigenous equipments like industrial shutters the local supplier has stopped manufacturing. We cannot afford to keep it idle as the shutters are responsible for goods movement. We also projected all time requirement of spares and addressed the ancillaries in industrial estate, but no one showed any interest. Drawings are never supplied by any original equipment manufacturers, and it is too difficult for us to develop drawing ourselves. The pattern of spares consumption is never steady and predictable by our maintenance crew. Spares such as axles, transmission parts, engines require replacement more than thrice during its economical life—in imported machines the situation is much worst.

The granting of import licence and dovetailing of requirements into the licence taking into account some estimates for the price escalation factors have been preeminent problem. All suppliers use the phrase—"price ruling at the time of supply will be charged". They quote labour index, inflationary index, budget deficit, fiscal deficit, monetary deficit, crude oil price, gold price, commodity prices etc., when pressed for details. The capital intensive fertiliser industry mostly imports the single stream equipments. The process is continuous involving round the clock working with high temperature and corrosive chemicals.

ROLE OF USER AND FINANCE

Crane Bedi continued: "We were importing special high pressure gaskets for our firm Santosh Hegde Pvt. Ltd. started manufacture of this gasket and we purchased our full requirement from them. After sometime Gaskets India, started the gaskets with 30% lower value approached and orders were placed after testing by technical and maintenance departments. After some time another firm started with 20% further reduction in price but the chief maintenance engineer threatened us saying that a single stream high pressure processing plant cannot go on working on a trial basis and he only is answerable for the shut down by being testing ground for all gasket suppliers.

The finance manager emphasised that the finance views gets rejected when he pointed out that each rotor we carry costs over ₹1 crore. When a rotor gets damaged it is moved to the shop, repaired in due course and returned to stores as spares while the insurance spare rotor in stores is drawn and fitted to the plant. In accounting jargon, the rotor in the plant is "fixed asset", while the rotor in the store is a current asset. It is quite possible that the damaged rotor can be used in the plant. After few cycles extensive records have to be kept with different pricing for 'plant rotor' and 'stores rotor'—we have to treat all of them as fixed assets and depreciate them at 10% under straight line method without any audit query. This helps us in the sense that financial charge is uniform every year and it is a better method. Since over the life of rotors, uniform debits are made, the burden is also smoothened every year as we depreciate 150,000 for each rotor. But these details are not discussed in every processing firm. Customs may treat a part as a regular item attracting higher duty. These are open to discussion. For instance the bearing block imported for our plant is a detachable or non detachable spare part—attracting of not only separate duties but fiscal account.

SPARES BANK

The MD—AHFL—called the senior consultant Prof. P. Gopalakrishnan to consider all the above problems and requested him to submit a detailed report within a month after discussing all offices of AFHL. Gopalakrishnan admitted these are inherent problems and limitations due to multitude of foreign collaborators and transfer of technologies. As a positive thinking senior consultant he found out a very costly catalyst working for about 10,000 hours and felt many process plants such as MFL, Coromandal, Fact, Zuari, Mangalore Fertilisers. Hindustan Fertilisers and other big fertilisers plants in public and private sectors also "may" be using such high value catalyst or rotors. He was thinking of meeting fertiliser association of India's office bearers to evolve a spares bank for packing bags—The fertiliser industry buys ₹500 crore worth of bags. This needs to standardized and brand printed later on—for bulk discounts. He was thinking of a spares part bank for rotors, compresses which will depend upon degree of standardisation of the mother equipment. Besides tackling AHFL problems, he was contacting the fertiliser association, before the preparation of final report—His team collected the details of products—

urea, Ammonia, NPK complex fertiliser, the capacity, capacity utilization, stagewise expansion programme, organisation charts, maintenance manuals, purchase and stores manuals, construction material, fuel oil, diesel, stock values, minutes of monthly meetings in AHFL etc. and started writing the report, which devotes special attention to total preventive maintenance. He is also wondering whether the standby on standing costly insurance parts can be operated in a common pool of fertiliser plants—as CHEP India Pvt. Ltd.—claims to be a global leader in equipment/pooling solutions committed to reduce environmental waste and protecting natural resources. CHEP located in Andheri Mumbai create pooling services, claim to help the entire supply chain, through reduce, reuse, and recycle by equipment pooling solutions.

Chapter 41

Vasanth—ASS Limited

COMPANY DETAILS

Vasanth Machinery Limited (VML) are the manufacturers of heavy earth-moving equipment in the country in collaboration agreements with a manufacturer of international repute. The company was the first to start manufacture of earth-moving equipment in the country. The equipment are considered to be versatile and have been supplied to large mining corporations and also to construction projects, irrigation projects, state electricity boards, etc. VML is interested in improving the image of its after sales service.

VML has four zonal offices which undertake the after-sales service (ASS) in the four regions of the country. Attached to the zonal offices are the qualified field staff who render after-sales service. VML's head office is located in the South from where the operations are coordinated.

VML is a large manufacturing unit employing over 15,000 work-force. The products manufactured are of wide range and are vital for the development of transportation and mining industry in India. The company is also thinking into paper bag and envelope making machines enclosed in Figures 41.1 and 41.2.

FAILURE OF ASS

The Chairman of the VML felt that with the various factors involved in the production and completion targets of projects, it is of utmost importance that correct type, size and a properly planned equipment schedule is made. The need for producing equipment of different capacities and types has been adding multifarious problems to the factors relating to spares and servicing which inhibit the performance of the equipment.

Figure 41.1 Envolope making machine.

Figure 41.2 Paper bag making machine

Mr. Malik, the Chairman of XLO Mining Corporation, was reviewing the performance of dumpers supplied by VML before meeting their officers. The review was done with the help of General Managers of various collieries of XLO Mining Corporation. XLO Mining Corporation is engaged in the coal production and comprises of a large number of mines and collieries spread over four different states. This corporation is considered to be one of the vital coal producing units and is looked upon as a potential unit for the development of thermal power station, cement industry in addition to being the major fuel suppliers to the Railways. The capital investment was ₹5,150 crore as per the latest balance sheets. The total manpower employed is of the order of 50,000 with a production of over 80,000,000 per annum. The production of coal in India has been stepped up from about 55 million tonnes in 1961 to about 180 million tonnes in 2011. This indicates a rising demand and potential prospects for the development of coal industries.

Increase in the demand for coal could only be met by increase in the mechanisation. This involves installation of sophisticated and expensive machinery which can serve the purpose and justify the cost only if the plant and equipment are properly maintained and kept to working at the optimum capacity. XLO company has ten open cast mines. In these mines in addition to the wide variety of coal mining machinery, dumpers of various makes are used for coal handling. Dumpers are equipments for handling earth, ore, rocks, etc., and are designed for use in quarries, mines, and construction work. In a coal industry, the smallest unit is a mine. Usually, the preventive maintenance task is carried out on the spot of operation. Breakdown repairs and major maintenance is carried out at the workshop.

AFTER SALES SERVICE POOR IMAGE

Mr. Malik said, "The different mining units of our corporation are continuing to face the idle time caused by non-availability of spares and parts and other facilities provided by VML. In my negotiations with the company I have sought to bring it to their notice that until and unless they also gear up their facilities to provide after-sales service and provide timely help of the supply of spares and parts it is an impossible task to make productive use of their vital equipments. The company's targeted production seems to be completely upset by frequent breakdowns of these equipments.

"As decided by the management further negotiations have been successfully carried out with VML for streamlining their service-after-sales", including annual maintenance contracts (AMC).

VML has supplied 100 dumpers so far and the future requirements will be about 75 dumpers in the next five years.

He indicated that by and large, in the past, equipment including the dumpers have been procured from different countries. Although standardisation has been achieved so far as the capacity is concerned, yet variety of equipments and designs supplied from different countries present problem with special reference to spares and parts. He also stated that indigenous manufacture of dumpers in the capacity range of 20/24 cu. yd. shovels which XLO has standardised is undertaken only by VML who are manufacturing a model of 23.3 cu. yd. capacity. The other available equipment in this capacity range is a dumper of 20 cu. yd. capacity.

Shri Kumar, the Supply Chain Manager of XLO Mining Corporation stated that the supplies of VML have not shown any improvement in quality and delivery time. Against committed deliveries of 57 dumpers during 2011 and commencing from May at 6/7 dumpers per month, no supplies have been made up to July and only 2 Nos. were reported to be ready for dispatch during August. Joint assessment made by technical officers of the company and VML in August last is that total supply during 2012–13 will be only 32 dumpers (shortfall of 25 dumpers). Mr. Kumar was very critical about the timely supplies and in his assessment VML will not be able to supply more than 24 dumpers during this period.

It may be noted that VML have been holding out promises of meeting the full requirements of XLO whereby import of dumpers by the corporation has been totally stopped. Mr. Kumar stated that VML can neither meet the full requirements nor keep up the delivery schedules. In most of the cases, they have even failed to meet the revised delivery schedules.

SUPPLY FAILURE

Mr. Baji Rao, the Maintenance Chief of XLO, highlighted the causes of failure of VML dumpers and broadly classified them into the following two groups:

1. *Failure on account of the problem relating to chassis:* (a) Transmission; (b) Oil Collar Assembly; (c) Self-starter; (d) Dynamo; (e) Suspension (Hydraulic Kit); (f) Hydraulic Pump and Seals; and (g) Final Drive.
2. *Failure on account of engine:* (a) Low Lub. Oil Pressure; (b) Camshaft Gear; (c) Crank-shaft and Gear; (d) Turbocharger (e) Inlet Valves; (f) Gear Lub. Pump; and (g) Radiator.

Mr. Baji Rao further explained that even after four years of commencement of manufacture the after-sales-service is unsatisfactory. He gave the following information to the members reviewing the performance.

PRICE DIFFERENCES

Of the dumpers so far supplied, 24 dumpers are out of commission due to warranty failure representing a locked up capital of ₹1.5 crore (approximately) and interest and depreciation charges at ₹2.5 crore per month.

The figure of average downtime from the statements furnished by Mr. Baji Rao (Column 4 of Table 41.1) indicate over 5 months of downtime in a warranty period of one year showing the low dependability of the dumpers.

Table 41.1 Downtime and Delays

Sl. No.	No. of Dumpers Supplied	Average Delay (after Revised Delivery Schedules)	Average Downtime for Failure During Warranty Period	Idle Hours After Warranty Period Due to Spare Parts
1	2	3	4	5
1.	30(2010)	3 1/2 months	1600 hours (5 months 10 days)	394 hours (2 months)
2.	12(2011)	18 days	1781 hours (5 months 20 days)	109 hours (15 days)
3.	41(2012)	60 days*	534 hours**	—

*This includes dumpers supplied incomplete, i.e., without tyres, oil filters, assembly, hydraulic system parts.

** These equipments have been commissioned recently and hence not completed their warranty period. Considering that the total hours availed of 1350 hours, the downtime of 594 hours (average) reflects 40% non-availability during the warranty period.

Mr. Kannan, the Financial Manager of XLO observed that price of VML dumpers is much higher than price of the imported equipment of similar capacity. The price details are given below:

Table 41.2 Price Detail of Imported Equipment

Price of 23.2 cu. yd. VML dumper Model-35	= ₹31.10 lakh (tentative)
Price of 20 cu. yd.	= ₹26.35 lakh (landed cost includes ₹55,000 worth of free spare parts)
Computed price of a 23.3 cu. yd. imported dumper	= ₹27.24 lakh
Allowing 15% price preference on the landed cost of equivalent imported dumper (price formula accepted by Government)	= ₹31.49

Even the tentative price of VML is ₹4.77 lakh higher than what should be paid according to accepted pricing formula. Mr. Kannan emphatically stated that within the extra amount for 80 dumpers XLO could have procured additional 46 to 50 imported dumpers but for the policy of banning and support to VML from the responsible quarters.

SUPPLY OF SPARES

The position of spare parts supply was explained by Mr. Kumar who was held responsible by the operations personnel for not providing the spare parts in time. Mr. Kumar while appreciating the problems of VML in their ardent desire to achieve import substitutions stated that, not only the supply of main equipment but even spares and parts were far from satisfactory. This he proved on the basis of the past experience. He gave the supply positions of spare and parts both of VML as well as SKODA* which are recorded as below.

Table 41.3 Spare Parts Supply of VML

Parts Supply	Total Value of Orders Placed on VML is ₹298.34 lakh
Orders placed in last year Chassis spares	= ₹139.36 lakh
Engine spares	= ₹112.21 lakh
Total spares	= ₹251.87 lakh
Supplies made	= Nil (₹)

Note: Except for some engine spares valued at 3.15 lakh all the balance items are being imported and there is yet no indication about possible date of supply.

In respect of SKODA* dumpers the purchase made during 1979 were 24. Orders placed within delivery dates—the equipments stated as above are under commissioning. The SKODA* dumpers have offered a delivery of yet another 12 numbers of dumpers before one year, if booked immediately. The company approached the Government for permission of release of foreign exchange for a group of 12 dumpers for the requirements for the first year and another group of 12 dumpers the likely shortage for the second year. The actual shortfall of supplies on the part of VML, Mr. Kumar indicated is 25. He has strongly felt that permission for import of 37 dumpers (12 + 25) would need to be accorded.

As regards the spares of SKODA* the following were noted:

(a) Orders placed directly on the foreign firm pending from 2007	= ₹80.41 lakh
Supplies made	= ₹61.95 lakh
Outstanding expected to be shipped before next October	= ₹18.46 lakh
(b) 15% for 25 dumpers (for SKODA*) supplied recently	= ₹71.62 lakh
Supply expected before next April (foreign exchange is expected shortly)	= ₹71.62 lakh
(c) Orders placed on M/s. Shri Ram & Co., an agent having rate contract with DGS&D	= ₹24.06 lakh
Orders executed	= Nil (no firm commitment yet)

(d) Orders placed on M/s. XYZ (Indian Agents of SKODA* dumpers) = ₹62.24 lakh
   Orders executed = ₹47.00 lakh
   Balance to be supplied = Within 4/6 months

*These are imported equipments from a country with trade agreements.

Operational shifts in open cast mines are generally two in number. The duration of the first shift is from 4.00 a.m. to 12.00 a.m. while the second shift is from 4.00 p.m. to 12.00 p.m. Such an agreement on the working of shifts allows a time of 8 hours for carrying out the maintenance tasks. There are various schedules laid down for maintenance of different equipments and every maintenance work is duly recorded and records maintained. The schedule of availability of time for maintenance of equipments on the surface (contrary to underground items) can improve scope of maintenance so far as the open cast mines are concerned; maintenance crew can reach the spot quickly which also facilitates inspection, movement of spares and tools are also rendered easy. The main problem for Mr. Baji Rao has been the non-supply of spares and parts. He further added, "I am sorry that I have to report adverse trends to this committee insofar as the VML's performance is concerned. However, I am confident that given time and the necessary feedback and the help required by the users it would be possible for VML to tide over the present difficulty and that it would be able to improve overall performance of the dumpers through improved after-sales-service and supply of spares and parts in due course of time."

SELF-RELIANCE TALK

Mr. Ramulu, representative of the Ministry of Finance, observed, "With the building up of various earth-moving equipment companies in India, we are also finding it difficult to justify the prices that we have been paying for the products of VML. Because it is a public sector undertaking we have been channeling orders in a big way to VML, but in the near future this might prove to be difficult as companies producing equally good products and selling these at reduced prices may place their position before the authorities. This would compel DGS & D who are our purchasing organisation, to divert such orders. VML management should be aware of these possibilities." VML should also practice T.P.M.

Mr. Malik, the Chairman, replied: "We have the built-up capacity and the competence for manufacturing good earth-moving equipments. Our country has achieved substantial progress in the field of import substitution. VML dumpers on an average have 30–45% F.E. content. The emphasis today is more and more on self-reliance. We are aware that the efforts of XLO Corporation to keep up the targets of increased coal demand have been greatly retarded by the non-availability of equipments and spares. We are aware also of the shortcoming on the part of our personnel policy in recruitment and training of equipment operators. Our staff on the maintenance, particularly the technicians, the supervisors and the departmental heads need to be trained in their respective fields of activity. We from the company are very seriously seized with the problem of VML, and we are making every effort to help VML in order that better services are provided in addition to improving the quality of the products. We still believe that given the facilities and the resources, the VML management should achieve improvement and such a collaborative arrangement should be the right step for both. The Commercial Director of VML observed that a short- and long-term policy of imports for VML is necessary. The policy maker must liberalise imports and reduce duties for the vital spares.

Chapter 42

Bhushan Refineries Limited (BRL)

COMPANY SCENARIO

Bhushan refineries was established in the Public Sector in 1990 and within a record period of 24 months it was erected and commissioned by an international consortium. The engineering management of the erection was done by an Indian Company. The plant has been designed to process about 3 million tonnes of crude annually. This is one of the most sophisticated refineries in the world. It handles high sulphur crude. The capital employed is around ₹400 crore which is divided equally between fixed assets and working capital. The paid-up capital is around ₹100 crore, and reserves and surpluses are around ₹150 crore. The borrowings are around ₹50 crore.

Liquid Petroleum Gas (LPG), Naphtha, Motor Spirit, Aviation-Turbine Fuel, Kerosene, High Speed and Low Speed diesels, Fuel oil, Lube oils, Residues, and Sulphur are the distillates obtained from the crude. Other than Naphtha which is pumped directly to a neighbouring fertiliser unit, almost all the other products are marketed through a government agency. Hence profitability depends on the international crude prices and the prices fixed for the products by the Government of India. For the past 4 years the company has been paying a 12% dividend to its shareholders and an average 8% to 9% Bonus/Exgratia payment to its employees. The financial strength of the company can be seen from the fact that the book value of the share is more than ten times its face value. The annual throughput is about 3 million tonnes of crude, which is processed into about 16% light distillates (LPG, Naphtha and Motor spirit), about 40% medium distillates (ATF, Kerosene, HS and LS diesels) and about 25% heavy distillates (Fuel oil and lubricants). Sulphur is also extracted. About 10% is consumed as fuel. The individual quantities are fixed by the government. There are 13 plants within the refineries out of which two are main units. Usually a two to three week annual turnaround shut down is planned for in the schedule.

The first plant is the crude distillation plant and consists of an atmospheric distillation section and a 2-stage vacuum distillation section. Light distillates and medium distillates are obtained from the atmospheric tower and the residue is fed to the vacuum tower from which heavy distillates are obtained.

Plant 2 has 2 sections: (i) a vapour recovery unit from which naphtha is obtained directly. It also produces the feedstock for plant 12 where sulphur is obtained and (ii) a merox treating plant (a proprietary trade name) which produces LPG and Petrol fractions and whose feedstock comes from the vapour recovery unit. The main feedstock for plant 2 is the light distillate from the atmospheric distillation tower.

Plant 3 is a unifiner and platforming tower which desulphurizes and blends to produce petrol. The feedstock comes from Plant 2.

Plant 4 is a hydro-treater for kerosene and produces kerosene and aviation fuel. The middle distillates from plant 1 is the feedstock. Plant 5 is a high speed diesel oil desulphurization unit. The feedstock is the middle distillate of the atmospheric distillation plant. Plant 6 is a thermal cracker unit which produces fuel oils. The heavy distillates from vacuum distillation unit either directly or through a desulphurizer (Plant 13) from the feedstock.

Plant 7 is fed with some heavy distillates and the residue of vacuum distillation plant and asphalt is obtained by air blowing.

Plants 8, 9 and 10 are connected in that order to the supply of heavy distillates from the vacuum distillation plant and produce lube base stocks. Plant 11 is a hydrogen manufacturing plant and supplies hydrogen wherever needed. As the refinery is processing high sulphur crude, the sulphur is removed by hydrosulphurisation. This is the only refinery in the country which has a hydrogen plant. Plants 12 and 13 are desulphurising plants. Plant 12 removes hydrogen-sulphide from the gas streams in Plant 2 and sulphur is recovered from the hydrogen-sulphide. Plant 13 is a vacuum (heavy) distillate desulphuriser and produces part of the feed-stock to Plant 6.

ORGANISATIONAL ASPECTS

There are executives in charge of Technical Services, Manufacturing, Maintenance and Construction, Finance and Personnel functions and they report to the Chief Executive.

There is no separate executive in BFL for the materials function which is looked after by the Maintenance and Construction executive. Other than the maintenance and construction heads, the head of warehouse and the head of foreign purchase also report to him. While the head of warehouse looks after local purchase, stores and material planning, the head of foreign purchase takes care of all foreign purchases.

It should be noted that crude is handled by the manufacturing division. This is because crude supply is on a long-term contract with a foreign supplier who is also a shareholder of the company. The Organisation Chart given in Figure 42.1 clearly indicates the structure, and the delegation of powers.

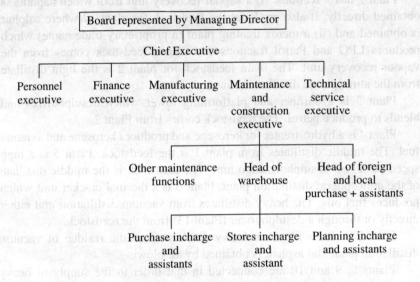

Figure 42.1 Organisation chart and powers of delegation.

BUSINESS LOGISTICS

The annual programme of processing about 3 million tonnes of crude is scheduled such that a 2 to 3 week annual shutdown is taken care of. This shutdown depends on the state of the plant, its past behaviour and the schedule of shutdowns in order refineries. Thus on an average 2,30,000 to 2,40,000 tonnes of crude is processed in a month. The refinery works 24 hours. The crude is shipped from a Gulf Port by the foreign agency. A national shipping company handles the crude. A six month firm requirement of crude and a year's forecast is given to the foreign agency and the shipping company. The shipping company has to be chased to detail tankers for this purpose.

Usually a tanker arrives every 10 to 15 days. Ships with 46 ft. draught in clear weather and 38 ft. draught during the monsoons are received at the oil jetty in the port. The oil jetty was specially constructed for this purpose. Ships with higher draughts have to unload into draughter ships in mid-sea.

The crude is piped directly from the ship to the refinery by a 10 km long pipeline of 75 cm diameter at the rate of about 3000 tonnes/hr. There are four storage tanks of about 35,000 tonnes capacity each. Usually one tank is kept empty for maintenance. They can store a total of 20 days crude supply. Though 6 tanks are usually needed (2 for settling, 2 for feeding, and 2 for maintenance/ spare etc.), by proper scheduling, the existing 4 are being utilized to the maximum.

All the processed products excepting asphalt, LPG and sulphur are pumped out of the refinery in bulk to various installations as per the requirements of the marketing agents. The pumps have sufficient capacity and each class of product has an individual line. Asphalt and LPG are filled in drums and cylinders respectively. Sulphur is either delivered in bulk into tank trucks or is flaked and supplied solid.

Naphtha is being pumped to a nearby fertiliser plant. All other products are marketed by the agency. Hence there is no marketing cell in the company.

The materials needed for the process can be broadly classified in Table 42.1.

Table 42.1 ABC Analysis and Financial Powers

Super A item	:	Crude: Board level
A items	:	Chemicals, catalysts etc. costly spares, pipe fittings etc.: Senior officers
B items	:	Consumables—soaps, rags small fittings, washers etc.: Officers
C items	:	Desirable spares: Assistants

Originally there were about 10,000 items and this has grown to about 25,000 items in recent times.

The financial performance of the company can be judged from the growth of shareholders' net worth per share (face value ₹1000). It has had an average annual rise of about ₹350 and is presently around ₹48,000. The finance department has prepared a report on splitting the shares and is awaiting government approval.

The value of sales does not represent a true picture of the performance because of the oil prices. Hence the growth in Net Worth gives a better indication of the company's performance.

Table 42.2 gives some of the important figures and their growth in the past few years. As stated earlier the total capital employed is around ₹300 crore, which is about equally divided between fixed assets and working capital. This is financed by about ₹100 crore of paid up capital, about ₹150 crore of reserves and surpluses, and ₹50 crore of loan. The collection period is about 15 to 20 days while payments are being made in 30 days.

Material Planning is very vital because the number of items is growing tremendously. In the past 5 years it has grown from 9000 to 15,000, because of the development of technology and other improvements. VED analysis for spare parts has been done by the technical services group. All the materials have been codified. The main groups which are 6 in number (1 digit in the code) are: (i) MRO items (maintenance, repairs and operations), (ii) Surplus, (iii) Incidentals, consumables, lubricants etc., (iv) Spares, (v) Direct charged items, (vi) Capital items.

The materials have been classified as follows:

Table 42.2   Materials Classification

Class	Type of Item (Basis for the Class)	Number of Items	% = Number	% Value of Consumption
Super A	Crude			This is the basic raw material, and the consumption figures are given elsewhere. This has not been taken in the calculation of value percentages as this alone accounts for 90%.
A	Chemicals, Costly spares, pipe fittings etc. Basis: (i) Consumption greater than ₹10,000 or (ii) Unit price greater than ₹50,000 or (iii) All vital items of spares:	1000	6.67%	70% (16 to 18 months stocks)
B	Consumables, smaller fittings, gaskets Basis: (i) Consumption between ₹2000 and ₹10,000 or (ii) Unit price between ₹500 and ₹5000	4000	26.67%	25% (9 to 10 months stock)
C	Desirable spares, bolts, nuts, washers etc. Basis: All other items which do not come in the above classes.	10,000	66.66%	5% (2 years stock)

Nil stock items: These form about 20% of C items—These are easily available and have no insurance value. There is also a fear of obsolescence in holding these materials—hence they are purchased on a need basis.

There are 2 digits for the sub groups. These are alphabetical characters e.g. automobiles etc. Within each sub group there is a 4 digit serial number. Each item has a 2 digit unit code in which its transactions take place.

Each item has a bin card. Postings are done with the help of computer. Monthly consumption statements are prepared in the computer and circulated.

REORDER LEVELS

Reorder levels, based on a 3-year history, have been fixed for 90% of the items. The material planner gets an exception report monthly and places his orders correspondingly. A minimum-maximum system is being operated and a minimum stock of three months is held for B and C items. Usually 'A' items are ordered on the basis of EOQ quantities, or as in the case of regular consumption materials (chemicals etc.) a contract order is placed and a delivery schedule is given. The deliveries may be daily, weekly or monthly depending upon the quantity consumed. For instance HCL acid is delivered daily.

For B&C class items, mini-max is followed more strictly, while as stated before 20% of C items are operated on nil-stock basis. Usually C items are ordered annually. All the indents, which are raised by any user department, come to the planning section. The indents are checked for specifications, substitutes, availability and then cleared for necessary purchase action. Capital items are not processed through materials planning section but directly by the process planners.

The purchase department calls for quotations. A limited tender policy is being followed and there is an approved list of suppliers. For orders above ₹500,000, a press advertisement is needed. The financial powers have been given in Table 42.1. Quotations are needed even for repeat orders. The orders are placed after financial concurrence is obtained, periodic chasing of the suppliers is kept up. Though the powers are limited, yet the purchase department has been able to manage because other than the foreign suppliers, about 80% of the suppliers are available within a distance of 20 km.

LOGISTICS WAREHOUSING

There is a main warehouse which has stocks and bins according to the classifications. An open steel yard is attached to it. There are separate sheds to store refractories, chemicals and lubricants because of the quantities involved. Material handling has been sub-contracted. The contractors move materials as per the work orders and collect their payments for which a piece rate system is in vogue. Only a general inspection (numbers, breakages etc.) is done by the warehouse men. The main inspection is done either by the user or the indentor as the case may be. Disposals are mainly mixed steel scrap, used compressors, sheet cuttings etc., accounting for about ₹1 crore per year. These have to be sold through press advertisements. As the parties collude and underquote financial concurrence is difficult to obtain under the plea that the quotations are below book value. At times when the prices fall, the tenderers are even willing to forego e.m.d. Hence there is a stockpiling of scrap.

Due to the constant rise in crude prices, the cost of sales (i.e. total manufacturing cost + overheads) to sales turnover has risen to 99%. About 90% of the cost of sales is accounted by crude alone. Correspondingly the return (before interest and taxes) on total capital employed has been decreasing from about 20% to 5%. The loan capital has been more or less steady around ₹50 crore. Hence the Chief Executive was wondering whether the loan amount could be reduced by reduction of stocks.

For the majority of A and B items (which were not on a contract order), the indigenous external lead time is about 4 months. For manufactured valves, this external lead time is about 6 months and for equipment this is about one year. Compared to this the internal lead time is about 2 months (materials planning clearance—2 weeks; quotations etc.—4 weeks; financial clearance—2 weeks). Hence if this could be reduced to around 2 weeks there can be a substantial reduction in holdings.

It can be seen that about 75% of the internal lead time is due to the low financial powers. The financial powers enjoyed are given in Table 42.1. The average ordering cost is around ₹500 per order. Hence for an argument an order for soap for say ₹750 has to wait for about 2 months. It incurs an order cost of ₹500 within the department, while the cost of the finance department which has to give its concurrence, has not yet been calculated.

The annual turnaround (annual shutdown) is a peculiarity of capital intensive chemical process plants. It lasts about 3 weeks. The plan is made well in advance and procurement is chased on the basis of a PERT network. A minimum 2 month slack is introduced between the availability of all materials and the beginning of turnaround. Materials are specially ordered and procured and these are not issued for regular consumption in the intervening period. The opposing costs in this case are the holding cost at 28% for about 2 to 3 months on spares worth about ₹50 lakh and extending the downtime of a ₹20 crore fixed asset by a few days. It may be noted that the plant has to be utilized annually for at least 80% of the time (inclusive of annual turnaround, failures, lack of materials) to break even. Hence a 99% (or even more) level of assurance of material availability is needed for this period. Even with the application of PERT the 3 weeks are very hectic and give rise to frayed tempers. The Chief Executive was wondering how to improve the situation. He was also wondering whether the 2 month slack time was really necessary or not.

Related to this annual turnaround is the development of indigenous sources. The average external lead time for imported items is about $1^1/_2$ years and imported spares form nearly 50% (by value) of all spares. Hence the stockpiling is more in these categories. Also, as the plant gets older and there is a technology leap in developed countries, these items will become more difficult to procure, because they would have become obsolete in developed nations.

But the major difficulties in indigenous development of the special type of valves of various internal diameters are: (a) non-availability of drawings and specifications of spares and materials required; (b) where the specifications are available the requisite raw materials may not be available, and hence equivalent substitutes have to be searched for; (c) due to the inadequacy of technical know-how the performance of substitutes is not as good as the original and thus causing a reluctance on the part of maintenance staff to use the substitutes; (d) the quantity required is too small for an indigenous supplier to manufacture economically.

Hence, the development of local sources has become one of the major activities. Where technology is lacking, it is usually provided by the Technical Services department. So far about 1000 items have been developed, out of a total number of 20,000 spares, 50% of which is imported.

An indigenous development committee has been established by all the refineries in the country and this is tackling the problem at the national level. But the inherent problem of such a committee is that the progress is not fast enough. The Chief Executive wants indigenization programmes to progress faster so that the capital locked up in imported inventories can be released (Table 42.3). Also, he wanted the maintenance system thoroughly examined by an external consultant and was wondering about the introduction of total productive maintenance.

(i) Purchases upto ₹2000 per order (proprietary items) and ₹5000 per order for others if the lowest of 3 quotations is chosen can be authorized by Maintenance and Construction executive upto an annual total of ₹4 lakhs.
(ii) Head of Warehouse is authorized upto ₹250.
(iii) Purchase Incharge is authorized upto ₹50.
(iv) For all other purchases financial concurrence must be obtained.
(v) For any value greater than ₹50,000, a press advertisement is needed.
(vi) Press advertisements are necessary for all disposals.

Table 42.3 Inventory Levels for the Recent 4 Years

	Year 1	Year 2	Year 3	Year 4
Crude oil	2.0	3.5	11.0	11.5
Chemicals and catalyst	0.4	0.5	0.6	0.6
Shares inventory	1.6	2.5	3.8	4.5
Finished products	2.0	2.0	6.0	6.0
W.I.P.	0.5	0.5	2.0	2.0
Sales	55.0	65.0	140.0	200.00
Total production in million tonnes	2.2	2.1	2.4	2.3
Capital employed	38.0	39.5	41.0	38.5
% of cost of sales to total sales	94.0	90.0	95.0	99.0
Spares consumption	0.4	1.1	1.8	2.0

The increase, is essentially due to hike in oil prices. (₹ 1000 crores)

Chapter 43

Middle East Air Transport

COMPANY BACKGROUND

Middle East Air Transport has hit global arms market to buy spares for its largely Russian fleet of aircrafts/helicopters. The move was triggered to overcome problems of spares from the original manufactures, that were not only expensive but supplies were erratic. Global bids have been invited for ageing transporters—An 32s to more recent Su-30 combat jets. The chairman mentioned that competitive bid would help in getting better price as relying only on the original equipment manufacturer was not always the best option. Only generalised spares are being routed from global vendors and for specialised parts, manufacturer would remain in the picture as they were not available outside.

SUPPLIERS FAULTS

The air head quarters has sought bids for a bulk of spares for almost the entire Russian float. It was the first time when such large scale purchasing of spares was being done at the same time. These items include tubeless tyres for IL-76 transporters to small equipment crucial for fleet maintenance. Most of the fleet is ageing and its maintenance cost has been criticised by every one.

Over dependence on a single source, mainly the Russian manufacturers had been a problem lending to several disputes over price against the contracted obligations. There was a tendency to charge exorbitant rates that resulted in a number of problems to keep the fleet fit. Maintenance has been a major challenge as the transport fleet was overstretched.

The air force has close to 100 An 32s which favour the backbone of the fleet, but at the moment they are barely able to meet daily operational requirements. The problem of spares was singled out as one of the reasons for frequent crashes of MIG Combat jets, the mainstay of the fleet. This is a continuous process and should ring alarm bells to the original supplier of aircraft. The defense services have complained about the maintenance support provided by the Russians for the equipment. A lot of companies outside Russia make generalised equipment that can be fitted over the aircraft and the airforce is trying to buy from them.

Air Chief Marshal Muhammed is not only the Chief of Mideast air forces, but also the Chairman of Middle East Air Transport. In the air transport company, he is assisted by Director of Operations, Director of Finance, Director of Engineering, Director of Commercial, and Director of Personnel. The Chairman after several meetings felt that the fleet utilisation must be improved urgently. He requested Prof. P. Gopalakrishnan, senior management consultant to study the problems of the airforce and the air transport company. Prof. P. Gopalakrishnan met senior officers of the airforce and also met the staff of air transport company and held detailed discussions. His tentative solution of airforce has been given in the opening paragraphs. He observed that private transport airlines have been operating with great profit margins at the expense of the government air transports which is not even able to pay the salary, perhaps due to involvement of some policy makers according to the pilot's association president. It was observed that the company consists of various models of aircraft, which has created a great strain on maintenance departments and spare parts management.

The maintenance chief informed facts, and the performance analysis of repair tasks has been increasing over the years. But the quantity of spares demanded was 100% more than delivered.

FORECASTING REPAIR TASK

It was also observed that the allowed repair time (in 2 months) was far better than the actual, which is 6 times more (12 months) waiting in queue for repairing was twice that of total allowed repair time (4 months), and actual repair was 4 times more i.e. 8 months. The manpower utilisation of all repair bases was 60% and could be improved if maintenance crew was further motivated. Since the maintenance shops also carry out repairs of rotable repairs which form an integral part of aircraft, the performance analysis of rotable repair shop indicates that the production to total task was only 50% spares not available and facilities not available constitute are the major reasons for this. The actual arisings were about 60% of the total projected. Prof. P. Gopalakrishnan questioned about the low arisings and was informed that the system of forecasting adopted in the company envisaged a certain level of utilisation of the aircraft and all task decisions and provisioning were based on this factor. If the utilisation is low, then arisings will fall below the anticipated level.

Prof. P. Gopalakrishnan found out that Mid East Air Transport Corporation has adopted the following procedure for forecasting of repair tasks for aircrafts, repair of rotables—basically modular units of instrumentation and other control devices which can be rotated between aircrafts and spares requirements. A five- year task is laid down giving breakdown for every year. This is reviewed and updated at the beginning of each year for subsequent 5 years.

The rotable task is issued for a 3-year period and is firm for the first year and tentative for the next two years.

Provisioning for spares is the responsibility of the materials planning and maintenance/mechanical engineering group. Since procurement lead time is for 24 months, spares requirement are to be forecast 2 years in advance.

For spares required for routine and line maintenance the following procedure is adopted—the current amount requirement based on past commention is multiplied by a fast factor to give an estimate of the requirements for two years hence. This forecast factor is the ratio of actual strength/effort during the past 12 months and the planned establishment effects during the authorised period of holding.

For spare querie for overhaul and major repair tasks the proportions are based on past consumption data expressed as units needed for overhauling $10°$ to $15°$. This requirement is multiplied by the overhaul of major repair task expectations to give the requirement forecast.

The provisioning decisions then takes into account dues in and dues out and stocks in hand before deciding on the net order quantity for the task period.

Prof. P. Gopalakrishnan asked the maintenance executives the reasons for the repair cycle time for the repair of aircrafts differ from the allowed-and the response was nonavailability of spares and rotables at the proper time and nonscheduled crossings leading to queueing delays. Whereas the arisings have been less than the predicted, the system should have generated more than 15% increase in spares holdings.

CONFLICTING VIEWS

In spite of this the maintenance engineer shows a shortfall of about 40% of the rotable repair task due to nonavailability of repairables pointed out Prof. P. Gopalakrishnan. Surely this situation calls for a detailed analysis as the company has taken a decision to improve the utilisation of aircrafts. For this, the maintenance delays have to be drastically curtailed and yet the inventory holding costs as well as reduced. He constituted a task force to solve the above controversies consisting of maintenance engineer, EDP manager, store purchase manager, materials planning engineer. Prof. P. Gopalakrishnan thought in the meanwhile that the above conflicting situation was further made chaotic by the Syria, Libya, Iraq, Afghanistan and in the middle east crisis, besides collapsing of many economies like Greece, lower ranking US dollar etc. in 2011 end.

The Canadian aircraft manufacturer BOMBARIER aerospace, $9 billion company headquartered in Montreal, has expressed its interest in selling 80-seater Q400 turboprops, 99-seater medium haul CRJ jets and the mainline 125 seater C series jets, for linking smaller cities and expressed that air transport business is cyclical in nature and Middle East Airport will turn the corner in the next 3 years.

INTERNATIONAL PRESSURES

The picture was confusing to the Chairman of middle east airways, who has to keep the future force/aircrafts/air transport ready in view of the worsening situations in Egypt, Libya, Iraq, Syria, Japanese tsunami/floods, Greece belligering, American economy downgraded by market observers, Chinese export of military hardware and balance of payment increasing, locking up of capital in Mid East Airforce/Airtransport, Indian Airlines merged with Air India, which is unable to pay salary to the pilots demanding heavy government help etc., and Russia unable to supply spares etc.

In this context the senior airforce officers thought that global tendering was the best way in Mid East Airforce, and scouting other Russian Airway uses for spares, cannibalisation of fighter aircrafts, buying of new aircrafts on soft lean basis and requesting them to help out in the Russian fighter aircrafts etc.

But the Chairman of Mid East Aircraft was anxious to collect more deformation particularly the operating of civilian aircraft in other countries. As per the suggestion of consultant Prof. P. Gopalakrishnan, attempts were made to pool the existing information from all sources of Mid East Air Transport. Some of the collected data is presented in the next few paragraphs.

ABC-FSN ANALYSIS

In airframe spares up to 1394 high value items have moved once in 10 years and 13361 items have not moved at all. Of the 'A' items 25% of items accounted for 95% consumption value while remaining 15% items accounted for 5% consumption. In the Aero engines 950 high value items have moved at least once and 985 items have not moved at all. Cost-criticality–availability combined MUSIC-3D has not been carried out. Variation in usage followed an informal distribution, while fluctuations in lead team followed a Poisson pattern.

The total lead time—provisioning review and draft indent preparation, approval of competent authority, preparation of final indent, supply lead time, transit lead time and inspection time took 5–6 months to 20 months. Data for each item and each part, summarised above is available.

DEMAND ANALYSIS

The demand pattern of spares—over the recent two years demand, risen out of major repair and overhaul of aircrafts, engines and rotables for A class items is given in Table 43.1.

Table 43.1 Performance Analysis of Repair Jobs

Period	Operating Balance	Input	Forecast	Performance	Close Balance
Year 1	45	14	30	22	37
Year 2	37	16	16	13	40

It was observed the total spares received was only 60% of the number demanded during the last three years in the repairs job—for all classes of aircrafts. The repair cycle time for all types of aircrafts was 25 months while the waiting time itself was over 8 months and actual repair time was 12 months thus adding up to 20 months! In the rotable repair shop for the last two years, only 50% of tasks were completed, while 10% of cases the spares was not available and in 5% of cases the facilities were not available.

Considering all the above the committee of directors asked the opinion of Prof. Gopalakrishnan whether periodic review system of fixing the replacement level 'M' is equivalent to total of 'B' buffer stock added to consumption (S) during lead time L plus review time $M = S(L + R) + B$.

Prof. Gopalakrishnan found that all past data are available within reasonable time. He is seeking your help to solve the mideast airforce and air transport problems, considering the socio-political, economic-political-macro/micro factors. He has promised to deliver his report on all aspects of maintenance, tpm inventory, purchase, supply, competition, supply problems, new suppliers, contractual obligations, overhauling, maintenance, repair facilities etc.

Chapter 44

Aruna Oil and Gas Limited (AOGL)

CORPORATE STRATEGY OF AOGL

The chairman of Aruna Oil and Gas Limited needs your help in introducing scientific inventory control on spare parts without affecting oil production and to develop suitable management information system. He would like to minimize obsolescence, standardize the parts, evolve procedures on stores and purchase, identify minimum/maximum limits, develop suitable training programmes for the officers and examine maintenance systems.

Aruna Oil and Gas Limited (AOGL) is a state organisation involved in planning, organising and implementing of oil resources in an Arab State. It started in a modest way about 15 years ago in collaboration with USA, UK, USSR, India and Rumania. AOGL has steadily improved the performance since 1993, and has matured into one of the few technologically self-sufficient oil companies in the developing nations, particularly in the areas of exploration, drilling and production. For this purpose, Institute of Petroleum Exploration, Institute of Reservoir Studies, Institute of Drilling Technology, have been established recently as part of AOGL. A sophisticated high speed fourth generation digital computer has been recently commissioned in AOGL for processing marine seismic data at a rate of nearly 10 km/h. Laptops have been provided to all senior officers.

Besides meeting the country's requirement of exports—50% of the total output—the policy of AOGL is to improve oil production by finding new sources in the country and also on off-shore. The company plans to double its production in the next four years. In this context, by the end of 2012, AOGL has geologically surveyed 65% country's area by detailed mapping, reconnaissance mapping, special studies and geomorphological mapping/neo-tectonic surveys. AOGL's geophysical parties have covered 75,348 km by seismic surveys and measured gravity magnetic data from 181,798 stations. The total number of structures taken for exploration both on shore and off-shore are 131 out of which 69 have been found to be oil and gas-bearing. Out of 1452 wells completed, 983 have proved to be oil-bearing and 198 gas-bearing. AOGL has extended its exploration activities to almost all the sedimentary basins of the country. Half of the total shelf area, up to 100 metres water depth has already been covered by marine seismic work. The crude output is handled by the government.

ORGANISATION STRUCTURE OF AOGL

The chairman is helped by five Director-Generals. They are incharge of (1) off-shore operations, (2) on-shore operations, (3) finance, (4) personnel and (5) materials and maintenance respectively. The whole country is divided into four zones—north, south, west and east and one General Manager reporting to the D.G. on-shore is incharge for the 'total operations' in each zone. Similarly, the D.G., off-shore, has two General Managers for two regions connected with off-shore, has two General Managers for two regions connected with off-shore drilling. Each General Manager has various functional heads including a materials manager under him. The D.G., Material and Maintenance, gives 'technical guidance' to the different materials managers. He also has legal, source development, vendor rating central purchasing, headquarter stores, inspection, capital equipment purchase, imports purchase, clearance, marine insurance and other service departments under him. The Materials Maintenance Managers reporting to the different General Managers, control the stores in the respective region and also purchase indigenous items worth 10,000 dollars per order. The indents are forwarded by Project Manager working under General Managers to the headquarters. The D.G., Materials Maintenance, is incharge of purchasing/hiring capital equipments, imported spares and miscellaneous items worth more than 10,000 dollars per order. The entire board is brought into the picture if the value of an item per order is more than one million dollars. AOGL has a Central Workshop under the control of D.G., on-shore. It has undertaken the manufacture of work over rigs, up to 50 tonnes capacity in the workshop. Through this workshop, and other organisations in the country, AOGL hopes to develop capability for manufacturing major equipments like rigs, drill bits, pumps, castings, drill pipes, platforms, supply vessels, seismographs and other equipments required for seismic work and well logging equipment. AOGL employs 50,000 people of all categories, 80% of which are Engineers, Geologists, Technicians and high skilled personnel. Four hundred officers, and 375 other category staff are directly involved in purchase/stores inspection activities.

ASSETS AND EQUIPMENTS

The equipments available in AOGL can be classified as for geological surveys, seismic survey vessels, drilling equipments, oil rigs, jack up rigs, drill ships, semi-submersible drilling unit, rack up platforms, supply vessels, helicopters, transport vehicles, construction equipments and all necessary machineries.

Geological survey equipments constitute 55% of the total assets, production and installation equipments to 28% and the remaining by other categories of machineries. Oil rig is the most important equipment in the above system and would cost anything from 5 million to 20 million dollars depending upon the specifications. The oil rig is installed to dig deep holes ranging up to 9000 metres to strike oil. These rigs installed at various sites have to be manned by engineers and the efficiency has got to be maintained by seeing that the right type of spares are available at the right time. The rig stands to a height of about 60 metres and is composed of various modules, namely, draw works, rotaries, travelling blocks, rig electrics, sound system, utilities and swivels. These modules have their own sub-systems and assemblies. In the above draw works and the rotary are the critical ones. The rig has a large number of transmission and mechanical components.

AOGL owns 105 rigs out of which 85 are in operation at various sites, spread evenly in all zones. Twelve rigs are under the process of assembly/dismantling at the sites and the remaining are not used due to lack of important spare parts. The number of productive days has 55% of the total available time. Non-availability of spare parts at the proper time has accounted for 80% of the idle time of the rigs.

Agewise analysis of rigs indicates that 20% of the rigs are over 15 years old, 40% between 5 to 15 years and only the remaining less than 5 years. The rigs are from USA, Russia, UK, France, India, Italy, Japan and Rumania. When the rigs are purchased, initially spare parts amounting to 8% of the value of rig are recommended and provided by the suppliers. Hence many spare parts in stock, have not been used at all even today. But, some critical spares for the same rigs are in perpetual shortages. In a few cases, the OEMs are unable to supply the spares as they have switched over the models.

SPARES DETAILS

The purchase budget of AOGL is about 1500 million dollars which includes 600 million dollars of capital equipment, 500 million dollars of hiring charges and the remaining is accounted by stores and spare items and 85% of the requirement in value is imported from foreign countries. AOGL maintains a list of 210,000 items. Only temporary codes have been allotted and hence there may be duplication in this as the item is inspected and used immediately.

The inventory value is 585 million dollars out of which 92 million dollars worth of items have been declared as obsolete by the D.G., Materials Maintenance. About 46 million dollars have been declared surplus and 20 million dollars as scrap. The inventory has been increasing in recent years indicating an upward trend in purchasing compared to the consumption.

The inventory carrying charges including the interest charges, to be paid to World Bank, Arab Development Fund, etc. have been estimated at 27%. The ordering cost per order has been estimated at 1000 dollars per order for imported items, and 125 dollars for indigenous items.

The annual consumption of stores and spares is about 125 million dollars. The chairman of AOGL is worried about the high inventory of around 60 months' consumption and wants it to be reduced to half. Analysis indicates that this figure is about 3 years for off-shore items, work for which was started only five years ago. Table 44.1 lists the typical stores, consumables and spares used by AOGL.

Table 44.1 List of Typical Stores and Spares

Sl. No.	Item
	Since the value is fluctuating, depends on country of origin theory have been omitted.
1.	Drilling pipes
2.	Casing pipes
3.	Other pipes and pipe fittings
4.	Drill bits
5.	Other drilling stores
6.	Electrical stores i.e. electrical fittings, cables, insulating materials, etc. including electrical instruments
7.	Building material and other civil engineering stores including timber
8.	Oil well cement
9.	Chemicals including mud chemicals
10.	POL e.g. oil grease and lubricating material etc.
11.	Metals e.g., bounds, bars, plates etc.
12.	General tools on stock
13.	Miscellaneous stores i.e. drawing material, bolts, nuts, rope, screw, tents, tarpaulin etc.
14.	Tubings, pipes and fittings
Spare Parts	
15.	Spares for turbodrills and connected items
16.	Spare parts for drilling equipment viz. rigs and diesel engine, mud pumps, air compressors etc.
	(i) Diesel engine
	(ii) Slush pump
	(iii) Draw works
	(iv) Crown block, travelling block, swivel and rotary table
	(v) Other spares

(Contd.)

Table 44.1 List of Typical Stores and Spares (*Contd.*)

Spare Parts
17. Spare parts for (i) production equipment and (ii) other spares
18. Spare parts for cementing unit
19. Spare parts for geological equipment
20. Spare parts for geophysical equipment
21. Spare parts for imported transport equipment
22. Spare parts for indigenous transport equipment
23. Spare parts for electrical plant and equipment
24. Spare parts for civil engineering plant and equipment
25. Spare parts for workshop plant and machinery and equipment
26. Other spares i.e. spares for other equipment not covered above, such as spare for instruments etc.
27. Spare parts for BOP elevators etc.

The material component has been estimated about 20% of the total turnover value. Since the price of oil per barrel has been fluctuating around 100 dollars per barrel, the material component has also been fluctuating. It may be mentioned that even for internal consumption, AOGL has been getting the market rate.

SPARES PROVISIONING

Five years' requirements, matching with the country's 5-year Plans, relate to the acquisition of capital assets and a hunch about spares requirement is calculated as a per cent—usually about 8%—of the value of the whole equipment. For procurement of indigenous spares, the powers for indenting are delegated to the Project Managers and procurement action initiated as per the powers of delegation. The requirement is based on current year's operation and the technology involved. However, in the case of indigenous consumable items, the recoupment is done on the basis of minimum/maximum levels. The quantity to be ordered usually is about six months' requirements. The annual indents for the imported items are prepared about 30 months as a lead-time by the concerned Projects. Later, these indents are screened by the Regional Material Planning Manager, where detailed examinations of the inventory control, consumption of the material and the budgetary sanctions etc. are made. After such screening, these indents are received at the HQ. The purchases of the imported materials are all centralized on land operations and for the off-shore operations. These indents are further scrutinised by the HQ. Material Planning and Provisioning Cell and the Inventory Control Cell. After these checks are made, the indents are now passed on to the concerned purchase section depending on the import of sources, say, Western countries, Russia, Rumania, India etc.

BID EVALUATION CRITERIA

Depending on the sanctioned amount and also on the nature of spares, say, where it is of proprietary nature or general spares, the purchase section decides to invite either open tenders or limited tenders. The quotations which are received are scrutinized by the Technical Department for the technical scrutiny. The technical scrutiny is done on the basis of past experience with similar materials, their proven efficiency in operations and their suitability.

Once the technical comments are prepared, based on the technical suitability the comparative chart of all the offers is prepared and a tender committee meeting consisting of an Officer from the Materials Department, Technical Department and Finance Department is called. According to the sanctioned amount, the level of tender committee officers is enhanced. The tender committee goes through the aspects of both technical and commercial and decides to place the offers on the suitable ones.

After the decision of the tender committee is received the supply orders are placed. Since these are imported spares, the requirement of foreign exchange is needed to be released by the Government. For this purpose, list of the requirement is furnished to the Government along with the application for the release of foreign exchange.

The HQ's Materials Maintenance Management Department keeps a continuous correspondence with the firms concerned to ensure that the materials are delivered as per the suppliers, are cleared from the customs and they are checked as per the specifications of the supply order. The spare parts are then transported to the various projects.

INVENTORY PROBLEMS

Since the machines are from different countries, AOGL is using all sorts of specifications like BSS, ASA, GOST, DIT, ISI etc. It has recently been stipulated that all critical items should have American Petroleum Institute specifications. The current classification is rig-wise and this would mean that spares for Russian rig could be used only on a particular Russian rig. Often people have been using both British and Metric Units as 2" × 6" by 4 metres. To mechanise inventory control system, the practice of giving codes to various items has been started recently. The temporary codes are based on usage and have 12 digits covering origin of the country, model, type of equipment etc. Often it is found that two codes exist for the same item kept in two zones.

Conventional 'ABC' analysis done in HQ has shown 850 'A' category items with 52 million dollars annual consumption. There are 3750 'B' category items worth 20 million dollars. The 'A' items include gear coupling, housings, shafts etc. of rigs as well as valve spares, transport spares, logging and geophysical spares. The classification is not uniform as flanges critically needed in Russian rigs have been classified in north zone whereas they have been classified as 'C' items for Italian rigs used in off-shore. The inventory policy is generally vague and stipulates that stockout should be avoided consistent with low stocks. The minimum and maximum have been fixed on a thumb rule basis and vary considerably from one zone to another. The VED classification has not been yet developed. The computer is used only occasionally for 'ABC' analysis and for determining obsolete items. In the annual stock verification discrepancies of great magnitude have been found out between book balance and physical balance. Items are stored on rig platform, region, on-shore head quarters off-shore head quarters and corporate head quarters.

The chairman of AOGL is wondering how to introduce scientific materials system without affecting oil production. He is also worried that AOGL has to pay 500 million dollars as hiring charges and was wondering whether they could be purchased at all. He was asking why TPM was not practised.

Chapter 45

Mahatma Gandhi Road Transport Corporation (MGRTC)

SCENARIO PAINTING

Mahatma Gandhi Road Transport Corporation (MGRTC) is a nationalised passenger transport organisation of a major state. By the RTC act of 1950, each state established its own RTC for providing adequate economic and coordinated transport services, particularly linking rural, tribal, hilly areas by mini buses. The board consists of 4 full time directors—MD cum CEO, directors incharge of finance, traffic and human relations, and 4 IAS officers from the State Secretariat and 6 non officials from different parts of the State. The board is keen to link up all villages/towns to the nearest railway station. Many private operators with high political-cum-bureaucratic connections have acquired through competitive bidding important and more profitable routes linking district headquarter as well as state headquarters and operate airconditioned sleepers, besides mini buses as well.

MGRTC has a fleet strength of 20,000 buses—mostly Telco, Ashok Leyland, and Volvo and Eicher—organised into 20 divisions. To enable better control of 8 adjacent divisions have Telco, 9 divisions with Leyland and 2 divisions with Volvo and 1 division with Eicher buses. Five per cent of the fleet is double décor buses operating within cities.

The Supply Chain Manager (SCM) and Logistics, reporting to the traffic director is helped by purchase officer, stores officer and chief mechanical engineer, who control stores, purchase, maintenance, spare parts and inventories in the divisions. Each division has 10 depots responsible to the divisional chief. The policies on all aspects have been covered on with powers of delegation cleared laidout, in the manual approved by the entire board. Minimum bus fare is kept at ₹5.00 per person.

DEPOT OPERATIONS

The objective is to serve the public in meeting the majority poor in a competitive way with the private sector. The bus conductor and driver are fully trained and 7% of collection are paid as incentives. For each depot there are nine checkers who ensure the collections are adequately returned. The average age of the bus is the same in all depots. Each depot manager has a stores assistant, collection accountant, maintenance foreman and traffic foreman. Vehicle maintenance is limited to regular preventive maintenance and minor repair of all the 90–120 buses/buses under its control. Typical depot operations in the table presented below. All activities at 5 selected depot level are indicated in Table 45.1.

Table 45.1 Select Indicators for 5 Depots

Indicators	Depots				
	1	2	3	4	5
No. of buses	80	119	85	95	105
No. on roads	70	110	80	80	95
No. of mechanics/assistants	25	32	27	29	30
Average km per day per bus	196	185	200	210	208
Failure per 10,000 km (average)	1.1	1.8	1.3	1.7	1.9
No. of type failures per year	110	130	109	120	110
Mechanical failures	165	215	180	225	204
Stores consumed in (000)	815	995	830	938	970
Kilometres lost (000 km)	95	180	129	111	121
Paise per km mechanical wages	52	71	63	59	68
Paise per km HSD oil	4.15	3.07	3.85	3.98	4.05
Paise per km engine oil	.11	.13	.12	.11	.10
Paise per km stores	.58	.52	.59	.57	.54
Paise per km tyres and tubes	.40	.40	.40	.40	.40
Paise per km depreciation	.81	.81	.81	.81	.81
Paise per km overheads	.57	.57	.57	.57	.57
Inventory (lakh ₹)	31	32	33	31	32
Obsolete inventory (000 ₹)	15	20	17	18	19
Consumption (lakh ₹)	14	20	15	16	18

MAINTENANCE POLICIES

Scientific maintenance policies have been prepared and approved and circulated in all departments, through the cost data of the table indicate variations. The depot manager is incharge of the fleet and authorised to provide buses as the traffic demands in the most efficient manner. The bus is considered as a capital investment and hence must be used to get the pay back in the shortest period. Hence the maintenance, playing a crucial supporting role, is done only when the buses are idle. The bus crew consider realisation of the target is more important than the maintenance.

If the breakdowns occur during the day time they are attended to immediately. But when they are reported during the late hours, all the spares cannot be (procured) supplied. "Emergency spares" are available to meet minor repairs. When a breakdown occurs enroute, a relief bus is arranged to carry the stranded passengers, if they are not accommodated in other buses passing in the route. If the reasons are major, then we send a pickup truck and tow the bus. If the causes are minor the mechanics are sent in another bus to repair the bus.

There have been occasional complaints on political pressures, drunken driving, traffic accidents, drivers complaining spurious parts, overloading, extra charges for luggage, the crew misusing free trips, concession for students, super senior citizens, private buses always overtaking MGRTC buses, buses not halting in recognised stops, but on the middle road, parallel bus stops in narrow road. Union protests stoning the buses by students, delays due to VIP movements etc. These are being looked into by the SCM.

Good preventive maintenance leads to performance, as "a stitch in time saves nine". Some crew often insisting change the spares more often although there may be some life still left, just to keep the vehicle in the best operational condition. So the SCM opined that at the depot level one should try to curtail the overall costs.

OVERHAULING PRACTICES

All the depot managers know a saving of one paise per kilometre results in about ₹2 lakh per annum. Hence the overhauling is done with this view. Normally overhauling is done after one lakh kilometres—about 2 years on the road. Overhauling is done at the headquarter control workshop, who are informed 10 days in advance. Reasonable amount of floats of 'engines and major assemblies' are kept in the central workshop. The SCM claims to apply PERT network is usually applied at overhauling and queuing principles and simulation are tried to calculate the floats in the workshop. Traditional ABC analysis has been done at central stores, division stores and depots. Usually 6 working days are lost in overhauling from pick up to delivery of the bus. The overhauling requirements are predicted with great accuracy. Fifteen days stock of fuel is usually kept and it is super 'A' item. Eight tyres per vehicle are spread in centre, divisional/depot levels, apart from the stepney and tool kits in the running buses. Lubricants, spares and consumables are other major items.

Even though 4 models of buses are maintained, there is always difficulty of getting original spare from OEM's some times due to the technological upgradation of the bus manufacturers. MGRTC scraps 500 vehicles, per annum, when they are 14 years old and sold by auction, while 600 new buses are added. There are over 30,000 items and consumption pattern vary widely, and SCM feels that only in these fastest moving spares and overhauling spares the prediction can be accurate. The depots/divisions send their estimates of their requirement (which is usually on the higher side to central office 18 months in advance of its use. But these are not regular in this process in spite of a weekly courses system. The indenting process is worked out assuming a total lead time of 1 year—which includes 1 month preparation of indents/estimates by depots, 1 month for setting the estimates, 2 months for consolidating, 5 months for purchasing activities and 3 months for the suppliers transport. Inter unit transfer serious emergencies, stock verification is carried out once a year and shortages beyond acceptable limits of .01% are investigated. Scrap and obsolete items are auctioned on an annual basis.

It is the depot managers who are questioned for bad performance at the end of each year.

ENTER THE CONSULTANT

The MD presented in the above information to Prof. P. Gopalakrishnan and asked him to suggest the suitability of decentralisation of materials, maintenance, operations, traffic and organisational details. Prof. Gopalakrishnan thought over pros and cons of the Tamil Nadu transport system of dividing the state fleet in the independent competing organisations which will practise TPM. He has prepared the report after meeting all persons at headquarters, division, depot and met some drivers/conductors. He also prompted driver cum conductor, with suitable incentives but without teas as in developed nations to some of union members. He also met the chief executives of some private bus companies. Then he presented the report to CEO.

In his interim report, Prof. P. Gopalakrishnan suggested the buying of procured tyre retreading machine priced about ₹22 lakh for which the detailed brochure of Gemini company was given. The MD and all officers thought it was a good idea, as tyres constute "Super A item in MGRTC". This offer has been found to be the best amongst 500 tyre retread machinery manufacturers in India with 5% commission on equipment. The manufacturer who gives one year warranty, informed that spares cost will be additional after the warranty period. The MD is nulling over the purchase of the tyre retread machinery and is awaiting the total formal report of problems in MGRTC from Prof. P. Gopalakrishnan.

GEMINI TYRE RETREADING MACHINERY DETAILS

This company started manufacturing of Precured Tyre Retreading Machinery in the year 1993. The range of machinery manufactured by the company includes Tyre Buffer, Tread Builder, Cyclon Filter, Automatic Inspection Spreader, Electric Curing Chambers of Capacity 2 tyre, 4 tyre, 8 tyre and 12 tyre and also steam operated Truck and LCV chambers. To date more than 100 sets of machinery have been produced and supplied to customers in India and abroad. Several Public Sector Road Transport Corporations in India have installed our Precured Tyre Retreading Machinery. The company so far has exported to countries such as Sri Lanka, Dubai, Nigeria, Sharjah, Madagascar, Mauritius, Muscut, Ghana and Egypt. The company has a team of technically trained personnel for installation and commissioning of machinery supplied by them and also impart on the job training. The features are given below:

TYRE BUFFER

The Gemini Inflated Tyre Buffer is designed for buffing tyres of sizes, ranging from 12" to 24".

Features

- Expandable hub with rims to ensure proper tyre holding, for various sizes of tyres.
- Specially designed turn table and rasp head to provide the right texture, ensuring quality of retreading.
- High speed dust collector ensures, clean working environment.

Technical Specification

Tyre range	:	4.50 × 12" to 10.00 × 20" (on option: 12.00 × 22.5" to 24")
Required air pressure	:	8 kg/cm^2
Power supply	:	7.5 kW/440V – 3Ph – 50 Hz (A.C. Motor)
Dimensions in mm	:	1500 × 1500 × 1700

TREAD BUILDER

The Tread Builder is designed for stitching the cushion gum and building the procured tread with the help of pressure, stitching and support rollers.

Features

- Expandable hub with rims to ensure proper tyre holding, for various sizes of tyres.

- Thyristor controlled drive with foot switch for inching and rotating tyre.
- Pneumatic treat cutter to ensure proper splicing.

Technical Specification

Tyre range	:	4.50 × 12" to 10.00 × 20" (on option: 12.00 × 22.5 to 24")
Required air pressure	:	8 kg/cm^2
Power supply	:	7.5 kW/220V – 1Ph – 50 Hz (D.C. Motor)
Dimensions in mm	:	1280 × 1100 × 1900

ELECTRIC CURING CHAMBER

The Electric Curing Chamber with both side dished ends ensure, proper bonding of tread to tyre. One year guarantee (Tools, accessories, electrical and machinery spares and priced separately.

Features

- Provided with safety inter locking system, prevents opening of the chamber under pressure.
- Independent Mono Rail arrangement for smooth loading and unloading of tyres.
- The chambers available with 2 tyre, 4 tyre, 8 tyre and 12 tyre capacity.

Technical Specifications

Chamber capacity	4 Tyres	8 Tyres	12 Tyres
Chamber dimensions			
Width in mm	2100	2100	2100
Length in mm	2200	3000	3800
Diameter in mm	1372	1372	1372
Temperature	125°C	125°C	125°C
Air Pressure in kg/cm^2 (chamber)	6	6	6
Air Pressure in kg/cm^2 (tyre)	8	8	8
Required air pressure	10	10	10
Connected load in amps	35	35	70
Power supply	440V/3 Ph. 50Hz		
Electric motor (Foot Mounted)	7.5Hp	10Hp	12.5Hp
RPM	1440	1440	1440
Heating elements	12kW	18kW	27kW
Type range	4.50 × 12" to 10.00 × 20"		
Price ₹(lakh)	15	19	22

Note: 1524 mm dia chamber option available for export.

MONO RAIL SYSTEM

The floor mounted Mono Rail system ensures smooth loading and unloading of tyres in and out of electric curing chamber.

Features

- Provided with pneumatic lifting arrangements.
- Well designed rollers ensures smooth and speedy loading and unloading of tyres, resulting in minimum energy loss.

INSPECTION SPREADER (MANUAL)

The Inspection Spreader is designed for close internal inspection of tyres ranging from 12" to 24".

Features

- Easy lifting of tyres.
- Sturdy pneumatic cylinders for easy indenting and spreading of tyres.
- Tyre rotatable in spread position.

Retread prices go upto 80% of new tyre

ENVELOPE EXPANDER

The Envelope Expander is suitable for tyre ranging from 20" to 24" rim sizes.

Features

- Rugged operation possible with the 8 pneumatic cylinders provided.
- Easy application and removal of envelope possible with no operator fatigue.
- Separate tyre trolley is provided for easy operation.

Section VI
Integrated Management Policy Cases

46. Ganapathy Ram Port Trust
47. Ganapathy Ram Steels Ltd.

Chapter 46

Ganapathy Ram Port Trust

ENTER THE CONSULTANT

Shri Ganapathy Ram, a reputed Senior Officer of the Indian Administrative Service and Chairman of Ganapathy Ram Port Trust, invited Prof. P. Gopalakrishnan to study the organisation in order to suggest improvements in the areas of maintenance, spare parts management, and materials handling. Professor Gopalakrishnan interviewed a large number of officers and his findings are as follows:

Ganapathy Ram Port Trust controls a major port located in Western India-accounting for over 20% of the country's total imports and exports. The functional heads reporting to the Chairman are the Chief Mechanical Engineer, Controller of Stores, Financial Advisor and Chief Accounts Officer and Chief Personnel Officer. These functional heads are duly helped by several executives, engineers and supervisors in discharging their duties.

The Chief Personnel Officer is incharge of human resource development, manpower estimation, advertisement, interview, recruiting, selection, induction, promotion, grievance handling, handling several unions, staff associations, labour welfare, contract labourers, casual labourers, canteen, charge sheeting, etc. The Chief Mechanical Engineer is incharge of the overall port operation, modernisation, automation, rationalisation, maintenance of equipments, workshop, tool room, planning the purchase of new machinery, movement of freight by containers, electrical/civil/electrical/mechanical/ instrument maintenance and safety engineering; he does liaison work with other ports and the Ministry of Shipping.

The Financial Advisor and Chief Accounts Officer are incharge of the total finance operations, like budgets, costs, variances, auditing, accounting, settlement of compensation, salary disposal, bill payment to suppliers, computer, collection of port charges, demurrage charges and cost control in port areas of operation. The procurement and stocking of spare parts for all the equipments owned by the Indian Port Trust is the responsibility of the Controller of Stores.

Shri Ganapathy Ram mentioned that for smooth working of the port operations, daily, weekly and monthly coordination meetings are held at different levels, but these meetings are converted into an occasion in which every department blames the other, unless it is presided over by the senior officer. He also emphasised that monthly coordination meetings are held with customs authorities, other ports, customers, the Ministry of Shipping and agencies to facilitate the smooth working of imports and exports of national importance. The activities of the port has been expanding over the years, in spite of new ports being started in western India.

Some maintenance executives interviewed by Prof. Gopalakrishnan described the current promotion policy of merit-cum-seniority adopted in Ganapathy Ram Port Trust as SWEAR BY CAT—subordination will be encouraged and rewarded but your competence at best will be tolerated! In this context, they were critical about the recruitment of MB As and retired defence officers. The Chief Personnel Officer, however, dismissed the criticism by explaining that Ganapathy Ram Port Trust adopts all modern management principles relating to participative management, mutual respect, team building, leadership by consensus, etc. He has promised to rectify any individual cases, if given in writing by the affected parties.

EQUIPMENT CATEGORIES

Ganapathy Ram Port Trust is over 100 years old. There are different types of equipment such as mobile cranes, EOT cranes, electrical shore cranes, ore handling plants, fertiliser handling equipment, conveyor system, container handling equipment, generating sets, dumpers, tugs, dredgers, floating cranes, steam locomotives, diesel locomotives, etc. The equipment are classified as A, B and C category plants. The A category plants are most vital for the operation of the port and any breakdown may result in stoppage of the operations, congestion at the port, diversion of freight to other ports, delays, loss of image, etc. The B category plants are of secondary importance, particularly those plants which have served their purpose to a great extent. The remaining equipment have been categorised in C class.

The spares for each plant are accounted under the head of that particular plant. The spares for the A and B category plants are stocked by the Controller of Stores reporting to AGM maintenance in the stocking depots in the port trust area. The spares for the C category plant are purchased as and when required on non-stock basis and handed over to the user without passing through the stocking points. The newly appointed industrial engineer has been preparing an alternative system to the above. He works under AGM maintenance reporting to the Chairman.

At present, the A and B category plants are 100 in number, while the category C plants are 25. Fifty per cent of the equipments have been imported from USA, the UK, Germany, Japan, Norway, Italy and Sweden. The machine utilisation of A and B categories is about 60%. The total number of spares individually amount to 75,000 in number. The internal audit and AG audit has criticised about the poor utilisation of machineries and wanted it to be increased to 90%. These have been audit enquiries about disposal of obsolete items. Attempts are being made to disperse manual transactions relating to spare parts through the recently acquired fourth generation computer and personal computers and laptops provided to senior officers. The industrial engineer wants to prepare computerised preventive maintenance programme instead of the present breakdown system.

SPARE PARTS IDENTIFICATION

Every A or B category equipment is assigned a plant code number which consists of an alphabet and three digits: for example, dredger goly is identified as Q 020, Q representing the stocking ward. The first spare part will be represented as Q 020–0001, the second item as Q 020–0002, and so on. The classification of spares has been done by using manufacturer's catalogue for recently procured machines. In the case of old machinery purchased long ago, where catalogues are not available, the classification has been done alphabetically and on the basis of location of machinery. Under this item, parts numbering 9999 can be accommodated for each plant, which is found to be more than adequate. It is also possible to identify each part quickly in the stores if the code number is correctly given irrespective of other mistakes in the indent or invoice. The assignment of plant number and code numbers is available with the Controller of Stores as well as the stores incharge who keep the user department informed of the same by periodical notifications. The newly appointed industrial engineer has been toying with the idea of a common codification of spare parts, irrespective of the machinery and its location. He is also thinking of a common codification for all equipments used in all port trusts in the country, ignoring suppliers part number and supplier profile. The AGM wants the supplier profile to be included.

STOCK LEVELS

Ganapathy Ram Port Trust has not so far fixed maximum, reorder point, reorder quantity and other parameters on stock levels on a scientific basis since in the service organisation the consumption pattern is extremely erratic. Hence the purchase of spares is generally arranged on an annual basis, based on specific requirement given by different maintenance branches. The Controller of Stores considers the stocks available, the internal administration lead time, manufacturing lead time, transporting lead time, inspection lead lime, import policies, availability of working capital, availability of the item, etc. The ordering cost has been estimated as ₹1,000 per order, while the inventory carrying charges, including the cost of capital, have been estimated at 30% per year. The average year-ending value of inventory of spare parts is nearly ₹500 crore, accounting for over four years consumption. This includes about ₹20 crore of stock not moved during last 10 years. The average total lead time for imported spares is about 40 months while the lead time for indigenous spares is almost 18 months. Only 20% of the items have been indigenised as the progress in import substitution has been hampered by the following factors: (a) non-availability of drawings; (b) maintenance people unwilling to take risk; (c) required quantity is small, resulting in higher price; (d) vendors insisting on higher price after initial approval; (e) poor after sales service even during warantee period and annual maintenance contract for some equipments and accessories.

PROCUREMENT

A mainframe fourth generation computer was purchased about two years ago and it is being used mainly for pay roll accounting. An on-line planning, transporting and material handling control system is also being procured from the same suppliers. There are plans to use computers in transportation logistics, handling, purchasing, stores, and inventory control, in the near future.

The main computer centre houses the computer system and the terminals have to be taken to different storage points. While four personal computers are in operations, quotations have again been called for the purchase of another 10 personal computers. In this aspect, the firm which supplied the first set of personal computers could not submit the quotation in time and hence it cannot be considered. But Ganapathy Ram Port Trust, which is satisfied with the performance of after sales service of the existing personal computers, is in a dilemma whether to cancel the quotation in order to recall this supplier. Even though the Port Trust is well within its jurisdiction to invite fresh quotations, it cannot keep the reasons without any leak and, being a Government department, it may lead to other complications, and the image will be adversely affected amongst the large number of suppliers. Computerised maintenance is only talked about.

ORGANISATION OF SPARES

The requirement for spare parts for A and B category plants are furnished in a consolidated form direct to the controller of stores for procurement action, The controller of stores analyses the requirement and fixes the quantity to be purchased after considering the stock on hand, pending previous orders and possible sources of supply. It has been noticed that about 50% of indents have been marked "most immediate" with two red "urgent" lags on the each tile. The indents are separated, depending on the source of purchase. Cost criticality-availability and MUSIC-3D are being started.

If items are proprietary in nature, a single tender is issued direct to the manufacturer or to accredited agents. For issue of such single tender for proprietary items, no monetary limits are fixed. The quotation of receipt is scrutinised by the user department and after financial concurrence and sanction of competent authority, procurement action is initiated. The user department approves the administrative sanction before the order is finally issued. The assistant controller of stores has the power of sanctioning up to ₹25,000 per indent, the deputy controller of stores up to ₹50,000, and the controller of stores up to ₹2,00,000 for non-proprietary items. The AGM maintenance upto ₹5 lakh. Representatives of users and finance at different levels are associated with the decision-making process. The chairman has purchase powers up to ₹15 lakh, and beyond this amount the sanction of the Ministry of Shipping is required. Orders are not usually split because of audit objections.

SUPPLIER PROFILE

For such of the items as are not proprietary and the value of items estimated exceeds ₹250,000, tenders are invited in the press by advertisement, insisting upon earnest money deposit and performance guarantee. When the tenders are received, the same are scrutinised by the user department. After necessary classification they are submitted to the tender committee and the minutes are recorded. Based on the minutes, purchase sanction is sought for from the competent authority, depending upon the value of purchase. For minor items with annual estimated cost less than ₹2,50,000, individual enquiries are issued to the likely suppliers. The purchase department has a system of registration of approved list of suppliers after visiting the suppliers plant to assess their capabilities. When a decision is taken to register a firm, it is required to remit a refundable security deposit of ₹50,000. Normally, enquiries are sent to all the firms registered under this approval supplier scheme, and the enquiries are consolidated and sent as weekly bulletins to the registered firms. Apart from these formalities, enquiries are also extended to important firms who may not have registered under this scheme. The aim of the purchase department is to have at least two suppliers for all critical items; however, the total number of suppliers should be less than 25,000.

PROCUREMENT OF IMPORTED SPARES

As pointed out earlier, over 75% of spares are imported in Ganapathy Ram Port Trust. Normally, imported spares are purchased on a single tender basis from the original equipment manufacturer who imposes stiff conditions on payment and using their own shipping. On receipt of the quotation from the suppliers, it is scrutinised by the user department, and the necessary technical and commercial classifications are obtained. Since the Government of India have laid down that the imported items should be procured only on FOB basis, it becomes necessary to arrange the shipment through the vessels belonging to the conference lines and marine freight paid in Indian rupees in India. Further, marine transit insurance premium is paid in Indian rupees to Indian insurance companies. To get the foreign firms to comply with these and other commercial conditions on guarantee of spares is indeed a Herculean task according to the Controller of Stores.

As a parallel action the Directorate General of Technical Development (DGTD) is requested to issue the requisite "No objection certificate" for import of required spare parts. The DGTD authorities normally raise some queries lo justify the import of spares, and the no objection certificate is issued. Thereafter, the Chairman corresponds with the Ministry of Shipping, Chief Controller of Imports and Exports, the Reserve Bank, and the Ministry of Finance for getting the requisite foreign exchange. This process takes three months.

On getting the foreign exchange release, the import licence is obtained from the Joint Chief Controller of Imports and Exports and thereafter the purchase order with import licence particulars is issued. Copies of the purchase order are sent to the Secretary, Shipping and Chartering, for necessary shipping arrangements. After issue of the purchase order, a letter of credit is issued in favour of the suppliers through the Stale Bank of India and made negotiable through the foreign supplier's bankers. The documents for negotiation are detailed in the letter of credit. The foreign supplier presents the document after shipping is effected to their bankers and realise the payment. Thereafter the documents are transmitted to the State Bank of India and Ganapathy Ram Port Trust account, debited by the corresponding rupee value amount. The documents are then handed over to the trust and the consignment cleared by paying the customs duty. The spares are inspected by the user department, and receipted in Ganapathy Ram Port Trust stores books. The spares, as and when required, are drawn by the user department from the stocking depot by placing a stock indent in the specified forms. Items which do not come in the banned or restricted category, can be imported without obtaining foreign exchange release and specific import license as per the import policy which changes every year. The Port Trust has also applied for duty drawback as it earns foreign exchange by handling foreign ships.

Some suppliers, when interviewed by the senior consultant Prof. Gopalakrishnan, have informed that the purchasing practices have not been fair in the Port Trust. They have used phrases such as kickbacks, corruption, cuts, commission, nepotism, favouritism, despotism, parochialism, political interference, interference from the ministry, punitive transfers, manipulation, manoeuvring, graft, winding-up charges, etc. in time with the corruption in India. While enquiries have been ordered, no one knows the fate of the results of the prolonged enquiries by which time the concerned officer is transferred leaves the organisation or departs from the world. But the Chairman defended and discussed about the fair practices prevalent in the organisation and promised to look into specific charges which are bound to exist in any organisation with a revenue account purchasing of over ₹75 crore per year and the capital account purchasing running into several hundred crore.

MAINTENANCE POLICIES

As pointed out earlier, the Assistant General Manager is incharge of the total operations, including maintenance. He is assisted by a Mechanical Engineer and superintendent engineers incharge of mechanical, electrical, civil, projects and electronic. There are 25 junior engineers while the foremen and fitters account to about 100. All the maintenance engineers are unaware of the latest plant engineering concepts like preventive, planned, productive, predictive, tpm, corrective, design out, terotechnology, signature analysis, vibration techniques, condition monitoring, safety engineering, non-destructive testing, energy conservation, lubrication, tribology, facilities investment, life cycle costing, reliability studies, bath-tub curve, mean time between failures, Weibull pattern, Poisson distribution, MUSIC-3D and XYZ analyses, work study, method study, snap study, etc. Adequate support facilities like tool room technical literature, testing equipment, inspection facilities, tool stores, workshop, computer facilities, and reconditioning facilities are available. The overall maintenance expenditure, including spare parts consumption, is about 3% of the total earnings of the Ganapathy Ram Port Trust. The official policy is for planned and predictive maintenance only with turnaround/preplanned overhaul scheduled as and when necessary. But a judicious combination of breakdown and preventive maintenance is the order of the day in Ganapathy Ram Port Trust. It has been noticed that equipment history cards are not maintained for some equipments.

MAINTENANCE: A–Z PROBLEMS

Prof. P. Gopalakrishnan interacted with the maintenance staff at all levels and he was appraised of the following problems by the maintenance staff:

(a) The maintenance engineers are never consulted before procuring or modifying the equipments which they are supposed to maintain.

(b) Since the operation and maintenance functions are handled by the mechanical engineers, preventive maintenance is often dispensed with, when large number of ships are awaiting to discharge freight.

(c) In order to increase the freight handled during the last quarter of the year, particularly during March, maintenance and overhauling are postponed and the plants are overworked, resulting in damages to equipments. In other words, long-term utility is sacrificed for short-term benefits to obtain increased tonnage for the year.

(d) Many plants have outlived their utility, some are even 100 years old and it is a big challenge to maintain these plants.

(e) With changes and upgradation of technology to catch up with local and foreign competitors, the original equipment manufacturer is not in a position to supply spare part after ten years, and the costly indigenous spares turn out to be spurious, resulting in more breakdowns.

(f) While importing replacement kits comprising a number of items, made of as diverse a material as rubber or stainless steel, the customs department insists on the break-up value for each item as duty for each varies. The problem has become more acute as the customs department adopts a tough attitude as it suspects that some workers of the port are hand in glove with the smugglers, criminals, and terrorists responsible for the periodic attacks in the city.

(g) The procedure for obtaining import licences, supplementary licences, clearance from various agencies, etc. takes a long time, making the quotations obtained from overseas agent as outdated, resulting in stockingpiling more items.

(h) High lead time of procurement, change of import policies and obsolescence are problems in imported spare parts.

(i) Often the purchase department is pleading for import substitution when drawings are not available and the price of local item is double that of the imported item with inferior quality.

(j) Facilities for rejecting spurious motors, bearings, compressors, pumps and assemblies are inadequate, resulting in repeated breakdowns. Due to shortage of stocks, maintenance is forced to accept spare parts that should be rejected in order to make the plant into operational stream.

(k) The fitters waste a lot of time in stores to identify the items as the stores are manned by non-technical staff and highly congested with material strewn all over the place.

(l) Improper coding, inadequate space, wrong binning, unreliable information, lack of communication between suppliers, purchase and maintenance staff are common daily problems for which solutions are found only in books.

(m) The pressure on maintenance, particularly on handling equipments is very large in the service sector as the ships awaiting berthing facilities are increasing every day.

(n) It is impossible to adhere to the maintenance budget as the original equipment manufacturers adopt market strategy for proprietary spares, and the local manufacturers increase the spare part price by blaming the inflationary spiral, manufacturing spares by diesel generators due to power cut, etc.

(o) The manufacturers are refusing to divulge the cost of manufacturing and use psychological pricing to extract the maximum price, with over 100% margin on spare parts.

(p) All foreign suppliers dump unwanted slow moving spares at the time of selling an equipment even though Ganapathy Ram Port Trust has paid a substantial amount for installation and commissioning the equipment. The company, has not entered into "buy back clauses" contract earlier and is now thinking of introducing the same for which the suppliers will be charging additional value, if they accept.

(q) After installation, the machine is covered by one year warranty during which period preventive maintenance has to be carried out as per contract, and all breakdown rectified by the supplier; but in actual practice, the foreign suppliers or their franchised agents are rarely available at the Port Trust during breakdowns even during the warranty period.

(r) On completion of the warranty, the seller usually offers "an all inclusive service contract" for one year or more; but in practice the after sales service is only 'ASS', because the service engineers are not available when contacted during breakdowns, forcing the port trust maintenance staff to do the maintenance jobs themselves.

(s) While computerised maintenance is being talked out, the source documents are inaccurate and software programs on preventive maintenance are not readily available for the peculiarities of the shipping company.

(t) The ministry has been urging the port trust to introduce robots for maintenance jobs but the maintenance staff have to liaise with the outdated transport and material handling equipments due to paucity of capital for replacement of old machineries and also the strong labour unions.

(u) The stores department insist on too much of clerical documentation like failure data, indent, issue, voucher, goods acceptance, note consumption norms, criticality analysis, rejections for each vendor, budgets, actual consumption for each machine, leaving little time for maintenance personnel to do the thinking job besides attending to regular maintenance work.

(v) Maintenance staff are forced to sign huge volumes of stores reconciliation vouchers at the year end as the book balance and physical balance widely differ in a large number of cases.

(w) The finance department is concerned more about maintenance budget, preaudits, financial concurrence, variables in actuals, too much money locked up in working capital, inaccurate estimates of quantity, wrong price forecasting and increased cost of maintenance, and do not care for the real maintenance problems.

(x) There is lack of motivation, reward, appreciation, recognition, incentives for good quality of maintenance; nevertheless, the maintenance man is remembered only when breakdown occurs.

(y) The plant capacity is overstretched to meet the annual targets by operating in unsafe conditions.

(z) The most important problem is non-availability of good quality reliable spares at the right time with a low cost from reputed manufacturers in order to carry out maintenance performance efficiently.

(z1) The fitters complain they have to spend a long time to identify the spares.

HIGH SPARES STOCK: A–Z CAUSES

The Controller of Stores listed the following reasons to justify the high level of spare parts inventories in Ganapathy Ram Port Trust:

(a) The maintenance staff blame the stores for any stockout, as revenue loss will be more than offset by the value of spare parts since the freight handled per day, including, imports and exports, may reach as high a figure as ₹150 crore.

(b) The long and uncontrollable lead time of imported spares.

(c) Documentation is not always complete as Titters remove the stores with a broken piece items without issue vouchers, resulting in more stock.

(d) Equipments are not standardised in all ports of the country and so the concept of spares bank or interchangeability or coordination is not applied inspite of the government raising in annual conferences.
(e) In the service industry operation affects not only the image of the Port Trust but the entire nation, when the ships are diverted to some other port like Karachi, Singapore or Colombo.
(f) In high tides and during the monsoon season, the efficiency is hampered, resulting in congestion of ships in the sea.
(g) Drawings are not available for imported spares.
(h) Some machinery is over 100 years old and spare parts are guaranteed by the original equipment manufacturers only for 10 years.
(i) About ₹20 crore worth of spares are non-moving but not declared as obsolete. Nobody wants to discuss this and is strewn all over.
(j) About ₹150 lakh worth imported spares are rejected and the stores become heavily congested and are lying in open.
(k) The Mechanical Engineer insists that spare parts should be readily available four months before overhauling the equipments and this would cost ₹20 crore in interest charges alone.
(l) Sometimes the overhauling is scheduled in the last minute, when dock and port workers go on lightning strikes, when parts have to be got at high prices.
(m) Since the original equipment manufacturer is not in a position to supply the spares for old machines, spurious parts made in Ulhasnagar Sindhi Association—USA—are used, resulting in repealed failures.
(n) There are difficulties in generation of data pertaining to failures, MTBF consumption, and forecasts as the demand is most erratic.
(o) Since the tempo of withdrawal cannot be predicted, FSN—fast moving, slow moving, non-moving—analysis does not give a clear indication of future movement. MUSIC 3D—cost criticality availability analysis has been started by the industrial engineer.
(p) Spare parts are not being identified as critical, essential, and desirable. Most of the spare parts are proprietary in nature,
(q) The procurement powers of delegation do not consider the widespread 20% inflation witnessed in the Indian economy, resulting in long lead time and high inventories.
(r) Past legacy persists or project surplus of the supplier dumping unwanted spares at the time of buying the original equipment and sometimes ordering parts when plants are phased out.
(s) There is no major thrust on indigenisation for all ports together. Difficulties in standardisation of equipments in all ports. Frequent modification by the foreign supplier to upgrade the technological challenges creates more problems.

(t) Purchase department overbuys in March as the funds lapse at the year end, and so sometimes unwanted items creep in leading to obsolescence.
(u) Suppliers send more materials than required at times hoping inspection will accept them.
(v) Maintenance and other departments only let off steam in case of stockouts without any responsibility for working capital.
(w) No mathematical and scientific models available to determine guidelines of stock levels in such erratic situations.
(x) Although the same part may be available in other depots, it is shown as "stockout" as each engineer is anxious to have his own quota of spares for maintenance work.
(y) Segregation of spares for overhauling, routine maintenance, rotable spares, insurance spares, proprietary etc. has to be done only in consultation with the user. Application of PERT has never been attempted in overhauling.
(z) Improper advance information and communication on long shutdown, short shutdown, major overhauling, minor repairs, etc., necessitating high spare parts inventory.
(z1) Attempts to make items in the workshop have not been successful as.
(z2) Due to Tsunami and high tides, the ships have difficulties.

In the context of high spares inventories, Shri Ganapathy Ram, Chairman of the Port Trust, has expressed to Prof. Gopalakrishnan a desire towards a scientific, systematic team approach for better working capital management.

Prof. P. Gopalakrishnan has started writing his report on structure, systems, skills, strategies, staff, styles, codification, identification of items, standardisation, corruption, value analysis, minimum-maximum levels, cannibalisation, overhauling, maintenance documentation, negotiation, categorisation of spares, maintenance policies, condition monitoring, reporting, maintenance systems, capital equipment buying, and other managerial aspects relevant to improving the corporate image for smoother operation of Ganapathy Ram Port Trust.

Chapter 47

Ganapathy Ram Steels Ltd.

CORPORATE BACKGROUND

Ganapathy Ram Steel Limited is one of the integrated steel plants in the public sector located in the eastern part of India. Dr. Ganapathy Ram, a well known metallurgical engineer from Banaras Hindu University, has been the Managing Director of the Company during the last two years. Prior to this, he was associated with the Tata Iron and Steel Company as a top senior executive for about two decades. The Ganapathy Ram Steel Plant started in the early sixties with foreign technical collaboration and the capacity was gradually expanded to 5 million tonnes. The present investment is about ₹30,000 crore and the company has plans to expand in the areas of special steels and super alloys. Dr. Ganapathy Ram has requested senior management consultant Prof. P. Gopalakrishnan to study the systems and procedures in the organisation, particularly those relating to maintenance HR, purchase, storage and inventory management. Prof. P. Gopalakrishnan has interviewed a large number of senior executives and has also gone through the files and production shops. His summary of findings after going through the records is presented now.

ORGANISATION STRUCTURE

The Managing Director, Dr. Ganapathy Ram, is the Chief Executive of the organisation responsible to the Board of Directors and to the Ministry of Steel and Mines. He has under him five general managers in charge of (a) works, (b) finance, (c) supply chain management, (d) personnel, and (e) projects. The sales are looked after by the corporate office which liases with the Steel and Mines Ministry. The General Manager, Works, is primarily responsible for the production. He has six deputy general managers under him, namely, (a) the Deputy General Manager, Mechanical Maintenance and Engineering

Shops; (b) Deputy General Manager, Electrical Maintenance and Power; (c) Deputy General Manager, Iron; (d) Deputy General Manager, Steel; (e) Deputy General Manager, Mills; and (f) Deputy General Manager, Safety. Each production shop is headed by a Chief Superintendent and a group of shops is under the control of an assistant general manager. Each Chief Superintendent in turn, is assisted by three superintendents for operations, mechanical and electrical aspects. These superintendents in turn, are assisted by managers and deputy managers. The first line executives are called junior managers and assistant managers, who report to the respective managers controlling different sections.

The General Manager, Projects and his subordinate staff are in charge of preparation of feasibility reports as well as detailed project reports for constant expansions and technological modifications. These reports supplemented by the financial viability, life cycle costing, capital machineries to be procured, sources of finance, foreign exchange content, additional manpower, marketing of finished goods and other details are submitted to the headquarters, which scrutinises and approves them on the basis of priorities of the corporation. India is likely to be a net exporter of steel by 2015.

TYPES OF AUDIT

The General Manager, Finance, aided by the Deputy General Manager, Assistant General Manager, managers, assistant managers, is incharge of finance, accounting, costing, audit, capital investment decisions, delegation of monetary powers for transactions, etc. He explained that the audits adopted in Ganapathy Ram Steel are classified into the following areas: (a) compliance audits which ensure compliance with procedures, policies, rules, regulations, statutory requirements such as financial audits; (b) efficiency audits which deal with evaluating the quality of performance against predetermined standards such as operational audit and performance audits; (c) effectiveness audit which ensures effectiveness in achieving results, including evaluating strategy such as management and social audits. The General Manager, Finance specially emphasised on introduction of management audit in the company with a view to appraising management's accomplishments of organisational objectives by providing indepth analysis of the strategic decisions in all functional areas. He is also trying to introduce managerial controls through which different managers assure that the resources are used effectively and efficiently in the accomplishment of the organisation's objectives. For this purpose, meetings are held to clarify the conception, the appropriate strategies, procedural systems, available skill-mix, staffing pattern, styles of management, and the goal itself. Corporate Social Responsibility and greenery are his pet subjects.

PRODUCTS

Ganapathy Ram Steel Plant has various mills, namely, blooming mill, billet mill, wire rod mill, rail mill, merchant mill, plate mill as well as steel melting shops from where the products are obtained. The various products rolled from the above mills include: steel ingots, slabs, blooms, billets, squares, rails, structures, wire rod of sizes 6 mm, 8 mm, 10 mm, and 12 mm, channels of different sizes, beams of different sizes, angles of different sizes, rounds of different sizes, ribbed bars of different sizes, and plates of varying thickness. Since expansion plans are being carried out to meet the consumer demands, the list of products will increase in the near future. The company also wants to produce stainless steel category, as it is more profitable.

Referring to the quality of products, Dr. Ganapathy Ram mentioned that the company's motto is: "Quality is the sea at high tide". "We start with quality and end with steel. Quality is an attitude of mind and a way of thinking to create the finest steel. We not only employ the most modern process but also the oldest mind. Nothing is as vital as the quality of the product in our plan."

PRODUCTION SHOPS

The old steel melting shop has a liquid steel capacity of 5 million tonnes per annum. The pig iron having high percentage of carbon, sulphur, phosphorus and silicon is again oxidised in the open hearth furnace in the form of coke oven gas and blast furnace gas. The steel having very low percentage of these impurities is sent to different mills where different products are manufactured.

The process of steel making in the new modem mill is done by the convenor and continuous casting process. This process is relatively cheaper as compared to the earlier one followed. The annual capacity of this steel melting shop is also about 2 million tonnes. Steel tapped from the melting mill goes to continuous casting shop to be cast in slabs and blooms. Slabs are then sent to the plate mill for rolling plates of different thickness, varying from 8 mm to 120 mm, and the blooms are sent to the rail mill for making rails.

Even though the average prices of saleable steel is about ₹1,00,000 per metric tonne, Ganapathy Ram Steel has not made any substantial profits in view of high material costs, overheads, depreciation, and interest charges.

The steel manufacturers are investing in a huge manner in fresh cold rolling capacities to meet the increasing auto demand.

HR AND PERSONNEL POLICIES

The General Manager, Personnel is incharge of manpower planning, advertisement, selection, recruitment, human resource development, training, management development, promotion policies, incentives, industrial relations, union activities, labour welfare, and so on. He emphasised about the progressive personnel policies adopted in the company. He mentioned that the workers were not up to the mark initially because the nominees of the land owners and local people were guaranteed jobs at the time of acquisition of the land. Due to their rural background, their adaptation to the industrial climate was slow in spite of training in industrial training institutes and on-the-job training and better supervision. However, a few maintenance, operation and materials staff criticise about the implementation of the progressive policies. They feel that a few MBAs' retired defence personnel and foreign trained technicians have been inducted at different levels, thereby affecting their promotion opportunities. These officers describe the merits-cum-promotion policy as SWEAR BY CAT, i.e. subordination will be encouraged and rewarded but your competence at best will be tolerated. They also say that political interference has resulted in top positions, particularly at the corporate office, being manned by people totally unqualified for the jobs, but chosen only for the right connections in Delhi. These policies have resulted in heart burning among a few who have migrated to the Middle East for better opportunities. However, the General Manager, Personnel points out that in a company with 60,000 employees with 5000 officers, there is bound to be some discontent among some employees. He has promised to look into specific cases, if brought to his attention. He defended the recruitment and promotion policy of the organisation in the larger national interest in the core sector.

MANAGEMENT DEVELOPMENT

The General Manager, Personnel, claims that the organisation treats human resources as the best asset and is one of the best professionally managed companies in the country committed to laudable objectives such as participative style for taking decisions, mutual respect, welfare schemes to meet the aspirations of the staff, self actualisation principles, and remunerative scales.

However, the impressions of a cross-section of senior executives, who have attended external management refresher courses, are not always encouraging. Their remarks about the training programmes are as follows:

(a) These programmes are not applicable to steel industry and totally irrelevant.
(b) The programmes are too much theoretical and not practical.
(c) The faculty is mostly incompetent and inexperienced and teach irrelevant topics, without even facing the audience.
(d) Emphasis is laid only on strategies, scenarios and broad policies.
(e) Lecturers seem to be confused in their thinking and are unable to answer questions; their communication abilities are poor.

(f) These lecturers speak without depth of knowledge; they have no dedication or commitment and concentrate on trivial or inconsequential matters.
(g) The same lecturer speaks on research as well as inventory control of spares.
(h) Training is not need based, but what could be offered by the faculty; it is intended only to show statistics of numbers trained to the ministry.
(i) There is improper mix of training through media, like audio, visual, and audio-visual (e.g. films) as well as through case studies.
(j) There is no proper infrastructure.
(k) There are more participants leaving the class during sessions or crosstalk.
(l) ABC analysis, which 100 years old is only discussed, without dealing cost-criticality-availability MUSIC 3D approach.

MAINTENANCE POLICIES

It was noted by Prof. P. Gopalakrishnan that Ganapathy Ram Steels has two types of maintenance organisation, taking into account the need for reliable maintenance in the steel industry: one under the shop headed by the respective shop superintendent and another by the central maintenance organisation under the Engineer, Maintenance. The shop maintenance head is assisted by managers, deputy managers, assistant managers and junior managers. Wherever necessary the shop maintenance is divided into subsections as hydraulics, lubrication, EOT cranes, ground equipment, and technical cells. The technical cell keeps all relevant data, history sheets, drawings of critical assemblies, equipment planning for the spares, preparation for major repairs, and arranging procurement at the right time. Coordination meetings are held at different levels of operations and maintenance personnel to sort out problems. For each major equipment, an equipment history card is maintained in the form of a book, where details of failure, change of item, down time, and manpower used are recorded. These documents are maintained in a centralised place in the shop. The shop maintenance observes and notes the performance of the equipment and does minor repairs needed during the operations. The central spare parts cell reporting to the Supply-chain Manager, is incharge of planning and controlling the spare parts.

The plant engineering policy is to carry out preventive maintenance as far as possible and to introduce computerised maintenance in the plant. The maintenance manual, including spare parts inventory and the necessary formats, is under preparation. The plant has 100 officers and 300 other categories of staff in order to carry out all functions of the maintenance. All maintenance engineers are well qualified and are aware of modern maintenance management principles like life cycle costing, tribology, condition monitoring, preventive, predictive, tpm productive, corrective, design out, terotechnology, and signature analysis. Constant updating of the technical skill is done by imparting training to maintenance staff in the required discipline. Ganapathy Ram Steel spends, on an average, about 2% of the total income on annual maintenance budget, 70% of which is accounted for by spare parts. This maintenance budget is apportioned to the various department activities after discussions. Whenever the actual maintenance expenditure is significantly more than the budgeted figure, explanations are called for.

While the maintenance policy is for planned preventive maintenance only, according to the maintenance personnel, breakdown cannot be avoided since some equipments are old. As the aim of maintenance is zero down time, the necessary jobs are attended on a crisis basis and the plant is handed over to operations. However, some maintenance personnel feel that during the months of February and March, planned maintenance is bypassed and neglected by production personnel to achieve the targets. It was also pointed out that the maintenance is not consulted at capital equipment buying stage. The feasibility of tpm usage is being examined.

OVERHAULING PRACTICES

All planned repairs, major repairs, capital repairs and relining blast furnaces are undertaken by the central maintenance organisation under the control of Manager, Maintenance. A modern engineering workshop is available which supplies about 40% of the spare parts needs of the organisation. The other departments under the control of central maintenance organisation include civil engineering, drawings, design, maintenance inspection, and bearings. The engineer wants to double the capacity utilisation of the workshop.

Planning for capital repair and overhauling is done at the headquarters corporate planning group, in consultation with the local operating and maintenance managers. The plant management is informed at least one year in advance to thoroughly identify the activities involved. Jobs which are to be done only in major shutdowns are also separately noted in a separate register. Whenever a shutdown is planned, the operating, maintenance and purchase staff meet together, plan for the shutdown, and inform the headquarters. The shutdowns are well planned through detailed network diagrams. For overhauling operations, the requirements of spares are worked out and action planned as per requirement. During the shutdown, major repairs, modifications, replacements, refurbishing and technological upgradation are also carried out in order to give a new lease of life to the equipment with minimum investments.

Professor Gopalakrishnan noticed that, in order to effectively carry out the overhauling job, the maintenance personnel lock up working capital in inventories worth ₹180 lakh of spares (which are exclusively meant for overhauling) at least three months in advance. The maintenance personnel feel that, otherwise, the overhauling of the plant will not be complete in time and fixed assets worth about ₹500 crore will remain idle.

PURCHASE POLICIES

The supply chain manager has the overall charge of purchase, stores, inventory control, source development, inspection of incoming materials, disposal of obsolete items, vendor rating, value analysis, cost reduction techniques, standardisation, and codification. Procurement of capital items is done on a global tender basis, after obtaining the sanction from the board. The materials department has 100 officers and 200 other categories of staff. Acquisition of stores, spares, refractories, consumables and miscellaneous items based on indents from user departments is done according to limited tenders, global tenders, open tenders, repeat orders as per the powers of delegation prescribed in the purchase manual developed about five years ago. Purchase committees for bid evaluation with user and finance have been formed as per the manual at different levels. The manual also provides emergency powers to the General Manager, Operations for ₹10,000 per item subject to a maximum of ₹10 lakh in a year. The aim of purchasing is to have at least two suppliers for most of the critical items. The system of purchasing is similar to any other public sector undertaking, like assessing the indents, inviting bids, comparison of quotations, getting approval of tender committees, placing orders, etc. Rate contracts are negotiated for repetitive items by the tender committees and included appropriately in these committees. Occasional queries are raised by audit on purchase violations.

The normal lead time of converting an indent to a purchase order is about four months for indigenous items and the total lead time is about one year, whereas the lead time for imported item goes up to three years. The current annual materials budget is about ₹135 crore, 10% of the materials are imported. However, some of the vendors complain about the practices adopted in the company, particularly rejecting items of acceptable quality on flimsy grounds. Phrases such as corruption, nepotism, kickbacks, parochialism, favouritism, manipulation, Machiavellian arts, manoeuvring, kickbacks, commissions, cuts, donations, graft, corruption, winding-up charges etc. are used by some suppliers, particularly against some offices. They also stress that when enquiries are ordered, no one knows the fate of the results of the prolonged enquiries, which are usually forgotten.

IMPORT SUBSTITUTION: A–Z PROBLEMS

The Managing Director, Dr. Ganapathy Ram, told Prof. Gopalakrishnan that the organisation has made rapid strides in indigenisation and import substitution. He has also stated that in the near future only capital machinery for steel making, that too in case of World Bank aided projects or tied credits, will be imported. However, according to the maintenance and operating technical personnel, the following problems are present in the field of import substitution:

(a) Inability to meet the precise demands of steel industry
(b) Using inferior quality raw material to produce the spares
(c) Lack of technical know-how on the part of vendors
(d) Need for foreign collaborator's approval before import substitution
(e) Vendors not interested in small quantity requirements of spares as it is not an economical batch size
(f) Not adhering to the specifications of approved sample
(g) Difficulty in getting drawing for intricate spare parts
(h) Inadequate infrastructure facilities with vendors for manufacturing
(i) Inadequate testing and inspection facilities with local vendors
(j) Repeated failures of the same part from the same vendor
(k) Increasing the price by more than 100% after obtaining orders
(l) Resorting to *force majeure* clauses and delaying the suppliers
(m) Complaining about difficulties in manufacturing with generator sets due to power cut
(n) Reliability of parts not specified
(o) Cost overruns and time overruns
(p) Poor after sales service
(q) Delays in replacement of rejected spare parts
(r) Non-removal of rejected parts even after three months
(s) Low salaried labour with lower motivation
(t) Insisting on advance for material
(u) Resorting to psychological pricing method
(v) Poor entrepreneurial quality of leadership and bringing political pressure
(w) Unsatisfactory organisation of vendor's purchases
(x) Vendor not using cost reduction methods like learning curve to reduce the price for repeat orders
(y) Wastage of too much time in price negotiations
(z) Lack of professional approach for quick response, prompt repair and least repeat failures.
(z1) He opined that steel forms may join to get together as a team to bargain imported cooking coal prices which is a key raw material as India consumes 40 million tonnes coking coal.

INVENTORY MANAGEMENT

Ganapathy Ram Steel has locked up over ₹500 crore in inventories in the form of raw materials, finished products, finished goods, refractories, and spare parts. Dr. Ganapathy Ram, is clearly worried about the high working capital commitment and its servicing, and is keen to reduce the total inventory. The inventory of refractories reaches a maximum of ₹20 crore. In this context, the spares inventory is about ₹130 crore, out of which ₹16 crore worth of spare parts have not moved during the last five years; but these have not been declared as obsolete. Imported rejected material of ₹5 crore is lying in the store and not included in the above. The inventory carrying charges, including the cost of capital, have been estimated at 20% and the ordering charges have been worked out at about ₹2000 per order. The purchase department has also estimated the acquisition cost as 0.7% of the total value of items procured.

All stock items numbering about 40,000 have been codified and listed. ABC analysis has been done with the help of a very sophisticated computer. Personal computers have not only been given to the shops but also to the stores and purchase. Criticality or VED analysis has not been done so far. Minimum-maximum levels have been worked out for all items based on availability, lead time, cost and criticality. Daily consumption report and balance report are prepared and circulated to the relevant departments. As mentioned earlier, the central spare parts cell headed by a Maintenance Manager is incharge of spare parts inventory control. Every body wants to start cost criticality—availability analysis—MUSIC 3D and treat each item reviewed by 3 dimension a two level basis.

MAINTENANCE PROBLEMS: A–Z ASPECTS

The maintenance personnel have admitted to Prof. Gopalakrishnan that by and large the maintenance system adopted in the plant is efficient. However, they were of the view that the following were problem areas:

(a) Quality of spare parts either manufactured within the workshop or bought from outside sources is not up to the mark and needs improvement.
(b) Lead time of indent to availability of reliable spare parts is very high.
(c) The breakdowns occur due to malfunctioning and overloading of the machine.
(d) The machines are not spared according to the maintenance schedule in time for inspection and checking as production gets priority particular in March.
(e) The Maintenance Manager attached to the shop is forced to accept the blames for delays in inspection.

(f) Inspection norms for purchased spares vary according to the whims and fancies of inspection staff.
(g) In view of tight schedules and pressure from the top management for production targets, quality repair is not possible for breakdowns, thereby leading to further breakdowns.
(h) There is lack of proper coordination between central maintenance and shop maintenance.
(i) There is no proper coordination between the different aspects of maintenance like electrical, mechanical, instrumentation, civil, etc.
(j) Coordination meetings are converted into forums for explaining difficulties only.
(k) There is a tendency to accept the lowest quotations in tender committee meetings; this affects the quality.
(l) Critical spares are not available on the shop floor as and when required in spite of adequate planning.
(m) Drawings are not available from the manufacturer or not developed in the company for all spare parts.
(n) In view of the humid, dusty and hot working environment, parts and instruments fail before their estimated economic life.
(o) The quality of repair and maintenance output is affected by unhealthy, dusty, hot, humid working environment.
(p) Non-availability of modern good quality maintenance, gadgets, tools, tackles, test facilities and inspection equipments affect the maintenance performance.
(q) Contract maintenance personnel do not carry out their obligations properly.
(r) Maintenance personnel are not consulted during the process of planning, buying, installing and commissioning the capital equipment.
(s) After sales.
(t) It takes a long time to communicate with the original equipment manufacturer for after sales service problems, due to poor communication facilities.
(u) The quality of highly priced indigenous spare parts varies and is inconsistent.
(v) Cannibalisation is resorted to, due to non-availability of spares.
(w) Spurious spare parts enter the system as good quality spares, resulting in increased cost and breakdown duration.
(x) Stores department gives nil stock advise against indents, resulting in stoppages.
(y) Mechanics spend a lot of time in identifying spare parts in stores as the stores personnel lack technical knowledge.

(z) A lot of reconciliation vouchers have to be signed at the year end to tally the physical stock with the book balance.
(z1) Some fitters store the item near the machine as they are accustomed to it over the year.

MATERIALS: A–Z PROBLEMS

According to the stores, purchase, inventory control, purchase intelligence, follow-up, source development and other materials executives of Ganapathy Ram Steel, the following difficulties we faced.

(a) The powers of delegation prescribed in the manuals are outmoded and do not consider the 100% inflation, resulting in open lender in a large number of cases, thus even minor decisions have to be taken at the higher level committees.
(b) Purchase staff are sometimes accused of receiving kickbacks and cuts even though decisions of purchasing are taken by purchase committees with the representatives from finance and user departments.
(c) The Chief Manager, Stores is held responsible for directly changed items that he has not even seen.
(d) The maintenance staff do not bother about documentation of indents or issue vouchers, saying that the down time of equipment is more important than paper work.
(e) The stores incharge is unable to know the quantity of spares lying in departmental stores as the entire lot of some category of critical spare parts is taken to the department as soon as it is inspected without completing documentation, thereby creating shortages.
(f) There are 10 temporary staff in the stores who are yet to be regularised.
(g) Sometimes the finance and audit insist that the stores staff must replace the broken items (glass) during handling in stores.
(h) Maintenance department always quotes only the part number of suppliers, making many items as proprietary items with its own delivery schedule and price.
(i) Specification in the indents is incomplete and improper.
(j) Ninety per cent of the indents are marked "urgent" or "immediate" with red tags.
(k) Indents mention as per the sample, where only (the broken item is attached without mentioning the internal codification.
(l) Dealings are directly done with the suppliers, ignoring the purchase and stores personnel.
(m) There is overshooting of budgeted consumption norms by 200%, thereby creating shortage in some items.

(n) Information on life of part and failure data for each vendor are not made available.
(o) Data on forecasts and estimated consumption are not made available, rendering the task of minimum-maximum inventory norms a difficult exercise.
(p) Good quality spares indented earlier, but not required later, are repeated, thus involving in disputes with the supplier.
(q) The concept of lead time is not appreciated; there is insistence that the material should be delivered the previous day.
(r) Notice for overhauling requirements is inadequate, thereby ignoring the procedures and lead time.
(s) There is no proper cooperation with the materials staff in effective implementation of cost reduction techniques like standardisation, codification, variety reduction, vendor development, vendor rating, value engineering, value analysis, obsolete control, and consumption control.
(t) The user department has not categorised the spare parts into various categories like maintenance/insurance/rotables/overhauling etc.
(u) The finance department has not given various costs elements like inventory carrying charges, ordering costs, stockout cost and overstocking cost in order to determine the service level.
(v) The finance department has not evolved a method to capitalise the slow moving costly insurance items.
(w) The finance department is not prepared to dispose of the obsolete items and insists on realisation of book value of the item.
(x) The bill passing section delays inordinately the vendor's bills, thus forcing to increase the price in subsequent lots.
(y) Maintenance staff refuse to contribute to the concept of spares bank for insurance spares of similar equipments operating in other steel plants in the country.
(z) There is lack of professional scientific team approach to the management in Ganapathy Ram Steel Limited.

Professor P. Gopalakrishnan has prepared the report on all aspects—particularly on maintenance and spare parts management and proceeded to meet Dr. Ganapathy Ram—the Chairman this evening at 1930 hr in Cosmopolitan Club.

Discussion Questions on Maintenance and Spares

In order to enable the readers understand the maintenance and spares management better and deal with problems in the day-to-day field in maintenance and spares management area, we have catalogued the major issues. Finding suitable answer will enable you understand the situation better and may help you in solving the cases in the following chapters, before you actually face maintenance and spares problems in your day-to-day life, with confidence. Whenever necessary you can refer to the text to find solutions. We have administered the questions to several maintenance/spares executives for writing in this book.

MAINTENANCE PERSPECTIVES

1. Identify the need of maintenance in organisations.
2. Explain the objectives of maintenance functions.
3. What is meant by down time and why does it occur?
4. Why maintenance is not recognized as a major profession?
5. Elucidate the maintenance functions?
6. What are the qualifications of maintenance engineer?
7. Identify the assets that are to be maintained.
8. Why does an equipment fail and what are its consequences?
9. What are the consequences of failures?
10. Link between productivity and maintenance.
11. Comment on the statement "insurance may meet your medical bill, but the pain has to be borne by you". Link this statement with maintenance problems and professionalisation of maintenance.
12. Discuss role of maintenance in the context of technological revolution.
13. Identify the problems faced by plant engineer.

MAINTENANCE ORGANISATION

14. What are the services provided by maintenance?
15. Discuss problems faced by a plant manager.
16. What are the adverse factors on maintenance management?
17. Discuss impact of automation on maintenance.
18. Identify the need of maintenance planning.
19. Enumerate the scheduling problems and discuss them.
20. What types of documents are needed for maintenance?
21. How do you control down time during maintenance?
22. Narrate the benefits of maintenance planning.
23. Comment "At the year end maintenance is not allowed in order to achieve production targets.
24. Discuss maintenance organisation—facilities needed.
25. What are the organisational perspectives?
26. Identify factors affecting the goal of maintenance.
27. Enumerate the types of organisation and reporting aspects.

MAINTENANCE SYSTEMS

28. Mention the types of maintenance.
29. What is meant by breakdown maintenance system?
30. Discuss routine maintenance.
31. Identify steps involved in planned maintenance.
32. Distinguish preventive maintenance from predictive maintenance.
33. When is corrective maintenance applied?
34. Describe design out maintenance.
35. Why do you adopt contract maintenance?
36. Narrate the steps for total productive maintenance.
37. Discuss the consequences of breakdown maintenance problems.

MAINTENANCE SYSTEMS DESIGN

38. Define machine criticality/part criticality.
39. How do you identify down time costs?
40. Identify standby availability.
41. Envisage the type of hazards in the event of failure.
42. Discuss maintenance relationship with age of plant.
43. What types of skills are needed for maintenance?
44. Enunciate the role of Bath tub curve in maintenance.
45. How do you optimise the systems design?
46. Give examples of vital, essential, desirable.
47. Why do you carryout VEIN analysis and explain significance?

CONDITION MONITORING

48. Explain role of condition monitoring.
49. What are the objectives of condition monitoring?
50. Enumerate the methods of condition monitoring.
51. When do you use vibration monitoring?
52. What are the priorities of equipments for condition monitoring?
53. Can you discuss the methods of condition monitoring?
54. What are the relevant costs in maintenance?
55. Narrate vibration monitoring systems.
56. What helps you need from OEM in maintenance?
57. Elaborate the role of NDT.
58. List down NDT Methodologies.
59. Detail general principles of trending.
60. Compare and contrast the diagnostic instruments.

TPM AND DOCUMENTATION

61. Strategise the TPM system components
62. Compare and contrast long range planning with short range.
63. Plot the strategy of facility register.
64. Profile the role of equipment record card in detail.
65. Explain maintenance scheduling process.
66. What are the principles of scheduling?
67. Find the role of maintenance records.
68. Categorise use of history record card.
69. Identify the importance of maintenance work order.
70. Design a sample work order.
71. Prepare a performance report card.
72. What type of information is needed in maintenance?
73. Describe all activities of maintenance control system.
74. Explain the control system for planned maintenance.

MAINTENANCE TURNAROUND AND NETWORK ANALYSIS

75. What is the meaning of outage management?
76. Deliberate on the term turnaround and its implications.
77. Explain the term opportunity maintenance.
78. Clarify the role of network analysis.
79. Explore the use of PEM/CPM in overhauling.
80. Can you treat maintenance work as a project?

INSPECTION LUBRICATION

81. What is meant by inspection of lubrication?
82. How do you prepare a check list for lubrication?
83. Which items are included in check list?
84. Why do you need lubrication?
85. How often is lubrication needed?
86. What do you do after lubrication?
87. Strategise the lubrication system.
88. Detail facilities for lubrication.
89. Determine frequency of lubrication.
90. Find out the equipments needed for lubrication for process industries.
91. Explain planning parameters of lubrication.
92. Elaborate problems in pooling together inspection facilities of similar equipments, plants.
93. What is the role of lubrication in maintenance?
94. Elaborate the need for survey of equipments for lubrication.
95. Compare and contrast lubrication with tribology.
96. List the documentation for lubrication.
97. Assess the quality of maintenance/lubrication.
98. How do you get zero defects maintenance?

HUMAN RELATIONS IN MAINTENANCE

99. Criticise the statement *"Hauthodi Wala"* a man with the hammer is maintenance manager!
100. Comment maintenance is not recognised as a profession.
101. "Maintenance is not recognised as a function at all in many firms and is remembered only during major breakdown". Clarify this. Can the maintenance engineer become the Chairman of the Corporation?
102. Maintenance is not treated as a subject in most universities/colleges/profession.
103. Compare and contrast maintenance with plant engineering.
104. Detail the difference in approach of maintenance of human beings—by high paid doctors in the costliest fashion—and maintenance of machine by low profile fitters.
105. Discuss role of HR manager in maintenance.
106. Strategise the need of selection, induction, training, promotion for levels training.
107. Explore the contents and methodologies in skill development in maintenance and safety field.

108. What are the limitations of internal training, training at suppliers' plant, training in outside courses?
109. Elaborate the scope of incentives in maintenance.
110. Deliberate organisational/reporting responsibilities of maintenance.
111. Find out role of statistical tools like activity sampling, Poisson distribution, etc.
112. How do you assess the work of maintenance engineers?
113. Comment maintenance comes into conflict with spares, materials, finance, production, HR departments.

CALIBRATION AND QUALITY

114. How is calibration related to maintenance?
115. Discuss calibration systems.
116. Clarify the documents in calibration, parameters and intervals.
117. Detail assessment of quality in maintenance.
118. List down steps in achieving total quality of maintenance.
119. Is there difference in approach between public sector and private sector in maintenance?
120. How do you magnify these points in defense applications?
121. Can you identify the maintenance oriented process companies and other manufacturing firms?

SAFETY ENGINEERING

122. Discuss—"Bhopal gas tragedy is man made and the culprits—Union Carbide now Dow Jone Chemicals Chairman allowed to leave the country by Congress Government without paying compensation.
123. Clarify reasons for accidents to happen in a plant.
124. Do you have a safety manual?
125. Have you any safety training programme?
126. Elaborate safety principles and guidelines.
127. Compare accidents and incidents, mistakes and sabotage.
128. Are you aware simulators being used for safety training like air forces?
129. Slipping, trifling, falling and burning are common incidents—elaborate and discuss.
130. Have you for hydrants in all places of working in shop floor?
131. Explore the realm of safety principles/guidelines.
132. Design a proforma for safety.
133. Why should cracker manufacturing plants/shops are always result in fatalities?

134. Explain role of fault free analysis in safety management.
135. How do you prevent flying sparks from a catastrophy on a shop floor?

COMPUTERS IN MAINTENANCE

136. What is a computer and how do you use it?
137. Extensively deal with "Garbage in Garbage out" and "Trash in Trash out" principles of computer.
138. Do you have a computersied maintenance software in your organisation?
139. What are the maintenance areas in which computers can be used?
140. Identify input data requirements for maintenance management needing computer assistance.
141. Clarify the computer output/screen image required by the maintenance manager.
142. How do you utilise computer for forecasting, spares control, analysis of past data?
143. Elaborate computer usage in past down time data, breakdowns, schedules, inspection, lubrication, per day works, etc.
144. How should the maintenance engineer be trained in computerisation of plant engineering?
145. Dwell on MIS from maintenance to corporate HQ.

MAINTENANCE COSTS/BUDGETS

146. Is there a maintenance budget provision for your organisation?
147. Identify the process of preparing from maintenance budget.
148. What is difference between actuals and budgeted?
149. Relate the total maintenance costs with overall output value.
150. Clarify the components of maintenance budget.
151. What happens when maintenance budget is exceeded or to the balance in budget provisioning?
152. What is proportion of components like manpower cost, spare part cost, overheads in maintenance budget?
153. Is the maintenance staff consulted in budget process?
154. Detail operating budget, performance budget, zero base budget in the context of maintenance?

PRODUCTIVITY AND MAINTENANCE

155. Signify maintenance, maintainability with productivity.
156. Find the utility of work study and time and motion study in maintenance field.

157. Clarify the step in work study in maintenance.
158. Deal with the objectives of work study in maintenance management.
159. Explain salient features of method study.
160. Relate productivity to method study and its contribution to the firm.
161. Detail work measurement process.
162. Explore the utility of method study and scope for application of work study in improving maintenance management.
163. Explore the feasibility of various recording techniques.
164. What are key facts in methods study?
165. Clarify the concept of ergonomics.
166. How does human factors contribute to maintenance?
167. Pictorise the areas of application.
168. Elaborate the terms mental load, physical load, perpetual load.
169. Relate the role of ergonomics with maintenance and work study.

ACTIVITY SAMPLING AND WORK MEASUREMENT

170. Clarify the meaning of activity sampling.
171. How does activity sampling help in work measurement?
172. Detail role of six sigma in using activity sampling.
173. Explain normal distribution in the context of snap reading.
174. Enumerate application areas of activity sampling.
175. Compare and contrast snap reading with time study in the context of maintenance.
176. Detail the differences between time study and activity sampling.
177. Probe relaxation allowance, contingency allowance, interference allowance, rest, personal allowances, process allowances special allowances, policy allowance, variable allowance and other allowances in snap reading.
178. How does activity sampling contribute to better maintenance?

ENERGY-STEAM-BOILER SYSTEMS

179. Detail the role of energy in any company.
180. Annotate—'cold air is leaking from air conditioners'.
181. Explore the feasibility of 'energy saving' in a company.
182. What infrastructure facilities are needed in energy saving?
183. Enunciate classification for concentration of energy savings.
184. Detail the energy saving measures for the building requirement, boiler, steam distribution, air compressor systems, high voltage systems, electrical system, and lighting system.
185. Detail maintenance/problems and steps in various systems.

186. Explain the steps of maintenance and energy in doors, windows, walls ceilings, roofs, boiler, steam distribution media, ventilation system.
187. Identify maintenance and energy loss problems of boilers.
188. Mention precautions of lighting systems, replacement of lamps and often light source efficiency.

FID AND LCC

189. Expand FID and relate to maintenance fields.
190. Elaborate LCC and its role in maintenance.
191. Analyse principles of asset management and process of increasing assets.
192. Dwell on decision variables for investment decisions.
193. Explain influencing factors on investment decisions.
194. Distinguish technical factors from environmental factors and list the human factors.
195. List the conditions for reconditioning and replacement.
196. What is meant by economic life?
197. Why does operating life differ from economic life?
198. Explain role of mathematical models in replacement theory.
199. In the context of FID, distinguish time value of money, capital recovery factor, ROI, NPV and pay back period?
200. Find out effects of tax and depreciation of assets.
201. Identify equipment discard policy.
202. What is the need of LCC?
203. Determine the objective of LCC.
204. Lucidly present the types of costs associated with LCC.
205. Find details of factors influencing LCC.
206. Can you work out an example of LCC and compare with different criteria by using alternative methods?

MAINTENANCE FUNCTION EVALUATION

207. Dwell on the challenges of maintenance function.
208. What are the skill men needed for maintenance?
209. Identify the parameters of expectations for maintenance evaluation.
210. Deliberate on managerial styles adopted by different companies.
211. Give examples of SWOT analysis in maintenance area.
212. Describe the process of MBO and identify HRA.
213. Articulate on the reports needed for objective evaluation.
214. Discuss the parameters of objective evaluation.

215. How do you use the subjective parameters for evaluation?
216. Discuss ratio analysis for evaluating maintenance function.
217. Dwell on the key ratio of evaluation of maintenance.
218. Lucidly present the ratios pertaining to human beings.
219. Which documents are needed for evaluation?
220. Clarify the ratios pertaining to machineries.
221. Find out the factors other than ratios used in rational evaluation process.
222. Make a presentation of futuristic role of maintenance function.
223. How does socio-political-economical-national-macro-micro policies including corruption affect maintenance?

SPARE PARTS DEFINITION

224. Define a spare part.
225. How do you distinguish spare part from consumables/components?
226. Describe the salient features of spares.
227. List the classification of spare part.
228. Explain the maintenance spares with examples.
229. Elucidate the concept of overhauling spare part.
230. When do you need overhauling spare part?
231. Give a comprehensive definition of insurance spares.
232. Enunciate the meaning of rotable spare part.
233. What is meant by mandatory spare parts and commissioning spare parts?
234. Examine the statement "Initial provisioning always consist of unwanted spare parts".
235. Elaborate the basis of subclassification of spares.

SPARES COST REDUCTION

236. Do you use part number (of suppliers) or internal codification?
237. How many digits is the codification?
238. Does the codification have a check digit.
239. Explore the possibility of variety reduction.
240. List the items which can be standardised.
241. Examine the items for value analysis/cost reduction.
242. Who are interested in spares management?
243. Clarify vein approach and its utility.
244. What is the total number of spare parts?
245. Do you have the total value of spare parts.
246. Express spare parts total value as a per cent of total machinery value.

247. What is the value of non moving/slow moving item?
248. Have you carried out obsolete items analysis?
249. Express ratio of total value of spare parts to value of consumption.
250. Do you have a workshop.
251. Can you make the item rather than buying from outside?
252. How much of your spare parts inventory is imported?

SPARES ORGANISATION

253. How should the spares function be organised?
254. Which departments are interested/involved in spares management?
255. Comment the spare part is identified/indented only with a broken piece.
256. Who should be responsible for control of spare parts?
257. What should be the qualification of spare parts manager at the user level/marketing level?
258. What is the role of material/maintenance spare parts in the control of spare parts?
259. Narrate the different departments involved in spares planning, buying, issuing, utilising and control of the parts.
260. Finance departments usually classify that too much working capital is tied up, while at the same time user departments show machines are stopped for want of spare part?
261. All unwanted/union staff are dumped in spare parts: Comment.
262. Only the filter is able to identify the spare part, examine this.
263. Indents, issues, ordering are only by part number—Examine this statement.
264. Who is incharge of inspection of spare parts?
265. Examine the communication system between different departments and spare parts.
266. Is it not common that operations department place orders for spares?
267. Elucidate the type of conflicts between different departments.
268. How and who resolve the conflict on spare parts?
269. Identify role of different sections in initial provision of specified parts.
270. Is it true that the supplier threatens guarantee will not work if mandatory spares are not lifted?
271. Explain the role of HR department in spares department—selection, induction, training, promotions, incentives, etc.
272. Enunciate the policy of technology upgradation in your firm.

SPARES CONTROL

273. What is meant by critical spare?
274. Explain the difference between critical and insurance spare part.
275. Define insurance spare part.
276. Enunciate steps for 'ABC' analysis for spare parts.
277. How is ABC analysis done?
278. Discuss the utility of ABC analysis in spares function.
279. Elucidate critical item.
280. Narrate the manner in which criticality is done.
281. Identify role of cost–criticality analysis.
282. How is service level for spares fixed?
283. Examine the utility of ABC/VED/SDE analysis of spare parts.
284. Identify the limitations of ABC/VED/SDE analysis for spare parts.
285. Explain MUSIC-3D analysis.
286. Clarify the role of MUSIC-3D in spares control.

INVENTORY POLICY

287. Have you categorised the spare parts?
288. What is meant by maintenance spare parts?
289. Define overhauling spare part.
290. Explain rotable spare part.
291. Elucidate insurance spare parts concept.
292. What is meant by capital spare parts?
293. Can you capitalise and depreciate the capital spare part?
294. Is it possible to start a spare parts bank for costly insurance spare parts?
295. Explain role of bath tub curve in spares control.
296. Identify the components of lead time.
297. How do you reduce the lead time?
298. Explain role of Poisson distribution in maintenance spare parts control.
299. How do you fix inventory levels for rotables?
300. Explain use of simulation in spare parts.
301. Identify the service level concept for fast moving spare parts.
302. Do you use part number for spare parts?
303. Explain role of standardisation, codification, value analysis in the case of spare parts.
304. Explain the concept of stockout cost in spares function.
305. Comment—The stockout cost is always higher than cost of spares—so hoard the spare part.

306. Have you experienced any production stoppage due to non availability of spare parts?
307. Do you have a fabrication workshop?
308. Who is controlling the workshop?
309. What is the capacity utilisation of workshop?
310. Do you manufacture any spare part in the workshop?
311. What is the reordering cost?
312. Have you estimated the inventory carrying charges.
313. Does your organisation need 100% protection against stockout, regarded of investments?
314. Elaborate the details X42 movement analysis.
315. Can you declare any nonmoving items as obsolete?
316. Have you declared any nonmoving insurance stock staying in the store for over 10 years—as obsolete?
317. Is there any procedure for disposing obsolete spare parts?
318. Enumerate the steps taken to reduce obsolescence.
319. Do you carry spare parts, when the original machinery has been changed?
320. Why traditional EOQ is not applicable for spare parts?
321. Explain the concept of service level.
322. Explain the concept of echelon system and stocks.

RELIABILITY AND QUALITY INSPECTION

323. Elucidate the term reliability and its relationship with quality of spares.
324. How do you determine the reliability of spare part?
325. Have you collected MTBF—mean time between failure for any spare part?
326. Dwell on the relevance of bath tub curve in spare parts management.
327. Explain in detail the term failure analysis.
328. Discuss how reliability and maintainability affect the equipment's availability.
329. Explain the scope of application of bath tub curve in understanding product reliability.
330. Identify the departments involved in improving product reliability.
331. How is reliability associated with quality of spare part?
332. Discuss the relationship between reliability of an item and inventory policy for the same item.

QUALITY AND INSPECTION

333. How do you assess the quality of spare parts?
334. Why is it necessary to inspect the spare parts?

335. Who does inspection and how is it carried out?
336. Detail the infrastructure needs of inspection.
337. Is it worthwhile inspecting imported item?
338. What is meant by inspecting on a random basis?
339. Do you measure the cost and benefit of inspection?
340. Do you resort to sampling, while inspecting spares?
341. Can you carry out stagewise inspection, while the item is being manufactured, at the suppliers' premises?
342. Identify the special needs for inspector.
343. What happens to rejected material in inspection?
344. How do you assess the quality of testing facilities?
345. Elaborate the accounting of shortages and damages.
346. Comment—Good quality items are rejected when you do not need and bad quality spares accepted when badly required.
347. Discuss merits and demerits of inspection by maintenance/materials/quality control/outside agencies/specialised in inspection personnel/shopkeepers.
348. What are the costs involved in inspection?
349. Do you get drawings from original suppliers?
350. Elucidate the policy for developing suppliers, requesting your own bright employees to supply spares.
351. Identify the procedures involved in importing items.
352. Is it difficult to get customs clearance?
353. Do you have clearing and forwarding agents to tackle corruption at clearing?

PROCUREMENT OF SPARES

354. Discuss the factors influencing spare parts buying.
355. Identify procedural difficulties in public sector purchase of spares.
356. Discuss problems involved in purchase of sensitive military harbour and its spare part.
357. The OEM does not supply spare parts as he has switched over to latest technology. Discuss this problem.
358. Elaborate ordering changes and its components.
359. How do you apply spares preservatives as they may not be consumed immediately?
360. Is it possible to introduce 'buy back' clause?
361. Comment—The supplier keeps fast moving items and dumps unwanted spares at initial procurement, to exploit the inexperience of the buyer.

362. What are the documents needed to procure spares?
363. Explain the right quantity, right quality, right price, right transport, handling, right supplier, right place of delivery, right handling and right spare part.
364. Do you negotiate on the above parameters?
365. Is it feasible to practice vendor managed inventory by keeping the ready stocks—at buyers' premises for spare parts?
366. Do you have a manual on procurement?
367. Discuss the powers of delegation for spare parts.
368. Discuss the type of contract/subcontract, you have with seller.
369. Elucidate the policy for developing alternate suppliers source development.
370. What is the payment system?
371. Do you borrow/take OD from banks for financing?
372. Discuss the market intelligence, material intelligence—on social economic-political-macro-micro-corruption scenario prevalent in your company.
373. Identify the documents—generated by computer systems—at the time of procuring.
374. Elaborate on warehousing charges, inventory carrying cost, ordering charges, same development cost inspection cost and other costs relevant to spares.
375. What is the information/documentation needed for procurement?
376. Is there surreptitious stock kept with fitters, tool room, near the machine, subscribers?
377. How do you consider the above costs in spares planning?
378. How do you avoid spurious spares in original pack?

MARKETING AND ASS

379. How do you assess the demand of spares of a category?
380. Lost the factors influencing price of spare part.
381. Do you have any priority allocation for the buyers?
382. Do you manufacture all spare parts or buy from subcontractors?
383. How do you ensure the quality of spare parts?
384. Do you face competition from spurious item manufacturers?
385. How much stock you determine in the channels of distribution?
386. Discuss legal aspects in selling at storage points in the multiechelon system.
387. Identify priority allocation if shortages occur.
388. Discuss management of issues/receipts and other documents when sales are executed.

389. How do you fix the price of any spare part? Do you indicate the reliability of the spares to clients?
390. Is it possible to determine the cost of manufacturing?
391. How do you forecast the quantity, demand and price of any spare part?
392. Do you provide supplier drawing of spare parts when you sell?
393. How do you handle the tough competition from spurious manufacturers as well as other firms supplying the spare part?
394. Explain relevance of RBI norms—Tandon/Chore committee norms for spare parts.
395. Detail the strategy of pricing at the time of initial selling and for replacement market.
396. Does the price vary from customer to customer, depending on their importance?
397. What is meant by ASS—after sales service?
398. Comment—no customer is satisfied with after sales service or AMC—Annual Maintenance Contract.
399. How do you build confidence in your user?
400. Is there a monetary levelling of stocks?
401. How often do you revise the price?
402. Elaborate the validity of your offer price for spare parts.
403. How often do you verify stocks at your warehouse?
404. Your price quotation includes spare part price, transport price, handling price.
405. Have you developed a list of equipment population supplied by you?
406. Do you advice your clients whenever you upgrade your technology?
407. Have you developed a database of failures of spares in your buyer's premises?
408. Are all your customers happy with your service/price delivery time/quality etc.?
409. What happens when you are unable to supply the spare parts promised?
410. For how many year, you as the supplier of equipment-guarantee the spare part.
411. How do you forecast the customer's requirement and plan the strategy?
412. Consider as the seller the statement—Price is not a consideration at all as the stockout cost is higher than cost of spares.
413. Identify the stocking levels at multiechelon levels.
414. Examine the proportion of manufacturing by subcontracting the spare parts—in relation to quality.

415. Have you analysed the customer's complaints?
416. Elaborate pricing strategy of spares.
417. How often you increase the price due to inflation?
418. Elaborate the influencing factors on pricing strategies.
419. What are the problems faced by the supplier due to competitive prices?
420. Discuss the policies of centralisation/decentralisation.
421. Have you analysed customers complaints and taken remedial measures?
422. How often organisations are forced to centralise other machines due to non availability of spare at right time?
423. Explain role of computer in marketing of spares.

EVALUATION OF SPARES FUNCTION

424. How are you satisfied as a spare parts service manager?
425. Explain the ratios that would help in evaluating spares function.
426. Explain the terms MBO/SWOT/KRA relations to spares evaluation.
427. Explain some of the audit in checking the performance of spare parts.
428. Discuss role of MTBF, MITR, and normal, Poisson distributions.
429. What is your policy relating to replacement vs. repair of a module of spare parts?
430. How do you evaluate the spares sales function and marketing forces?
431. How do you plan your time and allocate responsibilities?
432. Has your management audit team given a good certificate for your services of your department.
433. Explore the possibility of six sigma applications in maintenance.
434. Do you agree that spare parts management is a team effort to improve productivity and sales of organisation—following the motto "Sarva Jana Sukho Bhavanthi"?
435. Let all people—buyer/maintenance/spares manager and all be happy.
436. How do you improve the image of professional maintenance manager and spares function?

Bibliography

Aggarwal, S.C., *Maintenance Management*, Prabhu Book Service, Gurgaon, 1968.

Bagadia, K., *Microcomputer-aided Maintenance Management*, Marcel Dekker, Inc., New York, 1987.

Balachand, B.S., *Design and Management to Life Cycle Costing*, 1978.

Bell, C.F., *Information Needs for Effective Maintenance Management*, California, Rand Corporation, Santa Monica, 1965.

Ben-Daya, M., Duffuaa, S.O., Raouf, A. and Knezevic, J. (Eds.), *Handbook of Maintenance Management and Engineering*, Springer-Verlag, London, 2009.

Bullock, J.H., *Maintenance Planning and Control, National Association of Accountants*, New York, 1979.

Cooling, W.C., *Maintenance Management*, American Management Association, New York, 1974.

Corder, A.A., *Maintenance Management Techniques*, McGraw-Hill, New York, 1976.

Dunlop, C.L., *A Practical Guide to Maintenance Engineering*, Butterworths, London, 1990.

Get, K.H. and Bakh, I.C., *Models of Preventive Maintenance*, North Holland, Amsterdam, 1977.

Gopalakrishnan, P. and Narayanan, K.S., *Computers in India: An Overview*, Popular Prakashan, Bombay, 1975.

Gopalakrishnan, P. and Ramamurthy, V.G., *Project Management: Text and Cases*, Macmillan India, Bangalore, 2012.

Gopalakrishnan, P. and Sandilya, M.S., *Inventory Management*, Macmillan, New Delhi, 1978.

Gopalakrishnan, P. and Sandilya, M.S., *Management of Quality and Inspection*, Tata McGraw-Hill, New Delhi, 1979.

Gopalakrishnan, P. and Sandilya, M.S., *Purchasing Strategy*, Sterling, New Delhi, 1978.

Gopalakrishnan, P. and Sandilya, M.S., *Stores Management and Logistics*, S. Chand & Co., New Delhi, 1978.

Gopalakrishnan, P. and Sundaresan, N., *Materials Management—An Integrated Approach*, Prentice-Hall of India, New Delhi, 1989.

Gopalakrishnan, P., *Integrated Quality Management*, S. Chand & Co., New Delhi, 1988.

Gopalakrishnan, P., *Integrated Spares Management*, S. Chand & Co., New Delhi, 1984.

Gopalakrishnan, P., *Inventory and Working Capital Management Handbook*, Macmillan, New Delhi, 1991.

Gopalakrishnan, P., Janakiraman, *Artificial Intelligence and Expert Systems*, Macmillan India, 2012.

Gopalakrishnan, P., *Purchasing and Materials Management*, Tata McGraw-Hill, New Delhi, 1990.

Harriss, Etiya, M.J., *Management of Industrial Maintenance*, Butterworths, London, 1978.

Hartmann, E., *Maintenance Management*, IIE, USA, 1987.

Heinlzeiman, J.J., *Complete Handbook of Maintenance*, Prentice-Hall, Englewood Cliffs (N.J.), 1976.

Henley, E.J. and Hiromits Kumoto, *Reliability Engineering and Risk Management*, Prentice-Hall, Englewood Cliffs (N.J.), 1981.

Herbaty, F., *Cost-effective Maintenance Management: Productivity Improvement and Downtime Reduction*, Noyes Publication, New Jersey, 1983.

Hibi, S., *How to Measure Maintenance Performance*, Asian Productivity Organization, Tokyo, 1977.

Hibi, S., *How to Measure Maintenance Performance*, Asian Productivity Organization, Tokyo, 1977.

Higgins, L.R. and Morrow, L.C., *Maintenance Engineering Handbook*, 5th ed., McGraw-Hill, New York, 1985.

Husband, T.M., *Maintenance Management and Terotechnology*, Westmead-Saxon House, London, 1976.

Husband, T.M., *Maintenance Management and Terotechnology*, Saxon House, Westmead, 1976.

Kelly, A., *Maintenance Planning and Control*, Affiliated East-West Press, New Delhi, 1991.

Kelly, A., *Maintenance Management Auditing* [In search of maintenance management excellence), Industrial Press, New York, 2006.

Lewis, B.T. and Tow, L.M. (Eds.), *Readings in Maintenance Management*, Mass. Cahners Books, Boston, 1973.

Mann, L. (Jr.), *Maintenance Management, Health and Company*, Lexington, D.C. 1976 (658.58 L63 036574).

Mann, L., Jr., *Maintenance Management, Health and Company*, Lexington, D.C. UK, 1976.

Moss, M.A., *Designing for Minimal Maintenance Expense*, Marcel Dekker, New York, 1985.

National Productivity Council, *Maintenance Management*, Seminar Proceedings, New Delhi, 1977.

Newbrough, E.T., *Effective Maintenance Management: Organisation, Motivation and Control in Industrial Maintenance*, McGraw Hill, New York, 1967.

Niebel, B.W., *Engineering Maintenance Management*, Marcel Dekker, New York, 1985.

Patton, J.D., *Maintainability and Maintenance Management*, Instrument Society of America, USA, 1980.

Priel, V.Z., *Systematic Maintenance Organization*, Macdonald & Evans Ltd., 1974.

Saha, B.N., *Integrated Maintenance Management: Concept to Computerisation*, SBA Publication, New Delhi, 1995.

Scherer, E. (Ed.) *Shop: A Systems Perspective*, Springer, Berlin, 1998.

Suzuki, T., *New Directions for TPM*, Productivity Press, Cambridge, 1992.

United Nations Industrial Development Organization (UNIDO), *Introduction to Maintenance Planning in Manufacturing Establishments*, New York, 1975.

White, E.N., *Maintenance Planning Control and Documentation*, Gower Press, London, 1973.

Wolf, T.R., *Improving the Control of Maintenance Stores and Inventories, Industrial and Commercial Techniques*, London, 1974.

Index

ABC analysis, 266
Acquisition cost (*see also* ordering charges), 49, 276
Action limit curve, 67
Activity sampling, 171
After sales
  evaluation, 400
  service, 354
Air compressor, 186
Asset management, 194
  discard policy, 226, 248
Asset ownership cost (*see also* facilities and capital equipment), 218
Audit, 404
Automation, 11

Bath tub curve, 57, 324
Boiler maintenance, 185
Brain storming, 53
Breakdown maintenance, 37
Business logistics, 340

Calibration, 42, 122
Capacity utilisation, 237
Capital equipment, 247
Cash flow analysis, 204
Categorisation of spare parts, 250, 258
Codification of spare parts, 455

Computer application, areas of, 152, 394
Computer system, 151
Condition Based Maintenance (CBM), 71
Condition monitoring, 56, 64, 69,
Conflict
  external, 31
  internal, 31
  management, 30
  resolution, 132
Contract maintenance, 48–49, 397
Corporate scenario, 409
Corrective maintenance, 37, 45
Cost
  carrying, 276
  and consequences, 276
  down time, 296, 298
  of improvising, 15
  maintenance, 17, 154
  ordering, 260
  reduction, 261
Criticality analysis, 52, 266

Depreciation, 195, 203
Designing, 25, 51, 182
Design out maintenance, 46, 233
Diagnostic instruments, 77
Diagnostic study, 246
Disposal of obsolete spares, 336, 378
Distribution of spares, 350

Documentation, 17, 59, 84, 91
Down time reduction, 48
Drawings, development of, 229

Economic order quantity (EOQ), 278, 291
Economic analysis, 206
Economic life, 197, 198
Eddy current testing, 74
Energy conservation programme, 72
Energy saving, 181
Equipment
  availability, 44
  choice, 69
  types of, 154
Ergonomics, 164, 166
Evaluation
  challenges, 200
  criteria, 400
  of inventory, 403
  of maintenance, 223, 224
  need, 223
  parameters, 403
  process, 398
  of purchase, 399
  of spares, 400
  of stores, 402
  of training, 398

FSN analysis, 376
Facility(ies)
  condition of, 183
  investment decisions, factors influencing, 194–195
  management, 24
  register, 84
Failure
  analysis, 320, 321
  analysis concept, 5
  parameters, 321
Fault tree analysis, 143
Forecasting, 347

History card, 85, 86

Import
  of equipment, 251
  substitution, 254
Incentives, 133–135
Innovation, 125
Inspection, 113
Inspection of spares, 325
Inter-echelon interaction, 365
Insurance spares, 250
Inventory
  charges, 260
  control, 274
  evaluation, 403
  level, 280
  management, 473
Investment decisions, 194

Job description, 398

Lead time analysis, 251
Life cycle costing, 207, 218
Lighting system, 190
Load(s), 168
  mental, 168
  perceptual, 168
  physical, 168
Logistics management, 340, 341
Losses
  defect, 47
  down time, 47
  speed, 47
Lubrication
  planning for, 118
  programme, 118

Machine life, 14
Maintenance
  background, 226
  budget, 157
  challenges, 7
  concept, 8
  contract, 48
  control, 83